Leonid Baskin/Kjell Danell: Ecology of Ungulates

A Handbook of Species in Eastern Europe and Northern and Central Asia

Springer-Verlag Berlin Heidelberg GmbH

Leonid Baskin · Kjell Danell

Ecology of Ungulates

A Handbook of Species in Eastern Europe and Northern and Central Asia

With 61 Figures and 251 Tables

Springer

Professor Dr. LEONID BASKIN
Institute of Ecology and Evolution
Russian Academy of Sciences
Leninsky prospect 33
Moscow 119071
Russia

Professor Dr. KJELL DANELL
Department of Animal Ecology
Swedish University of Agricultural Sciences
SE-901 83 Umeå
Sweden

Cover illustration: Wild Goat, *Capra hircus* Linnaeus, 1758 (see Chap. 15). Drawing by A.N. Komarov

ISBN 978-3-642-07852-1 ISBN 978-3-662-06820-5 (eBook)
DOI 10.1007/978-3-662-06820-5

Library of Congress Cataloging-in-Publication Data
Baskin, L. (Leonid)
Ecology of ungulates : a handbook of species in Eastern Europe and Northern and Central Asia / Leonid Baskin, Kjell Danell.
p. cm.
Includes bibliographical references.

1. Ungulates - Ecology - Former Soviet republics. I. Danell, Kjell. II. Title.

QL737.U4 B366 2003
599.6'17'0947 - dc21

Originally published by Springer-Verlag Berlin Heidelberg in 2003.

Softcover reprint of the hardcover 1st edition 2003

Cover design: Design & Production, Heidelberg
Typesetting: SNP Best-set Typesetter Ltd., Hong Kong
31/3150 - 543210 - Printed on acid-free paper

Acknowledgements

We kindly thank the Swedish Institute, the Swedish University of Agricultural Sciences, and the Swedish Royal Academy of Sciences for financial support during preparation of this book.

Contents

Abbreviations

AN SSSR	Academy of Sciences of USSR
ASSR	Avtonomnaya Sovetskaya Socialisticheskaya Respublika (Autonomous Soviet Socialistic Republic)
RAN	Russian Academy of Sciences
RSFSR	Rossiyskaya Sovetskaya Federativnaya Socialisticheskaya Respublika (Russian Soviet Federative Socialist Republic)
SNG	Soyuz Nezavisimykh Gosudarstv (Union of Independent States)
SSR	Sovetskaya Socialisticheskaya Respublika (Soviet Socialist Republic)
SSSR	Soyuz Sovetskikh Socialisticheskikh Respublik (= USSR)
USSR	Union of Soviet Socialist Republics
VASKHNIL	Vsesoyuznaya Akademiya Selskokhozyaistvennykh Nauk imeni Lenina (The Lenin All-Union Academy of Agriculture Sciences)

Introduction

The knowledge on the ecology of ungulates (orders Perissodactyla and Artiodactyla) inhabiting eastern Europe and northern and central Asia is of special importance for those interested in zoology, ecology, nature conservation, hunting and management. There are 26 species of ungulates within the 22.4 million km^2 area, between 35–82°N and 20°E–169°W and they occupy several vegetation zones from arctic deserts to the subtropics.

In our opinion, the advancement of science can be retarded and general conclusions will be difficult to make, if the knowledge of the organisms inhabiting one sixth of the world's land surface, covered by this vast region, is not included. The language barrier, as well as the lack of international accessibility of local publications, makes it difficult to make use of the great volume of scientific information gathered within the territory of the former USSR. The only complete ecological review of ungulates of the Soviet Union, *Mammals of the Soviet Union, vol. 1. Ungulates*, was published by V.G. Heptner, A.A. Nasimovich, and A.G. Bannikov in 1961 (in Russian, Heptner and Naumov 1961) and 1989 (the English translation, Heptner and Naumov 1989). This excellent book does not, however, contain scientific contributions published after 1959. This is unfortunate, because during the following 40 years, intense and highly qualified research on the ecology and behavior of ungulates was performed.

One important reason for us to produce this review was the historical events of 1991, with the breakdown of the USSR, with its history of restoration and exploitation of the animal world. In 1917, many species of large mammals had low numbers and small ranges. Scientists and managers successfully replenished most of these species during the subsequent decades. Some tragic events in history resulted in the decline and disappearance of many human settlements over vast territories, and large strips of land along the borders were restricted for military use alone. However, these events were fortunate for the animals, as large areas of land were returned to wilderness. Since 1991, because of political and financial instability in some areas, the decline of many ungulate species has been observed. The time span reviewed by us includes a "cycle" from a phase with low population densities via a peak to another low phase.

The species with large ranges (wild boar, *Sus scrofa*, red deer, *Cervus elaphus*, moose, *Alces alces*, roe deer, *Capreolus capreolus*, *C. pygargus*, and reindeer, *Rangifer tarandus*) have their main distribution areas within the territory of the former Soviet Union. Here, they inhabit areas under abiotic and biotic conditions somewhat different to the conditions in other areas where the species are established. For many scientists, the ecology and behavior of these species in the former USSR territories are also interesting because here they live in virgin, or almost virgin, habitats. Information gained from studies carried out in such pristine environments can provide an important reference point, and can be used for conservation efforts in other areas.

After the dissolution of the Soviet Union, ungulates suffered from poaching, with the result that urgent measures for their protection are now necessary. These problems have now become a matter of international responsibility and activity. Many western specialists are involved in conservation activities, but also in determining how some of the ungulates within the territory of the former Soviet Union should be exploited. Conservation and sustainable use are especially important in central Russia, middle Asia and Caucasus, in the Ukraine and Moldova.

Aims, Definitions, Methods

Our aim is to present an updated review of the ecology and behavior of ungulates inhabiting eastern Europe and northern and central Asia. We have focused on quantitative data and have condensed presentations of the autoecology of the species, in order to facilitate comparisons between species, including data from several areas.

It is a challenge to present quantitative and semiquantitative data in a condensed way. We preferred to use tables for the presentation of data because it facilitates the extraction of information for different purposes. Original quantitative data are given as averages and lowest and highest values. Semiquantitative data, for example, conditional estimates, as well as verbal descriptions, are presented by using a relative scale. We have used +++ for >30% of the observations, ++ for 20–30%, + for 10–20%, and r for <10% of the observations; x is used when the behavior is noted, but no numerical value is given by the author. These estimates are applied to such variables as habitat use, diet, population composition, and predation. For food habits, we follow the scale developed by Vendland (1938) for the description of the diet of sika deer (*Cervus nippon*), because it was widely used in the USSR. We used +++ when a plant species is heavily grazed and constitutes one of the main sources of food in a season (5); ++ when the plant species are used often (4); + when plants are grazed occasionally (3), r for plants rarely eaten (2); we have not included situations where plants have been eaten by chance (1).

When covering such a large area, it is important to show the geographical variation in ecological features. For a common and updated classification system for ecoregions, we have used Bailey (1989, but see 1998 for more details) and, when possible, we have specified the information by ecoregions.

The species descriptions are structured in the following way. First, we give the scientific (Latin) name of the species, order and family/subfamily after the classification by Grubb (1993). However, in a few cases, we have adopted the common Russian nomenclature. The most important deviations are to include the saiga (*Saiga tatarica*) in the Antilopinae family, not Caprinae. Further, *Ovis ammon* was divided into three species (*O. ammon*, *O. vignei* and *O. aries*; Nadler et al. 1973); *Capreolus capreolus* was split into *C. capreolus* and *C. pygargus* (Sokolov et al. 1992), but the information for both species was kept within one chapter. We have given synonyms of species and previous Latin names (see Standley 1992), as well as the most common English names. We have used the drawings of the gifted artist A.N. Komarov, also published by Heptner and Naumov (1989).

The section "Status and Distribution" gives a brief description of the status and geographical distribution of the species/subspecies, mostly as described by Heptner (1989). Updated information on the conservation status is given. The subsection "Distribution" includes a rough estimate of the range size together with a schematic map of the occupied areas, where some of the main localities are indicated. The occurrence of the species in different ecoregions is also given.

In the section "Measurements" we give estimates (usually separately for the two sexes) on body mass, body + head length and length of the tail. The development of size, of young animals over time, is presented in graphs if enough data are available.

The section "Habitat" gives characteristics of habitat features important for the existence and survival of the species under natural conditions. We have noted if human activities have forced the species to occupy suboptimal habitats. We have given descriptions of terrain forms, altitude and have listed the occupied habitat/biotope types. Further, we have included a small section on the snow conditions and temperatures, which are of importance to the animal. Also included are the general needs of access to water during different seasons, and information has been given on the tolerance to salt water.

The section "Diet and Feeding Habits" gives information on the diet, plant species included in the diet, degree of herbivory, utilization of different plant forms and plant parts as well as a description of the feeding niche. We have also given information on the amount of food consumed.

The section "Social Organization and Behavior" includes information on the social life of the species, its home range size, dispersal/migration, diurnal activity, and some important features resulting from the activity of the animal, such as trails, constant areas of defecation and the choice of resting places. To describe animal group composition we have used a classification based on the character of individual relationships between group members. We have described types of movements (Hildebrand 1988) and some special activities, such as digging in the snow and soil, and moving over rocks. We have also given information on the speed of movement, and any special behavior engaged in when escaping predators.

In the section "Reproduction" we have given data on species fecundity, age of participation in breeding, sexual behavior, maternal and offspring behavior. We describe reproductive behavior, the extent of seasonal breeding events and maximum length of life.

The section "Mortality" includes mortality factors, role of predators, parasites and diseases.

The section "Population Structure and Densities" gives gender and age composition of the populations in different ecoregions, summarized information on population dynamics, density as well as resources of species in different parts of their range. We have given population data by year to show population trends in different areas. For some species, there is only fragmentary data and in such cases we have chosen to present such figures, even if they only cover small areas and a short time.

The last section "Interactions with Man" gives productivity indicators and harvest data. The influences of forestry, agriculture and other land uses are discussed. In the last two short chapters, we have included information on the Przewalski horse, which lives only in captivity, and the Mongolian gazelle, which sometimes visits areas along the southern border of the area covered by us.

In the appendices, we provide lists with genera of plants eaten and lists of helminths and ectoparasites reported for the different ungulate species.

In addition to a full list of references used, we have listed some additional books and papers giving valuable, but more specialized, information on the ecology and behavior of the species described here. We hope that our book will assist orientation in the ocean of Russian zoological literature and will stimulate further use of it.

1 Asian Wild Ass, Kulan – *Equus hemionus*

Equus hemionus Pallas, 1775.
Perissodactyla, Equidae

1.1 Status and Distribution (Fig. 1.1)

1.1.1 Taxonomic Remarks

Only one subspecies, *E. h. onager* Boddaert 1785, exists in the modern fauna of the former USSR. *E. h. hemionus* Pallas 1775 inhabited the Transbaikal part of the range and has been extinct since the latter part of the 18th century. *E. h. finschii* Matschie 1911 inhabited the Kazakhstan part of the range and became extinct in 1936–1937 (Heptner 1989; Blank and Dzhanyspaev 1990).

1.1.2 Status

Vulnerable (VU–A2c) (Hilton-Taylor 2000).

1.1.3 Distribution (Fig. 1.2)

Distribution Area. Ca. 4,268,000 km^2 (estimated historical range).

Occurrence in Ecoregions. Temperate steppe (332), temperate desert (342, 343).

1.2 Measurements (Fig. 1.3)

1.3 Habitat (Table 1.1)

Semi-deserts and deserts of uplands and foothills. During winter mostly in deserts, during summer in semi-deserts and more seldom in steppes (Solomatin 1973).

Fig. 1.1. Kulan

Terrain Form. Flat and hilly uplands; foothills of mountains.

Altitude. In upland at 800 m a.s.l. (Badkhyz population).

Snow. Snow is important as water source. At more than 40 cm snow depth animals follow each another. Snow depth more than 60–70 cm is limiting (Sludsky 1963; Solomatin 1973).

Temperature. In the past, kulans inhabited areas with a temperature range from −50 to +45 °C.

Water. Need watering except during early spring when food plants are rich in water. On Barsakelmes Island, they can drink salt water (<20 g salt/l, except pregnant females <10 g salt/l). Evacuation of kulans from Barsakelmes Island started due to increasing saltiness of the Aral Sea in 1986 (Blank and Dghanyspaev 1990).

1.3.1 Biotope (Table 1.1)

1.4 Diet and Feeding Habits

1.4.1 Diet (Table 1.2 and Appendix 1, Table 1)

In winter, when snow is absent or not too deep, green parts of sedges, cereals, wormgrass and *Salsola* are the main diet; large herbs and shrubs

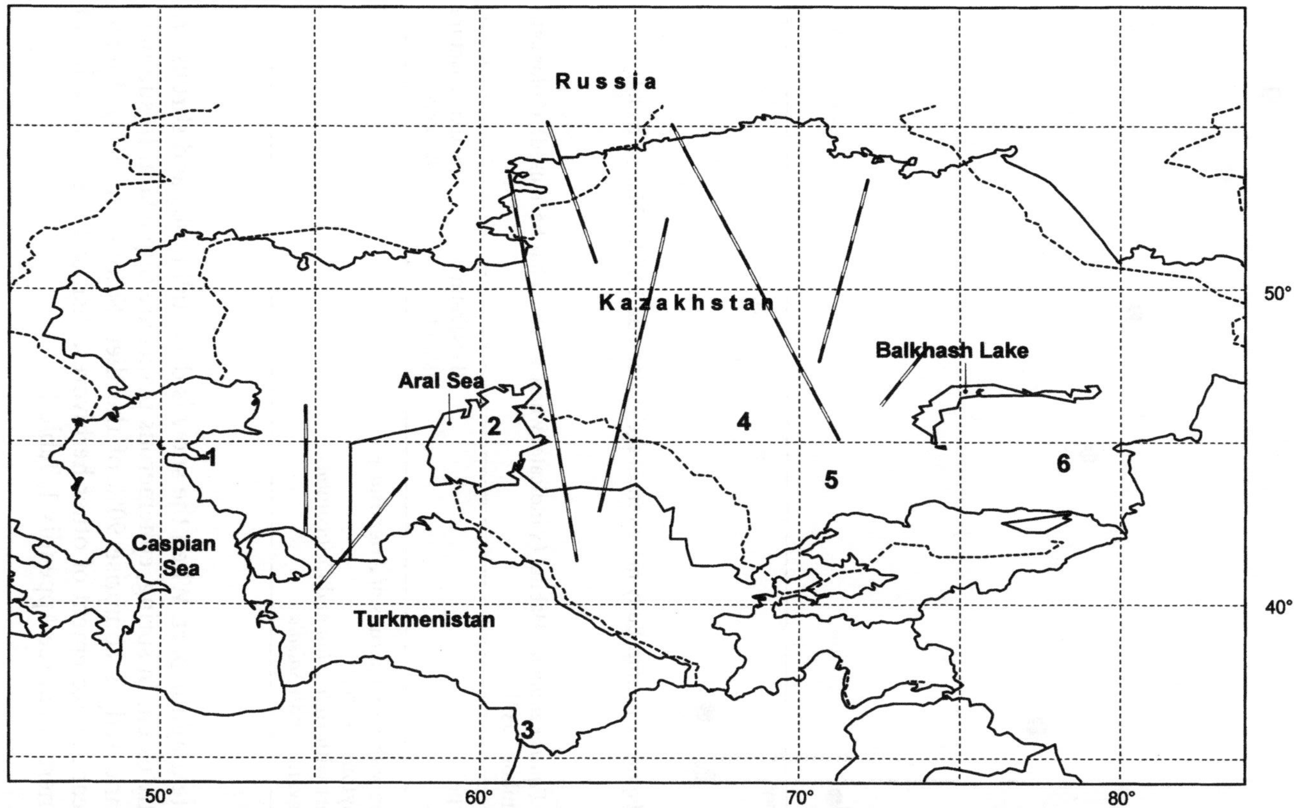

Fig. 1.2. *Equus hemionus*. Distribution map and migration routes (Blank and Dzhanyspaev 1990). *1* Aktau-Bazatchi; *2* Barsakelmes Island; *3* Badkhyz Nature Reserve; *4* Betpaqdala semi-desert; *5* Andasaysky Game Husbandry; *6* Qapshaghay Game Husbandry. *Thick lines* represent routes of migration in the past

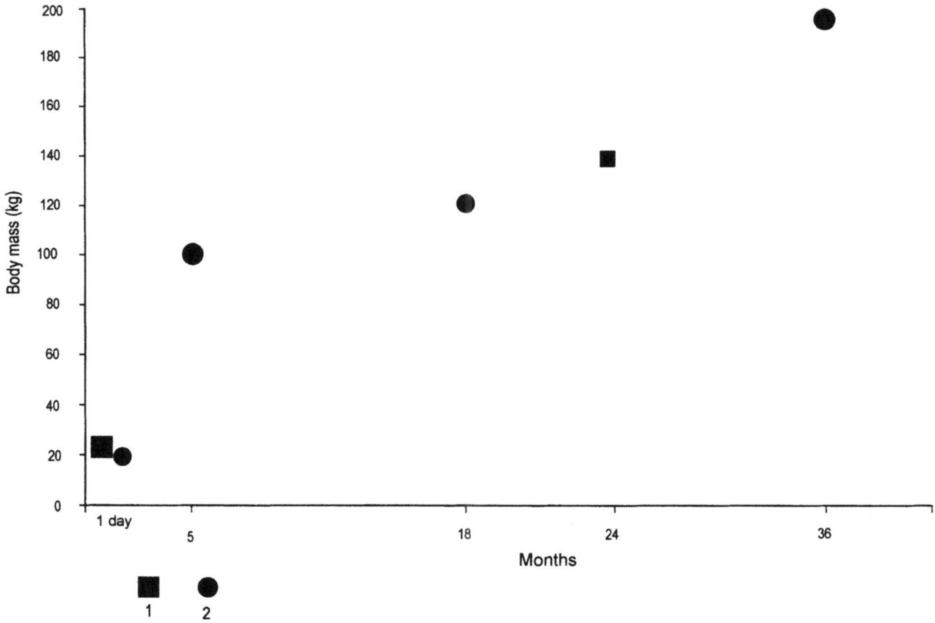

Fig. 1.3. *Equus hemionus.* Body mass versus age (Solomatin 1973). *1* Male; *2* female

Table 1.1. *Equus hemionus.* Habitat (x indicates where a habitat has been used). (Solomatin 1973; Bannikov 1989)

Habitat type	Barsakelmes Island	Badkhyz Nature Reserve
Clay desert with dwarf shrubs of *Artemisia* + nanophytons + cereals	x	x
Sand desert with *Haloxylon* and *Calligonum*	x	
Pistacia savanna with *Salsola*		x

constitute a minor part. When the snow is deep, tall herbs and shrubs are frequently eaten. In spring, ephemerous plants are important. In summer, cereals and sedges comprise 90% of the diet (Solomatin 1973). Kulans usually eat the excrement of other herbivores. Visit salt licks. For a list of plant genera eaten see Appendix 1, Table 1.

1.4.2
Life Forms/Plant Parts Eaten (Table 1.3)

Consumption of Food/Day. Ca. 5 kg of dry food (calculated by Solomatin 1973).

Table 1.2. *Equus hemionus.* Seasonal changes in utilisation of shrubs and grass in Barsakelmes Island (% of time used for feeding). (Rashek 1967; Bekenov and Fadeev 1984)

Plants	Plant type	Early spring (%)	Late spring (%)	Summer (%)	Autumn (%)	Winter (%)
Anabasis	Shrub	48–70	0–33	0–50	0–43	21–67
Artemisia	Herb	30–34	0	0	10–30	19–78
Alhagi	Herb	10–20	0	0	0	0
Cereals (total)	Grass	0–52	67–100	87–100	51–87	1–26
Agropyron	Grass	0–12	13–50	1–65	0–71	12–26
Eremopyrum	Grass	0–52	13–50	16–28	0	1–25
Stipa	Grass	0	4–66	2–42	0–40	0

Table 1.3. *Equus hemionus.* Life forms/plant parts eaten (x indicates that plant type has been used)

Life forms	Plant parts eaten						
	Leaves	Stems/twigs	Bark	Roots/bulbs	Flowers	Fruits	Seeds
Trees	x	x					x
Shrubs		x					
Dwarf shrubs		x					
Herbs	x	x			x	x	x
Grasses/sedges	x	x					x

Feeding Niche. In field and shrub layers.

1.5 Social Organization and Behavior

1.5.1 Social Life (Table 1.4)

There are permanent groups which consist of a stallion, some mares and some young animals. However, the stallion can leave the group for a few days. Average size of groups is 11–13. In winter the groups can form concentrations of up to 150–200 animals (Solomatin 1973; Bannikov 1981; Table 1.4).

A mature female leads the group during movements and selects the direction and rhythm of pasturing. The stallion is dominant and brings

Table 1.4. *Equus hemionus.* Group size (% of observed groups) in Badkhyz Nature Reserve. (Solomatin 1973)

Season	Number of animals per group					
	1	2–10	11–20	21–50	51–100	>100
Winter	14	42	20	21	3	<1
Spring	25	41	19	12	3	<1
Summer	16	52	15	15	2	<1
Autumn	8	57	24	9	2	0

members of the group together, protects them against predators and drives away other males, keeping them at some distance. When the density of kulans increases, the hierarchy among stallions becomes more obvious, with the strongest stallions being found in the centre and the weak stallions at the periphery. During rut, there are harems, consisting of a male and some females with foals and yearlings and bachelor herds, consisting of young males expelled from harems by dominant males.

1.5.2 Home Range

In winter, kulans can live in a small area if there is enough food, e.g. herds of 100–250 animals can live within 5-15 km^2 (Solomatin 1973). A network of trails develops in heavily used areas. Also, there are places for bathing in dust, lavatories with excrement can be deposited over several years (Bannikov 1981; Solomatin 1973). In summer, animals live within 10–15 km from water. During other seasons, when there is no shortage of water, large areas are used.

1.5.3 Mode of Movement

Capable of running 85 km/h (Gambaryan, citing Solomatin 1973). Jumps up to 6.5 m in length and 1.8 m in height. Climb slopes up to 70°, and can jump from heights of 2.5 m. Dig into snow to find food.

1.5.4 Dispersal/Migrations

In the past, regular and distant migrations between NW Kazakhstan (summer pastures) to Betpaqdala (winter pastures) and along the N Balkhash lake were described (Fig. 1.2; Gern 1891; Slovtsov 1897; Pevtsov 1951). Also, non-regular migrations into the south during snow-rich

Table 1.5. *Equus hemionus.* Diurnal activity (shadowed parts denote activity). (Solomatin 1973)

Month	Time of day			
	0–7	7–13	13–19	19–24
February				
April				
June				

winters, and summer migrations to find water are known. Sudden movements were initiated when vegetation or snow was coated by ice, after steppe fires, burns or occupation of watering sources by domestic herds (Solomatin 1973).

1.5.5 Diurnal Activity (Table 1.5)

Closely connected with watering regime. In winter and spring kulans are mostly active during the day. In summer, especially during hot days, they are active preferably at night in order to decrease water losses.

1.5.6 Shelters

In windy weather kulans form very dense gatherings; young animals are in the front (in the downwind area) (Solomatin 1973).

1.5.7 Escape Behavior

Wolves can chase kulans at speeds of 45–58 km/h, which is much less than the running speed of kulans. Kulans also demonstrate great endurance. Kulans show great vigilance when approaching watering places; they always come to a water source through an open area and use very steep slopes. Leaders, especially stallions, are very cautious and keep group members together (Solomatin 1973). Females protect their foals against dogs.

1.6 Reproduction (Tables 1.6, 1.7 and 1.8)

From winter to foaling time, males stay 100–150 m away from family groups; they are very excited and battles often occur. Females and yearlings live in groups of 30–100 animals. Many times a day stallions burst into the female groups and try to mate. After foaling harems increase in size, the stallion (owner of harem) mates with the females in heat. Males participate in breeding up to an age of 9–10 years, after this age they are expelled from harems by stronger competitors (Bekenov and Fadeev 1984). Female heat occurs 6–7 days after foaling, and again after 17–28 days (Rashek 1966; Solomatin 1973). Females prefer to foal in small depressions on the open plateau. After 2–3 h the foal can follow its mother, but during the first 2 days it spends most of the time lying down (Table 1.7).

1.6.1 Seasonal Breeding Events (Table 1.8)

Table 1.6. *Equus hemionus.* Breeding parameters (Rashek 1966; Solomatin 1973; Bannikov et al. 1976)

Age at sexual maturity (years)		Participation in first breeding (years)		Pregnancy (months)	Barrenness (%)	Lactation (months)	Litter (size)
Male	Female	Male	Female				
2–3	2–3	3–4	2–3	12	14–29	11–12	1

Table 1.7. Female *Equus hemionus* fecundity in different age classes. (Rashek 1966; Solomatin 1973)

Age (years)	Number of foals/female (including barren animals)
3–4	0.33
5–6	1.0
8–9	0.75–1.0
10–15	0.33–1.0

Table 1.8. *Equus hemionus.* Seasonal breeding events

Locality[a]	Rutting season			Foaling season		
	Preparation for rut	Basic rut	Last demonstrations	First foals	Main foaling	Last foals
Badkhyz Nature Reserve[1]	January–February	End January–end June		8 April	10 April–end June	28 July
Barsakelmes Island[2]	10 April	Mid May–end June	28 December		7 May–end June	

[a]The numbers denote the following references: 1, Solomatin (1973); 2, Rashek (1966).

1.7 Mortality

In the Badkhyz population, after foaling, 12–20% of the population consisted of foals. In August 14% were foals, but by the following April, only 9% were foals (Solomatin 1973).

Table 1.9. *Equus hemionus.* Gender and age composition (%) of the Badkhyz population in 1959. (Solomatin 1973)

Males			Females		
Adult (%)	2–3 years (%)	Yearling (%)	Adult (%)	2–3 years (%)	Yearling (%)
10	23	11	45	4	7

Table 1.10. *Equus hemionus.* Population size. (Solomatin 1973; Blank and Dzhanyspaev 1990; Flint et al. 1991; Bekenov 1999; Gorelov and Ishanov 1999; Pereladova, pers. comm.)

Locality	Year	Population size
Badkhyz Nature Reserve	1941	300
	1952	200
	1967	700
	1975	980
	1976	1,254
	1980	2,000
	1995	5,000
	1999	1,500
	2000	900
Barsakelmes Island	1953	8
	1958	13
	1963	47
	1968	64
	1973	97
	1978	175
	1983	228
Qapshaghay	1982	23
	1987	83
Andasaysky	1987	55
Total	1991	3,000

1.7.1
Predators

Wolves, golden eagles and dogs are probably the main predators. Wolf predation is most likely to be the least important.

1.7.2
Diseases, Helminths, and Ectoparasites (Appendix 1, Table 1)

1.8
Population Structure and Densities (Table 1.9)

Gender ratio of kulans (M/F) in the Badkhyz: newborns 1:1; yearlings 1.6:1; adult 1:4.5 (Solomatin 1973); in the Barsakelmes (1956–1964): newborns 1:1.27; adult 1:1.83 (Rashek 1966).

1.8.1
Population Density and Size (Table 1.10)

In 1980 the density of the Badkhyz population reached 25 animals per 10 km^2 (Bannikov 1989).

1.9
Relations to Man

Since 1953, reintroduction of kulans into new areas with appropriate conditions was arranged: in Kazakhstan (Barsa-Kelmes Island, Kapchagay Game Husbandry, Andasay Reserve, Aktau-Buzachi Reserve) (Kovshar and Kargopolov 1982; Bannikov 1989; Blank and Dzhanyspaev 1990; Plakhov 1997), in Turkmenistan (W Kopet-Dag) (Lukarevsky 1999).

2 Wild Boar – *Sus scrofa*

Sus scrofa Linnaeus, 1758.
Artiodactyla, Suidae, Suinae

2.1 Status and Distribution (Fig. 2.1)

2.1.1 Taxonomic Remarks

Five subspecies occur: *S. s. scrofa* Linnaeus, 1758 (European part of Russia); *S. s. attila* Thomas, 1912 (Caucasus, Carpathian Mountains); *S. s. nigripes* Blanford, 1875 (Middle Asia, Kazakhstan); *S. s. sibiricus* Staffe, 1922 (Transbaikal); *S. s. ussuricus* Heude, 1888 (Far East) (Adlerberg 1950). During 1935–1988, 9285 animals were transferred between different parts of the range (Danilkin 2002).

2.1.2 Status

Numerous, hunted (Sokolov and Lebedeva 1989; Smirnov 1994; Danilkin 2002).

2.1.3 Distribution (Fig. 2.2)

In 1960s, the species quickly enlarged its range in the European part of Russia, and in 1970 it had reached N Karelia (Fadeev 1975a). Although the re-established population in W Siberia disappeared after the very severe winter of 1913/1914; the population was again restored in 1980 by re-introduction (Azarov and Klimov 1990). The expanding range in Altai has reached 52°N (Sobansky 1992).

Distribution area. Ca. 10,057,000 km^2.

Occurrence in ecoregions. Subarctic (134), Subarctic Regime Mountains (M134), Warm Continental (212), Warm Continental Regime Mountains (M212), Subtropical (231), Subtropical Regime Mountains (M232), Prairie

Fig. 2.1. Wild boar

(252, 253), Prairie Regime Mountains (M251, M252), Mediterranean Regime Mountains (M262), Tropical/Subtropical Steppe (315), Tropical/Subtropical Desert (321), Tropical/Subtropical Desert Regime Mountains (M321), Temperate Steppe (331, 332), Temperate Desert (342, 343), Temperate Desert Regime Mountains (M341).

2.2 Measurements (Table 2.1, Fig. 2.3)

2.3 Habitat

Almost eurytopic. Shrubs, tall grass or dense forest stands are important for shelter. Adult males can inhabit harsh environment, e.g. subalpine habitats with deep snow (Lebedeva 1956).

Terrain Form. Shores of fresh water; flood plains; plains (ravines), valleys; foothills, slopes of mountains (Donaurov and Teplov 1938; Nasimovich 1939, 1955; Sludsky 1956a; Kozlo 1975; Vatolin 1980; Bromley and

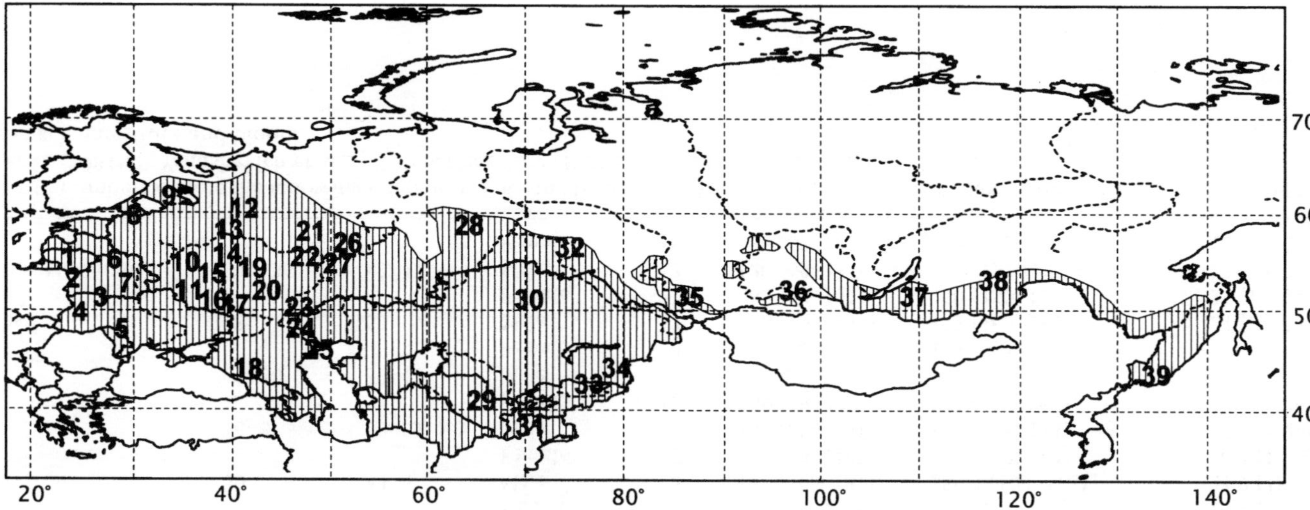

Fig. 2.2. *Sus scrofa.* Distribution map (Fadeev 1980a, 1988; Ovsyukova 1996). *1* Lithuania; *2* Belowezhskaya Forest Nature Reserve; *3* Volyn; *4* Carpathian Mountains; *5* Moldova; *6* Berezinsky Nature Reserve, Belarus; *7* Polesye; *8* Leningrad oblast; *9* Karelia; *10* Smolensk oblast; *11* Bryansk oblast; *12* Vologda oblast; *13* Darwinsky Nature Reserve; *14* Tver oblast, Moscow oblast; *15* Kaluga oblast; *16* Orel oblast, Kursk oblast, Belgorod oblast; *17* Lipetsk oblast, Voronezh oblast; *18* Kavkazsky Nature Reserve; *19* Ryazan oblast, Oksky Nature Reserve; *20* Tambov oblast; *21* Kostroma oblast; *22* Nizhny Novgorod oblast; *23* Saratov oblast; *24* Volgograd oblast; *25* Astrakhansky Nature Reserve; *26* Kirov oblast; *27* Tatarstan; *28* Ural Mountains; *29* Uzbekistan; *30* plains of Kazakhstan; *31* Tajikistan; *32* W. Siberia; *33* Kyrgyzstan; *34* Kazakhstan Mountains; *35* Altai Mountains; *36* S Siberian Mountains; *37* Transbaikal; *38* E Siberia; *39* Sikhote-Alin Mountains, Lazovsky Nature Reserve

Table 2.1. *Sus scrofa*. Measurements

Locality[a]	Longitude (°E)	Latitude (°N)	Gender	*n*	Body mass (kg)	Head and body length (cm)	Tail length (cm)
Belovezhskaya Forest[1]	24	53	Male				
			5 years	21	113.7 ± 3.1	158.9 ± 1.4	29.4 ± 0.7
			6–7 years	9	137.2 ± 5.2	160.2 ± 2.3	27.2 ± 1.0
			8–9 years	5	148.1 ± 7.4	167.4 ± 2.3	30.3 ± 1.5
			Female				
			5 years	12	80 ± 4.3	141.4 ± 1.2	26.4 ± 0.8
			6–7 years	10	85 ± 3.4	147.5 ± 2.3	27.8 ± 1.7
			8–9 years	8	80.7 ± 3.3	146 ± 2.2	28.2 ± 0.7
Voronezh oblast[2]	38–43	49–52	Male	126	118 (62–285)	162 (132–202)	24.6 (17–31)
			Female	137	110 (71–186)	156 (132–179)	23.4 (17–35)
Kavkazsky[3]	40	44	Male	6	144 (105–181)	156.3 (109–176)	24.9 (18–30)
			Female	7	117 (83–148)	152 (142–160)	20.4 (14–25)
Kazakhstan[4]	46–87	40–55	Male	11	106 (80–183)[b]	160 (148–174)	25 (20–29)
			Female	12	69 (49–80)[b]	144 (127–157)	22 (19–26)
Kyrgyzstan[5]	69–80	39–43	Male	5	130–243	182–217	
			Female	4	90–120	149–160	
Irkutsk oblast[6]	108	53	Male	12	118 (103–139)	173 (163–182)	24 (22–28)
			Female	21	85 (64–104)	160 (150–169)	21 (18–24)
S Sikhote-Alin Mountains[7]	134–136	43–45	Male	17	120 (69–238)	171 (137–217)	25 (20–31)
			Female	16	86 (53–131)	149 (135–173)	22 (21–24)

[a]The superscript numbers denote the following references: 1, Kozlo (1975); 2, Dyakova (1989); 3, Durov and Alexandrov (1968); 4, Sludsky (1956a); Fedosenko et al. (1984); 5, Yanushevich et al. (1972); 6, Nedzelsky (1994); 7, Bromley and Kucherenko (1983).
[b]Without stomach, intestines and blood.

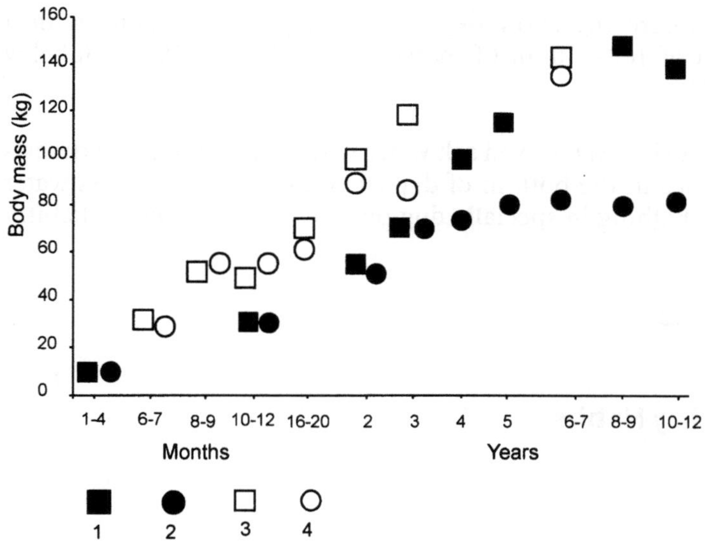

Fig. 2.3. *Sus scrofa.* Body mass versus age in Belowezhskaya Forest Nature Reserve (*1, 2*) (Kozlo 1975) and Sikhote-Alin Mountains (*3, 4*) (Bromley 1964). *1, 3* male mass; *2, 4* female mass

Kucherenko 1983; Sopina 1983; Chegorka 1984; Tupitsyna 1988; Muntyanu et al. 1992; Fadeev 1988, 1993; Nedzelsky 1994; Prostakov 1996).

Soil Texture. Capacity to dig in stony soil; frozen soil is more difficult.

Altitude. 50–2400 m a.s.l. (Kazakhstan, Kyrgyzstan, Tajikistan, Uzbekistan); 1,000–2,000 m a.s.l. (Altai); 200–1,000 m a.s.l. (E Siberia); 200–2,000 m a.s.l. (Sikhote-Alin) (Bromley 1964; Sapozhnikov 1975; Grachev and Smirnova 1980; Bromley and Kucherenko 1983; Sopina 1983; Fedosenko et al. 1984; Sobansky 1992; Sokov 1993).

Temperature. Sensitive to low temperature. Temperature lower than −15 °C is limiting for newborn animals, although they can spend several days in the nest. Mutual warming in family groups is important (Sludsky 1956a; Timofeeva 1975; Fedosenko et al. 1984).

Snow. Snow depth >30–40 cm limiting according to Nasimovich (1955), but Bromley (1959) suggested that adult boars only suffer in snow >60 cm. Weight per square centimeter of hoof is 184 g/cm^2 for piglets, 450 g/cm^2 for yearlings and 446 g/cm^2 for adults. Because of growth of hooves over winter, weight per square centimeter decreases from autumn to spring by 1.5 times (Bromley 1964). However, contrary to earlier predictions (Nasimovich 1955; Bannikov 1989), the species has, during recent decades, occupied areas where the snow depth on average is 60–70 cm, with a maximum depth of 100–110 cm (Timofeeva 1975; Fadeev 1980a; Sokolov and Baskin 1984).

Water. Regularly drink water, even salt water with up to 8 g salt/l. In deserts they dig deep holes in the bottom of dry lakes and rivers to access water (Sludsky 1956a). Bathing in specially dug puddles is a well-known habit.

2.3.1 Biotope (Table 2.2)

2.4 Diet and Feeding Habits

2.4.1 Diet

Omnivorous, but in some areas they feed only on a few species of plants (e.g. roots of *Phragmites* and *Typha*). Soil invertebrates, acorns, ground-living fungi, bird's eggs, rodents, fish, newborn mammals and carrion form the main diet in many areas (Lavrovsky 1952; Sludsky 1956a; Kozlo 1975; Rusakov 1979; Bromley and Kucherenko 1983; Fedosenko et al. 1984). In Sikhote-Alin, *Equisetum* is often an important food. During the most severe winters in N Russia, Transbaikal, W Siberia, they survive by feeding on bark, twigs and dry grass. Resistant to some poisonous plants: *Aconitum, Anemone, Calla, Caltha, Ferula,* and *Pteridium.* For a list of plant genera eaten see Appendix 1, Table 2.

Degree of Herbivory. When feeding in soil it eats about 60% invertebrates, selecting mostly large ones (Rusakov 1979). In general, animal and/or invertebrate food contributes to a small part of the diet (1.5–8%), but when abundant, animal food can make up to 100% of the diet (e.g. locusts, salmon, fish under ice in shallow lakes, and mollusks on shorelines).

2.4.2 Life Forms/Plant Parts Eaten (Table 2.3)

Consumption of Food/Day. In captivity, yearlings: 2.2 kg fresh food, adult male: 4.3 kg, adult female: 3.0 kg (Donaurov and Teplov 1938).

Feeding niche. In field and shrub layers, also below ground.

Table 2.2. *Sus scrofa*. Habitat[a]

Habitat type	Heading A[b] 1 Darwinsky[1], 2 NW Russia[2], 3 Lithuania[3]			Heading B[b] 1 Belarus[4], 2 Bryansk, Moscow[5], 3 Tatarstan[6], 4 Saratov[7]				Heading C[b] 1 Kazakhstan (plains)[8], 2 Carpathians, Caucasus[9], 3 Central Asia (mountains)[10]			Heading D[b] 1 Altai[11], 2 Pri-Baikal[12], 3 Sikhote-Alin[13]		
	1	2	3	1	2	3	4	1	2	3	1	2	3
(Ecoregions)	134	134, 212	212	212	212	252	252	342, 343	M252	M341	M251	M134	M212
Mature forest													
Coniferous	++	++	+++	++	x	r	+	x	++	x	x	x[e]	
Mixed deciduous-coniferous	r	+	r	+++	x	+++	++		+++	++[d]	x	x[e]	x
Broadleaf		+		+			++	x	+++	++[d]	x	x[e]	x
Young forest													
Coniferous				++		+							
Mixed deciduous-coniferous				r									
Deciduous						+							
Forest (all ages)							+						
Coniferous					x					x			
Broadleaf					x		++			++	x		
Savanna, *Pistacia*										x			
Fresh water shore		+											
Tall grass	+++	+		r	x		+	x	+++	+++			
Shrub		r		r	x	++		x		+++	x		
Broadleaf forest		r	r	++[c]	x					x			
Flood plain						++							
Broadleaf forest							+++						x

Table 2.2. *Continued*

Habitat type	Heading A[b] 1 Darwinsky[1], 2 NW Russia[2], 3 Lithuania[3]			Heading B[b] 1 Belarus[4], 2 Bryansk, Moscow[5], 3 Tatarstan[6], 4 Saratov[7]				Heading C[b] 1 Kazakhstan (plains)[8], 2 Carpathians, Caucasus[9], 3 Central Asia (mountains)[10]			Heading D[b] 1 Altai[11], 2 Pri-Baikal[12], 3 Sikhote-Alin[13]		
	1	2	3	1	2	3	4	1	2	3	1	2	3
Bog	r	+	++										
Burnt forest					x								
Cut forest				r	x								x
Shrubland, *Juniperus*		r		+				x		+++			
Grassland				r									
Meadows near forest		+	++										
Mountain steppe											x		
Subalpine meadows									+				
Desert								x					
Arable lands	+	+	++	+	x				x				

[a]Habitat used in >30% (+++); 20–30% (++): 10–20% (+); <10% (r); x indicates habitat use noted, no numerical data reported.
[b]The superscript numbers denote the following references: 1, Tupitsyna (1988); 2, Rusakov (1979); Rusakov and Timofeeva (1984); 3, Balĉhauskas (1988); 4, Sablina (1955); Kozlo (1975); Gatikh (1975); 5, Fadeev (1975a); Ivanova (1975); Vatolin (1980); 6, Gorshkov (1988, 1989); 7, Fadeev (1993); 8, Sludsky (1956a); Ishunin (1961); Yanushevich et al. (1972); Fedosenko et al. (1984); 9, Donaurov and Teplov (1938); Nasimovich (1939; 1955); Chegorka (1984); Alekperov and Kuliev (1987); Muntyanu et al. (1992); 10, Sludsky (1956a); Yanushevich et al. (1972); Sokov (1993); 11, Sobansky (1992); 12, Nedzelsky (1994); Sopina (1983); 13, Abramov (1963a); Bromley (1964); Antonov (1999).
[c]Alder.
[d]In subalpine zone.
[e]With *Rhododendron* shrubs.

Table 2.3. *Sus scrofa.* Life forms/plant parts eaten[a]

Life forms	Plant parts eaten							
	Whole plant	Leaves	Stems/twigs	Bark	Roots/bulbs	Flowers	Fruits	Seeds
Trees		+		+	+		+++	+++
Shrubs		+			+		+++	++
Herbs		++	+++		+++	+		
Grasses/sedges		++			++			
Horsetails	+++							
Mosses	r							
Lichens	r							
Fungi	+							

[a]Food type used: heavily and is one of main foods (+++); often (++); occasionally (+); rarely (r).

2.5 Social Organization and Behavior

2.5.1 Social Life (Tables 2.4 and 2.5)

The family consists of female + piglets. The prolongation of social connections between a mother and her daughters leads to formation of a family group that exists for about 3 years. Females of the same group can feed their piglets jointly (Fedosenko and Zhiryakov 1984; Durov 1986; Tsarev 1989; Prostakov 1996). Male-yearlings abandon their families and live alone, or in small groups. Adult males live alone.

2.5.2 Home Range

Within the home range there are trails, lying places, bathing puddles and scratching places. In winter, the home range of a female group in NW Russia was 5–15 ha (Tsarev 1980); in the Oksky Nature Reserve the home range in early winter was 44.6 ha, in late winter 25.7 ha (Chervonnyi 1980, 1990). During the hunting season in C Russia, female groups range over 15,000–20,000 ha (Voronin 1980); they return to the same pasture after 1–2 weeks. Distance covered daily: 10–20 km (females) and 4.2 km (males). In Tajikistan, the family home range was 3–15 km^2 and boars in some areas traveled 16–18 km daily to visit feeding areas outside the home range (Sokov 1993).

2.5.3 Mode of Movement

Walk, run, jump, swim. Maximum speed 40 km/h; height of jump 140–150 cm. Can dig with the nose; capable of digging 8–10 cm into frozen ground and can turn stones weighing 40–50 kg (Bromley and Kucherenko 1983). Because of sharp edged hooves, they are good runners on smooth ice (Sludsky 1956a).

2.5.4 Dispersal/Migration

In flat areas, boars often search for habitats with good food and snow conditions (Rusakov 1979). In mountainous areas, such as the Caucasus and M Asia, seasonal vertical migrations (in autumn - down, in spring - up) to escape deep snow and to find food rich areas are usual and can involve migrations of up to 20–50 km (Nasimovich 1939; Sludsky 1956a; Durov

Table 2.4. *Sus scrofa.* Group size

Locality[a]	Belovezhskaya Forest[1]	Carpathians[2]	NW Russia[3]	Voronezh[4]	Caucasus[5]	Kazakhstan (plains)[6]	Pri-Baikal[7]	Altai[8]
Group size	3.3–5.1	4–5, max. 27	3.8–6.7	3–6	6 (3–10)	6 (3–8)	5 (4–6)	5–8 (max. 14)

[a]The superscript numbers denote the following references: 1, Kozlo (1975); 2, Chegorka (1984); 3, Timofeeva (1975); 4, Prostakov (1996); 5, Donaurov and Teplov (1938); 6, Annenkov (1989); 7, Nedzelsky (1994); 8, Sobansky (1992).

Table 2.5. *Sus scrofa.* Gender and age composition of differently sized groups

Region[a]	Number of animals/group												
	1	2	3	4	5	6	7	8	9	10	11–15	16–20	21–30
Lipetsk oblast[1] (% of all observed groups)	31	12	12	12	6	6	4	6	2	3	6		
NW Russia[2] (% of all observed groups)	26.5	7.3	7.4	6.6	7.7	9	7.8	6	4.3	4.9	8	2.9	1.6
Lipetsk oblast[1] Group composition (%)													
Males, adult	39	0	34	20	7	0	0	0	0	0	0		
Yearlings and piglets	19	34	14	25	8	0	0	0	0	0	0		
Females + juveniles	0	5	8	6	14	6	13		14	17	17		
Males + females + juveniles	0	0	0	0	0	42	0	0	36	22	0		

[a]The superscript numbers denote the following references: 1, Antonets (1993); 2, Rusakov and Timofeeva (1984).

1973; Sokov 1993). Over a 24-h period, they are able to walk 30–40 km and over 10–15 days they can cover 200–300 km (Bromley and Kucherenko 1983; Sopina 1983; Nedzelsky 1994). Migrations occur near the Caspian Sea to escape floods and in Kazakhstan (up to 250–300 km) to escape deep snow (Lavrovsky 1952; Sludsky 1956a; Kotov and Ryabov 1963).

2.5.5 Diurnal Activity (Table 2.6)

2.5.6 Shelters

Resting places are often situated near streams, in swamp forest, in tall grass or shrub thickets. Shelters are used under spruce, with bedding. When disturbed by insects, boars cover themselves with soil and pine needles. Shelters are occupied by whole family groups, with adult males lying separately (Sablina 1955; Bromley 1964). Boars never defecate in a shelter. In NW Russia, shelters contain insulating warming material such as spruce branches and dry hay.

2.5.7 Nests

The pregnant sow prepares a special nest for parturition with a roof constructed from branches and bedding material of shrubs and dry grass: length 0.9–2.5 m; height up to 1.2 m. The nest protects piglets from snow, rain, sunlight, and crows. In more northern areas, nests are made from dry grass (Sludsky 1956a; Bromley 1964; Rusakov 1979; Prostakov 1996).

2.5.8 Escape Behavior

Good smelling and hearing, but poor vision. Cannot recognize a standing human at 10–15 m. In groups, and after starvation, they are less alert (Bromley 1964). When disturbed by a tiger, the group scatters in all directions, often successfully. Large males sometimes protect themselves and their group from predators and humans. Sows will protect their litter when piglets are in the nest.

2.6 Reproduction (Tables 2.7 and 2.8)

In rutting season very fierce fights can occur between males. A male will follow and mate with 5–10 females.

Table 2.6. *Sus scrofa*. Diurnal activity (shaded parts denote feeding or walking)

Locality	Season	Time of day						
		Midnight	Late night	Sunrise	Day	Sunset	Twilight	Early night
Plain Kazakhstan[1]								
Females with piglets	Spring							
Males, Yearlings								
	Summer							
	Winter							
Mountain Kazakhstan[2]								
	Spring							
	Summer							
	Winter							
Tadijkistan[3]	Summer							
	Winter							

[a]The superscript numbers denote the following references: 1, Sludsky (1956a); Fedosenko et al. (1984); 2, Fedosenko and Zhiryakov (1984); Grachev (1988); 3, Sokov (1993).

Table 2.7. *Sus scrofa*. Breeding parameters

Locality[a]	Age at sexual maturity (months)		Participation in first breeding (years)		Pregnancy (days)	Embryos/female	Litter size	Lactation period (months)
	Male	Female	Male	Female				
Belovezhskaya Forest[1]	19–20			1.5–2.5		5.7	5.5 (5.2–5.8)	
Berezinsky Nature Reserve[2]						6.1		
NW Russia[3]					135	6.6	4.6	3
Voronezh oblast[4]	18		4.5	1.5	128 (128–134)			
Astrakhansky Nature Reserve[5]							6.0 (4–10)	
Caucasus[6]							7.6 (6–10)	3
Kazakhstan[7]	18–20		5–7		140		5.9 (3–9)	
Tajikistan[7]	18	18	4–6			4.3		
Kyrgyzstan[8]					125–130	4–8		
Pri-Baikal[9]	31–32	19–20			120–130		5.4–7.3	
S Sikhote-Alin[10]			4–5			6.5	6.4 (2–9)	3–3.5

[a]The superscript numbers denote the following references: 1, 2, Kozlo (1975); 3, Rusakov (1979); Rusakov and Timofeeva (1984); 4, Prostakov (1996); 5, Lavrovsky (1952); 6, Dinnik (1910); Durov (1987); 7, Sludsky (1956a); Annenkov (1989); 8, Yanushevich et al. (1972); 9, Nedzelsky (1994); 10, Bromley (1964).

Table 2.8. *Sus scrofa.* Fertility of females (blank cells indicate that breeding females were not found, or that data were absent)

Age (years)	Breeding females (%)					Number of embryos				
	2	3	4–5	6–7	8–10<	2	3	4–5	6–7	≥8–10
Region[a] NW Russia[1]	40	55	64–68			5.5	6.0	(6.8–7.0)	7.7	6.5
Belovezhskaya Forest[2]	31	55	69	78	67	4.2	5.5	6.2	6.6	6.3
Moscow[3]			69 (50–90)			2.3 (1–4)	4.5 (3–6)	7.3 (4–12)	0	0
Kyrgyzstan[4]				50–55						
Pri-Baikal[5]				70–75						
Sikhote-Alin[6]						4–5		7–9		

[a]The superscript numbers denote the following references: 1, Rusakov (1979); 2, Kozlo (1975); 3, Ivanova (1975, 1985); 4, Yanushevich et al. (1972); 5, Nedzelsky (1994); 6, Bromley and Kucherenko (1983).

2.6.1
Seasonal Breeding Events (Table 2.9)

Males show rutting activity 1 month before actual breeding starts.

Maximum Length of Life. More than 12 years (Kozlo 1975); in Tajikistan <9–10 years (Sokov 1993). In captivity, 20 years (Rusakov 1979).

2.7
Mortality (Table 2.15)

High mortality of piglets is usual; in areas with mild climate 8–20% of piglets die during their first year of life (Moscow, Belovezhskaya Forest); in Caucasus, Carpathians, Kazakhstan and Sikhote-Alin, first year mortality reaches 50–60% (Sludsky 1956a; Ivanova 1975; Kozlo 1975; Rusakov 1979; Bromley and Kucherenko 1983; Chegorka 1984; Durov 1987; Annenkov 1989; Glushkov et al. 1989). Ognev (1951) suggested that there are two periods of high mortality for piglets: in the first 3 months of life when piglets are very fragile and during the seventh month of life when sows reduce their care of the litters. Animals lose 30–50% of their weight during winter (Fadeev 1980a). Severe winters with deep snow and strong frost severely reduced boar numbers in Kazakhstan, N European Russia and the Far East. In W Siberia, boars were extinct after the very severe winter of 1913/1914. In Sikhote-Alin high mortality of boars was observed every 18 years during the period 1928–1978 (Bromley and Kucherenko 1983). Boars drown when crossing rivers on thin ice. Floods in the delta of the Volga River can kill up to 25% of the population (Donaurov and Teplov 1938; Kiselev 1975). In Belovezhskaya Forest (Poland), wild boar have suffered mostly from disease/starvation (61%) and wolf predation (16%). A heavy acorn crop led to an increase in the numbers of wild boar, many of which died the following year (Okarma et al. 1995). In the Tambov oblast, 30% of animals found dead were wounded by hunters, 8% were killed by cars and trains, 5% drowned and 8% were killed by wolves (Prostakov 1996). In European Russia from a total of 1389 animals that were found dead, 24.6% died from starvation, 21.4% were killed by predators, 27.9% were killed by poachers, 7% were drowned, 9.8% were killed during traffic accidents, 5.7% were poisoned by fertilizers and 3.6% died from various diseases (official data, Ivanova 1985).

2.7.1
Predators

The list of predators includes: wolf, tiger, snow leopard, leopard, jungle cat, wild cat, lynx, brown bear, wolverine, yellow-throated marten (Sludsky

Table 2.9. *Sus scrofa.* Seasonal breeding events[a]

Locality[b]	January		February		March		April		May		June		July	Aug	Sept	October		November		December	
Belovezhskaya Forest[1]						N	N											R	R	R	
NW Russia[2]	R						N	N	N	N	N						R	R	R	R	R
Voronezh[3]					N	N	N												R	R	R
Caucasus[4]	R						N	N	N									R	R	R	R
Kazakhstan[5]	R					N	N	N									R	R	R	R	R
Kyrgyzstan[6]	R					N	N	N											R	R	R
Uzbekistan[7]			N	N	N	N												R	R	R	R
Pri-Baikal[8]							N	N	N										R	R	R
S Sikhote-Alin[9]						N	N	N											R	R	R

[a]R, Rutting season; N, newborns.
[b]The superscript numbers denote the following references: 1, Kozlo (1975); 2, Rusakov (1979); Timofeeva (1975); 3, Prostakov (1996); 4, Donaurov and Teplov (1938); 5, Sludsky (1956a); Fedosenko and Zhiryakov (1984); Annenkov (1989); 6, Yanuschevich et al. (1972); 7, Ishunin (1961); 8, Nedzelsky (1994); 9, Bromley and Kucherenko (1983).

Table 2.10. *Sus scrofa*. Gender ratio in populations (males : females)

Region[a]	NW Russia[1]	1 Belovezhskaya Forest[2], 2 Kaluga[2], 3 Bryansk[3], 4 Oksky[4]				1 Astrakhansky[5], 2 Kazakhstan[6]		1 Carpathians[7], 2 Caucasus[8]		Kyrgyzstan[9]
(Ecoregion)	134, 212	212				343		M252		M341
		1	2	3	4	1	2	1	2	
Embryos	1:2					1:1.9	1:1.7		1:2.2	
Adults	1.44:1	1:1.6–1:1.3	1.9:1–2:1	1.9:1	1.4:1–1:1.2	1.1:1	1:1	1:1.8	1:1.1	1:1.23

[a]The superscript numbers denote the following references: 1, Rusakov (1979); 2, Severtsov and Sablina (1953); 3, Voronin and Zaikin (1989); 4, Chervonnyi (1980); 5, Lavrovsky (1952); 6, Sludsky (1956a); 7, Chegorka (1984); 8, Durov (1980); 9, Yanushevich et al. (1972).

1956a; Ivanova 1975; Kozlo 1975; Kudaktin 1981, 1998; Bromley and Kucherenko 1983; Fedosenko 1986; Durov 1987; Tupitsyna 1988; Annenkov 1989; Nedzelsky 1994). Predators have the greatest impact on the mortality of piglets and yearlings. The wolf is an important predator in all areas except in the Far East, where the tiger is the main predator. In Transcaucasia, boar comprises 67% of the wolf diet, while in the Voronezh oblast, 13% (Ryabov et al. 1994). In the Berezinsky Nature Reserve, wolves kill 60–70% of piglets and in the Caucasus they take 16% of all boars, while bears take 7% (Kudaktin 1981, 1998). At Sikhote-Alin one tiger can kill 30–34 boars per year; one adult brown bear can kill 3–8 boars per month (Kucherenko 1977). In 2.5% of lynx stomachs remnants of boars were found (Kozlo 1975). In the Caucasus, 10–37% of wolf scats included boar remnants (Kudaktin 1981; Durov 1987), in Kopet-dag, 27% of leopard scats included boar remnants (Lukarevsky 1995). However, adult males can defend themselves against wolves and even bears (Sludsky 1956a; Bromley and Kucherenko 1983).

Table 2.11. *Sus scrofa.* Gender and age composition of populations (%)

Localities[a]	Adult		Yearlings	Piglets
	Males	Females		
European Russia[1]	27–47		12–27	34–56
NW Russia[2]	33		16	51
Belovezhskaya Forest[3]	20	12	9	59
Darwinsky[4]	55–58		4–8	
Kaluga[5]	82		13	5
Kazakhstan[6]	18	19	14	49
Carpathians[7]	31		24	45
Caucasus[8]	24	28	15	33
Altai[9]	16	15	19	50

[a]The superscript numbers denote the following references: 1, Ivanova (1985); 2, Rusakov (1979); 3, Severtsov and Sablina (1953); 4, Zavyalova (1997); 5, Voronin and Zaikin (1989); 6, Sludsky (1956a); 7, Chegorka (1984); 8, Durov (1980); 9, Sobansky (1992).

Table 2.12. Age structure of a boar population in NW Russia in 1971–1975. (Rusakov 1979)

Age (years)	Piglets	1	2	3	4	5	6	≥7
Proportion in population (%)	5	15	21	22	18	12	4	3

Table 2.13. *Sus scrofa*. Population density and size in typical habitats in different areas

Locality[a]	Year	Population density (animals/10 km^2)	Total number of animals (×1,000)
Darwinsky[1]	1987	2 (1–3)	
	1992	5.3	0.3
	1993	4.4	0.18
Belovezhskaya Forest[2]	1890–1945	2–19	
Polesye[3]		8, max. 17	
S C Russia[b,4,5]	1965	0.3	0.3
	1970	1.7	2.0
	1975	5.8	6.4
	1980	7.0	7.3
	1985	6.6	8.5
	1990	6.6	8.8
	1995	4.4	6.2
Volyn[6]		17 (16–18)	
Tatarstan[7]	1988	5–46	4.7
Volgograd[8]	1995	10	1.9
Saratov[9,14]	1991	6–8	5
Tajikistan[10]	1887–1937		Increase
	1938–1958	50	Rapid increase
	1959–1968	9	Stabilization
	1973–1987	3.2 (1.1–10)	Stable–decline
	1987		7.0
S Siberian mountains[11]	1992	1–4	
Pri-Baikal[12]	1992	(0.1–0.3), max. 0.9	
Sikhote-Alin[13]	1980, 1993	1–3	

[a]The superscript numbers denote the following references: 1, Tupitsyna (1988); Zavyalova (1997); 2, Kozlo (1975); 3, Serzhanin et al. (1975); 4, Vatolin (1980); 5, Ivanova (1980); Prostakov (1996); 6, Kolisnik (1990); 7, Gorshkov (1988); 8, Kiselev (1975); 9, Fadeev (1993); 10, Sokov (1993); 11, Smirnov (1994); 12, Nedzelsky (1994); 13, Bromley and Kucherenko (1983); Zaumyslova (1998); 14, Ovsyukova (1996).
[b]Voronezh+Lipetsk+Belgorod+Kursk+Tambov oblast'.

2.7.2
Diseases, Helminths and Ectoparasites (Appendix 2, Table 2)

Anthrax, rabies, *Picorna virus apthae*, *Leptospira*, pig plague (Belovezhskaya Forest, Caucasus, eastern Siberia, and Far East) with mortality up to 90%, *Erysipelotrix insidiosa* (Litvinov 1980, 1981; Gorelov 1994).

Table 2.14. *Sus scrofa.* Total number of animals (official data)

Locality[a]	Population size (×1,000)		
Year	1990	1995	1999
Belarus[1]		32–35	
Kazakhstan[2]	10.0	10	10
Plain	6		
Mountain	4		
Russian Federation[3]	275.2	160.7	173.6
N European Russia	13.6	4.9	3.2
NW European Russia	24.3	13	8.8
C European Russia	67.3	25.3	37.3
Volgo-Vyatka region	8.8	4	5.0
S European Russia	11	7.1	9.6
Volga region	16.8	12.2	18.1
Ural region	20.4	15.8	13.5
N Caucasus	25.7	15.3	18.6
W Siberia	4.2	3.7	3.6
E Siberia	45.4	18.7	21.5
Far East	46.8	38	31.8

[a]The superscript numbers denote the following references: 1, Kozlo (1995); 2, Bekenov et al. (1992); Bekenov (1999); 3, Ovsyukova (1996); Lomanov (2000).

Table 2.15. *Sus scrofa.* Mortality, Russian Federation, 1986–1988, official data. (Glushkov et al. 1989)

Locality	Killed by hunters (spring number, %)	Poached[a] (animals/10 km²)	Killed by predators and other means
Vologda	14	0.1	0.2
NW Russia	11	0.2	0.6
Kirov, Kostroma	27	0.3	0.2
Moscow, Smolensk	22	0.6	0.6
Tambov, Voronezh	25	0.2	0
Saratov, Volgograd	44	0.2	0.2
Ural	17	0.1	0.3
W Siberia	5	0.1	0
E Siberia	2.5	1	0.2
Far East	2.2	0.1	0.6

[a]Found poached.

2.8
Population Structure and Densities (Tables 2.10, 2.11 and 2.12)

2.8.1
Population Density and Size (Tables 2.13 and 2.14)

A large capacity to increase in numbers and high mortality after both severe winters and diseases, are responsible for the large fluctuations observed in population sizes between years. In the Lazovsky Nature Reserve the population fluctuated 27-fold, in Belovezhskaya Forest between 1990–1966, 6-fold (Kozlo 1975). After occupation of a new area population growth occurs rapidly, then becomes stable. In 1969, boar invaded Karelia and in 1990 the population had reached 2800 animals. In forest-steppe areas of C Russia, where there are excellent conditions for wild boar, the fluctuations in population size were within 10–26% (Prostakov 1996; Pobedinsky 1997).

2.8.2
Impact on Vegetation

In oak forests, boars dig up 10–30% of the surface area (Timofeeva 1978), in subalpine meadows up to 51% (Fedosenko and Zhiryakov 1984). This extensive digging alters the distribution of plant species and soil chemistry (Kozlo 1975; Dvornikov and Dvornikova 1989).

Table 2.16. *Sus scrofa*. Harvest in C Russia, 1988–1997. (Ovsyukova 1996; Lomanov 2000)

Locality	1988		1998	
	Total number (×1,000)	Legal harvest (×1,000)	Total number (×1,000)	Legal harvest (×1,000)
Bryansk oblast	3.5	0.44	1.8	0.15
Vladimir oblast	2.0	0.98	1.6	0.17
Ivanovo oblast	3.8	1.42	1.2	0.12
Kaluga oblast	3.7	1.22	3.0	0.41
Kostroma oblast	2.5	0.52	1.8	0.05
Moscow	7.2	3.33	4.6	0.63
Oryol oblast	1.5	0.4	1.2	0.05
Ryazan oblast	2.1	0.51	1.6	0.07
Smolensk oblast	8.7	1.7	4.7	0.5
Tver oblast	8.2	1.29	5.9	0.37
Tula oblast	1.9	0.88	2.2	0.48
Yaroslavl oblast	3.5	0.26	3.2	0.78

2.9
Relations to Man (Tables 2.15 and 2.16)

Important hunting species, with both meat and tusks of males taken as trophies (Danilkin 2002).

Have an impact on agriculture by damaging crops (Varnakov and Mosheva 1975). Damage to melons, potatoes, oats, barley and wheat are significant (Sokov 1993). They harm wildlife by taking bird's eggs, and young animals (Shiryaev 1975; Rusakov 1979). Before 1917, about 50,000 boars were legally killed per year, mostly in the Caucasus, on the shores of the Caspian Sea in the Far East. In the 1950s, 60,000–70,000 per year and 39,000 in 1973 (Rusakov 1979).

3 Musk Deer – *Moschus moschiferus*

Moschus moschiferus Linnaeus, 1758.
Artiodactyla, Moschidae

3.1 Status and Distribution (Fig. 3.1)

3.1.1 Taxonomic Remarks

Three subspecies occur: *M. m. moschiferus* Linnaeus, 1758 (Siberia except the Amur Basin and Sakhalin Island); *M. m. parvipes* Hol. 1911 (Far East); *M. m. sachalinensis* Flerov, 1929 (Sakhalin) (Heptner 1989).

3.1.2 Status

Vulnerable (VU – A1acd) (Hilton-Taylor 2000). CITES I. In Russia *M. m. sachalinensis* is endangered (Rozhnov 2001).

3.1.3 Distribution (Fig. 3.2)

Distribution Area. Ca. 7,647,000 km^2.

Occurrence in Ecoregions. Subarctic (133, 136), Subarctic Regime Mountains (M133, M134),), Warm Continental (212), Warm Continental Regime Mountains (M212).

3.2 Measurements (Table 3.1)

Body mass of newborns is 440–470 g (Bromley and Kucherenko 1983).

Fig. 3.1. Musk deer

3.3
Habitat (Tables 3.2 and 3.3)

Mountains, preferably coniferous forest, mostly dark, often with high abundance of tree lichens, rich moss cover on ground; with rocks that give shelter from predators.

Terrain Form. Terraces, stream valleys, foothills, valleys and slopes in mountains.

Altitude. In Kazakhstan, 600–1,000 m a.s.l. (winter); 400–1400 m a.s.l. (summer) (Baidavletov 1984).

Snow. Weight on hoof is 80–120 g cm^2 (Dulkeit, cited in Nasimovich 1955). Because of small body weight and low hoof pressure, snow that has been compressed by wind is favorable (Ustinov 1969a). However, snow depth more than 60–70 cm is limiting (Table 3.2). Can dig in snow up to 40 cm deep. Snow is often eaten(Bromley and Kucherenko 1983).

Water. Regularly visit watering places in summer.

3.3.1
Biotope (Table 3.3)

Visibility in the best habitats of musk deer: 20–100 m.

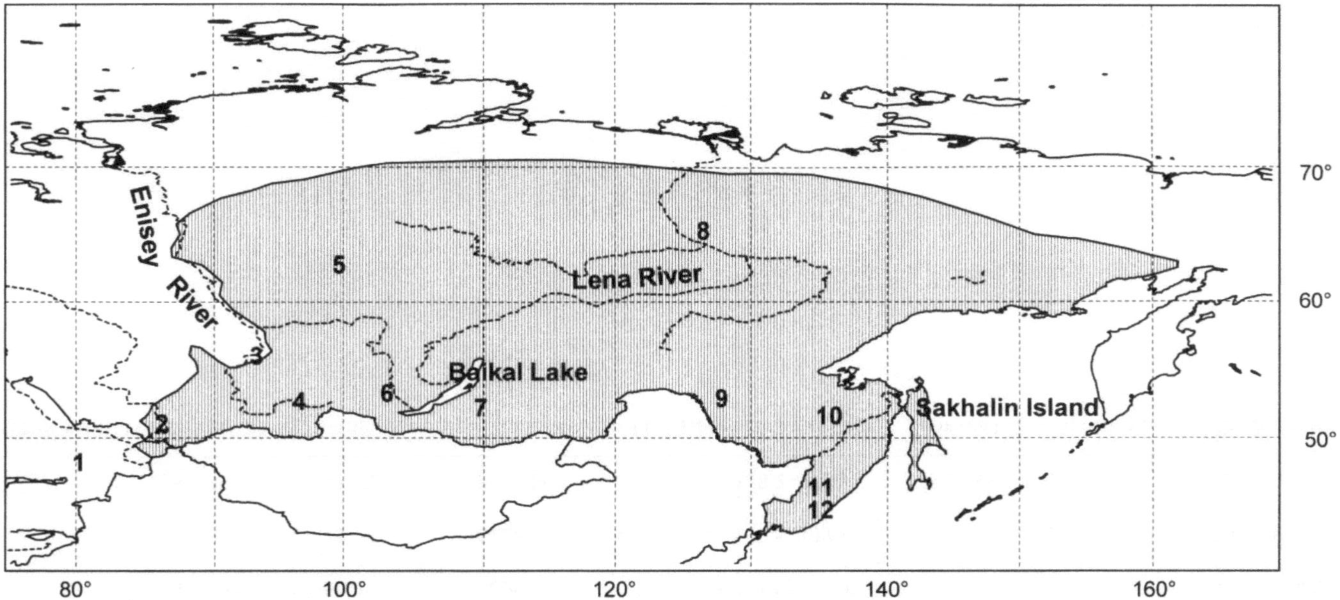

Fig. 3.2. *Moschus moschiferus.* Distribution map (Heptner 1989). *1* Kazakhstan; *2* Altai Mountains; *3* Stolby Nature Reserve; *4* E Sayan Mountains; *5* C Siberia; *6* Irkutsk oblast; *7* Transbaikal; *8* Yakutia; *9* Amur oblast; *10* Khabarovsk kray; *11* Primorsky kray; *12* Sikhote-Alin Range, Sikhote-Alinsky Nature Reserve

Table 3.1. *Moschus moschiferus.* Measurements

Locality[a]	Longitude (°E)	Latitude (°N)	Gender	*n*	Body mass (kg)	Head and body length (cm)	Tail length (cm)
Altai, Kazakhstan[1]	84–87	49–50	Male	3	14.9 (14–16)	96.0 (93–98)	4.7 (4.5–5.1)
Altai, Kazakhstan[1]	84–87	49–50	Female	8	12.6 (11–15.8)	92.1 (83–100)	4.5 (3.6–6.0)
Yakutia[2]	106–152	56–67	Male	6	11.1 (9.0–12.0)		3.9 (3.5–4.0)
Yakutia[2]	106–152	56–67	Female	7	13.7 (12.5–17.0)		3.6 (3.0–4.0)
Sikhote-Alin[3]	133–139	43–52	Male	8	13.4 (8.5–18.6)	83.1 (76.3–92.6)	4.1 (4.0–4.5)
Sikhote-Alin[3]	133–139	43–52	Female	9	12.9 (8.2–17.4)	88.8 (85.3–93.1)	4.5 (4.0–5.0)

[a]The superscript numbers denote the following references: 1, Baidavletov (1984); Sobansky (1992); 2, Egorov (1971); 3, Bromley and Kucherenko (1983).

Table 3.2. *Moschus moschiferus.* Sinking depth of animals in snow in Sayan Mountains. (Bannikov et al. 1978)

Snow depth (cm)	35	41	43	52	50
Sinking depth (cm)	26	27	23	21	0

Table 3.3. *Moschus moschiferus.* Habitat[a] (W winter, S summer)

Locality[b]		Altai[1]	Stolby Nature Reserve[2]	Yakutia[3]	Sikhote-Alin Mountains[4]
Dark coniferous (spruce, pine, fir) with abundant lichen cover	W	+++		x	
Light coniferous (larch) with abundant lichen cover	W	+			+
Dark coniferous in mountains and mountain valleys with abundant overgrowth, moss and lichen cover	W	+++	+++	x	+++
	S	+++	+++	x	+++
Coniferous-broadleaf forest with good undergrowth in mountains	W	+++			
	S	+++			+
Broadleaf forest	W			x	
	S			x	
Shrubs on slopes	WS			x	

[a]Habitat used in >30% (+++); 20–30% (++); 10–20% (+); <10% (r); x indicates habitat use noted, no numerical data reported.
[b]The superscript numbers denote the following references: 1, Baidavletov (1984); Sobansky (1992); 2, Zyryanov (1975); 3, Egorov (1971); 4, Bromley and Kucherenko (1983).

3.4 Diet and Feeding Habits

3.4.1 Diet (Table 3.4)

Lichens (especially tree lichens) comprise more than 80% of the diet in winter and significantly improve digestion of green plants (leaves, flowers) in summer (Ustinov 1969b; Kholodova and Prikhodko 1984; Prikhodko 1992). Other important foods are fir needles. In areas where fir is absent needles of pine and larch, berries of *Vaccinium*, *Equisetum*, dry cereals, and mushrooms are consumed. For list of plant species eaten see Table 3.4;

Table 3.4. *Moschus moschiferus.* The main food sources in winter and the quantity (%) of each type of food consumed at each feed. (Zaitcev 1981)

Plants	Plant parts eaten	Quantity consumed (%)
Abies	Needles, twigs	15
Larix	Needles	3
Acer	Leaves, twigs	3
Quercus	Leaves, twigs	3
Corylus	Leaves	1
Rhododendron	Leaves, twigs	2
Gramineae sp.	Leaves	1
Mushroom		2
Lichens		
Usnea		69
Cladonia		1

Appendix 1, Table 3. A small proportion of the available food is taken at any one time, thus the same feeding patch can be returned to many times. Food is taken from the snow surface (up to 80%), under snow (5–8%), from tree branches (12–45%) (Altai, Sikhote-Alin) (Baidavletov 1984; Zaitcev 1991).

3.4.2 Life Forms/Plant Parts Eaten (Table 3.5)

Consumption of Food/Day. In summer, 160–265 g dry weight and 208–1,000 g fresh weight(Kholodova 1989). In captivity, daily consumption is ca. 800 g. In the wild (Sayan region), food consumption is as follows: fir needles 3–17 g, mushrooms up to 20 g, herbs up to 18 g, tree lichens 62–78 g; total 135 g) (Bannikov et al. 1978; Prikhodko 1992).

Feeding Niche. In field and shrub layers, sometimes also below ground.

3.5 Social Organization and Behavior

3.5.1 Social Life

Musk deer are usually solitary. However, a community of 5–7 acquainted neighbors can exist where male territories overlap female (or females with

Table 3.5. *Moschus moschiferus*. Life forms/plant parts eaten[a]

Life forms	Plant parts eaten						
	Whole plant	Leaves	Stems/ twigs	Bark	Roots/ bulbs	Flowers	Fruits	Seeds
Trees		++[c]	+					
Shrubs		++	r			+		
Dwarf shrubs		+	r					
Herbs		r			r			
Grasses/sedges		r						
Horsetails	r							
Mosses	r							
Lichens[b]	+++							
Fungi	r							

[a]Plants used in >30% (+++); 20–30% (++): 10–20% (+); <10% (r); x indicates plant use noted, no numerical data reported.
[b]Tree lichens.
[c]Needles of fir, larch, and pine especially.

fawn) territories. Animals with highest status occupy areas near the center of the community area where the conditions are best. Borders between the individual territories are determined by habitat characteristics and new inhabitants use the same territories as their predecessors (Zaitcev 1991). Males mark their territory all year-round and not only in the rutting season. Marking can be achieved with a fluid from the nose, a secretion from the tail gland, urine, pellets and by scratching the ground with the hooves. Permanent latrines are used, usually situated on trails, the crest of a ridge, or on plots without grass. There are up to 10 pellet groups in each latrine at any one time and 7–10 latrines in a home range are used over a period of several months (Ustinov 1965; Zaitcev 1991).

3.5.2 Home Range

Daily home range is 2–10 ha (Bromley and Kucherenko 1983). It includes feeding places, watering places, and small patches (about 40–20 cm) on rocks where the animal can remain motionless and inaccessible to predators; other types of shelter include piles of wood and shrub thickets. During winter, animals can move on average 2,000 m (between 300–3,500 m). This route can include 5–23 feeding patches, 4–7 resting points, 3–8 defecation points, and 1–4 urination points. Year-round home range size: males, 870 ha in coniferous-deciduous forest in Altai, 40–50 ha in Sikhote-Alin; females, 210 ha in coniferous-deciduous forest in Altai, 200–300 ha

Table 3.6. *Moschus moschiferus*. Diurnal activity (shaded parts denote activity)

Locality	Season	Time of day							
		0–3	3–6	6–9	9–12	12–15	15–18	18–21	21–24
Sayan Mountains[1]	Winter								
	Early spring								
Enclosure[2]	Spring, summer								
	Autumn (male)								

[a]The superscript numbers denote the following references: 1, Bannikov et al. (1978); 2, Prikhodko (1992).

in Sikhote-Alin (Zyryanov 1975; Bannikov et al. 1978; Bromley and Kucherenko 1983; Baidavletov 1984; Zaitcev 1991; Sobansky 1992).

3.5.3
Mode of Movement

Very good at jumping (up to 5–6 m); landing on all four legs; capable of turning up to 90° during a jump. Never run, will only walk and jump. Can dig into snow. Can stretch to reach lichens at 140 cm. Climbs on sloping trees.

3.5.4
Dispersal/Migration

Settled way of life. Difference in altitude between winter and summer habitats of 2,000–3,000 m. During rut, males make 3–4 km trips in search of females (Baidavletov 1984). Bad snow conditions can force animals to move 8–35 km in Yakutia (Egorov 1971); 4–8 km in Sikhote-Alin (Bromley and Kucherenko 1983).

3.5.5
Diurnal Activity (Table 3.6)

Mostly active in twilight and at night (Table 3.6). During 24 h, 4–12 peaks of feeding activity were observed.

3.5.6
Shelters

Resting places are mostly situated where the animals have a good view: ridge tops and tops of ant hills. Hiding is also important: under sloping

Table 3.7. *Moschus moschiferus*. Breeding parameters

Locality	Age at sexual maturity (months)		Pregnancy (days)	Number of embryos/female (%)		
	Male	Female		1	2	3
E Sayan[1]	18	15–17		31	56	13
Yakutia[2]				50	50	0
Stolby Nature Reserve[3]				69	31	0
Sikhote-Alin Mountains[3]			185–195	30	70	Very rare

[a]The superscript numbers denote the following references: 1, Lobanov (1975); Sokolov and Prikhodko (1992); 2, Egorov (1965); 3, Salmin (1972); Bromley and Kucherenko (1983).

trees, and in piles of tree branches. In bad weather, musk deer select dry places under branched trees, rocks, or sloping trees.

3.5.7 Escape Behavior

Very vigilant; 55% of feeding time is spent listening (Ustinov 1971). A very characteristic behavior is flight to "otstoi" (an area of about 40–20 cm on rocks where the animal will stay motionless and inaccessible to predators). Often there is only one "otstoi" for all musk deer in an area inhabiting a community of acquainted neighbors. Disturbed musk deer "freeze", or escape. When escaping from predators, musk deer mix up trails, running twice along the same trail, then will jump to the side and hide. Flight distance in Kazakhstan is 30–70 m, in Sikhote-Alin about 55 m (Zaitcev 1983; Baidavletov 1984). When chased for 5 h by dogs, musk deer stayed within an area of 50 ha (Zyryanov 1975).

3.6 Reproduction (Table 3.7)

During the rut, animals gather in groups of 3–4. Battles between males occur, although they are not fierce. Females can be on heat for 1–2 days and will mate 5–7 times. Birth of calf occurs in a hidden place. Calves first suckle after 25–30 min; females will suckle their young for 3–4 months; near the end of suckling a calf is fed once every 1–5 days (Sokolov and Prikhodko 1992). Calves less than 3 months old do not follow their mother, but are hidden 100–150 m away (Bannikov et al. 1978).

Table 3.8. *Moschus moschiferus.* Seasonal breeding events

Locality[a]	Rutting season			Calving season	
	Preparation to rut	Basic rut	Last appearances of rut	First calves	Basic calving
Kazakhstan[1]		Late November–early December	Middle–end December		
Yakutia[2]	Late November		First half of January	10 June	
Sea of Okhotsk shore[3]		December			June
Sikhote-Alin[3]		November–December			Late April–middle of May

[a]The superscript numbers denote the following references: 1, Baidavletov (1984); 2, Egorov (1971); 3, Bromley and Kucherenko (1983).

3.6.1 Seasonal Breeding Events (Table 3.8)

Maximum Length of Life. 10–12 years.

3.7 Mortality

In 1919, mass mortality from an unknown disease was recorded (Bannikov et al. 1978).

3.7.1 Predators

In Sikhote-Alin, 8–12% of the musk deer population was killed by yellow-throated marten (Matyushkin 1974), which hunt in packs of 1-4 individuals. Success of the marten's hunting reaches 90% (Salmin 1972; Zaitsev 1991). Musk deer remnants are found in 2–10% of wolf excrement, 2–50% of wolverine excrement, about 10% of red fox excrement and 18–43% of lynx excrements. Musk deer are rarely preyed upon by tigers or brown bears (Egorov 1965; Zyryanov 1975; Bromley and Kucherenko 1983).

3.7.2 Diseases, Helminths and Ectoparasites (Appendix 2, Table 3)

Table 3.9. *Moschus moschiferus*. Gender and age composition of the E Sayan Mountain population (Lobanov 1975)

Adult male : female ratio	1.08 : 1–1.33 : 1				
Age (years)	Newborn	1	2	3	⩾4
Age composition of group (%)	19	14	32	16	19

Table 3.10. *Moschus moschiferus*. Population density, in typical habitats, and population size

Species[a]	Locality[a]	Year	Population density/ (number/10 km²)	Population size (×1,000)
M. m. moschiferus[1,a]	Total	1997		47–52
	Kazakhstan (Altai)[2]	1977–1981	2–7	
	Kazakhstan (Altai)[2]	1992		(1.0–1.2)
	Altaisky kray[3]	1997–1999		1.5
	E Sayan[4]	1966–1974	(25–40)–(65–70)	
	E Siberia[4]	1978		27–30
	Buryatia[3]	1997		18.9
	Buryatia[3]	1999		12.6
	Irkutsk oblast[3]	1997		15.2
	Irkutsk oblast[3]	1999		20.5
	Krasnoyarsk kray[3]	1997		10.5
	Krasnoyarsk kray[3]	1999		12
	Khakassiya[3]	1999		2.6
	Tuva[3]	1999		10.5
	Chita oblast[3]	1997–1999		22.9
	Yakutia[5]	1965	2–3, max. 19	50
	Yakutia[3]	1997–1999		13
M. m. parvipes[1,b]	Total	1997		4–5
	Amur oblast, larch forest[6]	1975	2–8	
	Amur oblast[6]	1980		8–10
	Amur oblast[3]	1999		13
	Khabarovsk kray[6]	1980		30
	Khabarovsk kray[3]	1997		20.7
	Khabarovsk kray[3]	1999		24
	Primorsky kray[6]	1980		14–16
	Primorsky kray[3]	1997		17.7
	Primorsky kray[3]	1999		21
	Sikhote-Alinsky Nature Reserve[6]	1975	(15–20)–(30–40)	
M. m. sachalinensis	Total	1997[1]		0.3–0.4
		1997[3]		1.1
		1999[3]		1.6
All subspecies	Former USSR territory[4]	1978		100
	Russian Federation[3]	1997		146
	Russian Federation[3]	1999		156

[a]The superscript numbers denote the following references: 1, Prikhodko (1997); 2, Baidavletov (1984); 3, Mosheva (official data) (2000); 4, Bannikov et al. (1978); 5, Egorov (1965); 6, Kucherenko (1977); Bromley and Kucherenko (1983).
[b]Prikhodko's estimation is extremely conservative.

Table 3.11. *Moschus moschiferus.* Population density in different habitats of Transbaikal. (Ustinov 1965)

Habitat	Number of "otstoi"[a] (per 10 km of the observers route)	Population density (animals/10 km^2)
Dark coniferous forest on slopes with good undergrowth, abundant windfall, plenty of fallen/rotten trees and branches, plenty of tree lichens; visibility 20–100 m; snow depth 55–60 cm	7	44
Dark coniferous forest in mountains with good overgrowth, abundant of wooden trash, tree lichen resources are large; visibility 50–80 m; snow depth 50–70 cm	1	15
Light coniferous forest in smoothed mountains with good overgrowth, tree lichens resources are small; visibility 100–150 m; snow depth 40–50 cm	0	4

[a]"Otstoi" – small area on a rock, where musk deer remain motionless and are inaccessible to predators.

Table 3.12. *Moschus moschiferus.* Harvest of animals and glands. (Silantev 1898; Bannikov et al. 1978; Mosheva 2000)

Locality	Year	Harvest (×1,000)	
		Animals	Glands
Irkutsk	1850 (for 3 years)		3.4
All Siberia	1855		81.2
Transbaikal	1960–1970	1.4–1.8	
Yakutia	1960–1970	1	
Far East	1960–1970	1.2	
Former USSR territory	1960–1970	5	
Russian Federation	1999	2.8	

3.8 Population Structure and Densities (Table 3.9)

3.8.1 Population Density and Size (Tables 3.10 and 3.11)

Hunters stress that the great fluctuations of musk deer numbers which occurred during the end of the 19th–early 20th centuries were not the result of hunting (Bromley and Kucherenko 1983). Prikhodko (1997) reported a significant decline of musk deer in all areas.

3.9 Relations to Man (Table 3.12)

Meat, skin and especially musk gland of males are valuable (Table 3.12).

4 Red Deer – *Cervus elaphus*

Cervus elaphus Linnaeus, 1758.
Artiodactyla, Cervidae, Cervinae

4.1 Status and Distribution (Fig. 4.1)

4.1.1 Taxonomic Remarks

Eight subspecies occur: *C. e. hippelaphus* Erxleben, 1792 (European part of USSR, exclusive the Carpathians); *C. e. montanus* Botezat, 1903 (Carpathian Mountains); *C. e. maral* Ogilby, 1840 (Caucasus Mountains); *C. e. brauneri* Charlemagne, 1920 (Crimean Peninsula); *C. e. sibiricus* Severtzov, 1872 (Altai Mountains, Sayan Mountains, Pri-Baikal region); *C. e. xanthopygos* Milne-Edwards, 1860 (Trans-Baikal, Far East); *C. e. songaricus* Severtzov, 1872 (Tien Shan Mountains, Dzungarian Alatau Range); *C. e. bactrianus* Lydekker, 1900 (Kazakhstan, Tajikistan, Turkmenistan, Uzbekistan) (Heptner 1989).

4.1.2 Status

C. e. bactrianus is vulnerable (VU – D1) (Hilton-Taylor 2000). CITES II. *C. e. hippelaphus* is hunted mostly in well-protected and well-managed game husbandry areas. Other subspecies are numerous and, while hunted, are protected by hunting regulations (Bannikov and Pivovarova 1983; Bromley and Kucherenko 1983; Sobansky 1992; Smirnov 1994; Prostakov 1996). Populations on the plains of European Russia and the Ural became extinct around 1750 (Barabash-Nikiforov 1957) and have been reestablished by man. In the second part of the 19th century, some deer from Germany were used to establish a population for hunting near Voronezh. Their seven descendants survived the Civil War, and became founders of the new European Russian population. This population increased and about 3,200 animals were then distributed over the European part of the USSR. Also, 800 marals (*C. e. sibiricus*) were acclimatized there. The last-mentioned subspecies easily adapts to new conditions and survived better in the more severe climate. As a result, southern and central parts of European Russia

Fig. 4.1. Maral

are inhabited mostly by *C. e. hippelaphus*, northern and northwestern parts and Bashkortostan are inhabited mostly by marals, while in Moldova, a hybrid form of deer from Askaniya-Nova was acclimatized (Lavrov and Pavlov 1975; Fadeev 1990). In Yakutia, red deer have features intermediate between *C. e. sibiricus* and *C. e. xanthopygos*. This variant appeared in S Yakutia in the last part of the 19th century and nowadays occupies an area of 60,000 km^2 along tributaries of the Lena and Aldan Rivers (Stepanova 2002).

4.1.3 Distribution (Fig. 4.2)

Distribution Area. Ca. 7,297,000 km^2 - estimated historical range.

Occurrence in Ecoregions. Subarctic (133), Subarctic Regime Mountains (M133, M134), Warm Continental (212), Warm Continental Regime Mountains (M212), Subtropical (231), Prairie (252, 253), Prairie Regime Mountains (M251, M252,), Mediterranean Regime Mountains (M262), Tropical/Subtropical Desert Regime Mountains (M321), Steppe (331, 332), Temperate Desert (342, 343), Temperate Desert Regime Mountains (M341).

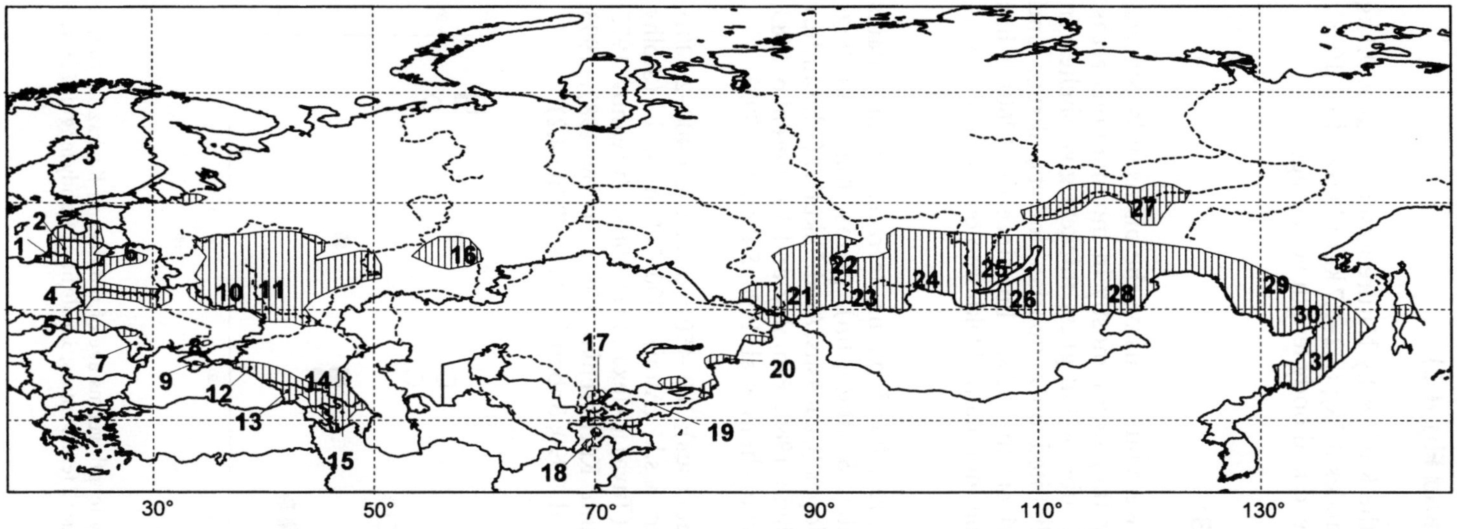

Fig. 4.2. *Cervus elaphus*. Distribution map (Fadeev 1990; Sokolov and Tembotov 1993; Mirutenko 1996). *1* Kalinigrad oblast; *2* Lithuania; *3* Latvia; *4* Belovezhskaya Forest Nature Reserve; *5* Carpathian Mountains; *6* Belarus; *7* Moldova; *8* Askania-Nova Nature Reserve; *9* Crimean Peninsula, Crymsky Nature Reserve; *10* Kursk oblast, Belgorod oblast; *11* Lipetsk oblast, Tambov oblast, Voronezh oblast, Voronezhsky Nature Reserve; *12* W Caucasus, Kavkazsky Nature Reserve; *13* Georgia, Borzhomi Nature Reserve; *14* Great Caucasus Range; *15* Azerbaijan; *16* Bashkortostan, Bashkirsky Nature Reserve; *17* Talas Alatau Range; *18* Tajikistan, Ramit Nature Reserve, Tigrovaya Balka Nature Reserve; *19* Kyrgyzstan; *20* Dzungarian Alatau Range; *21* Altai Mountains, Altaisky Nature Reserve, Teletskoe Lake; *22* Krasnoyarsk kray, Stolby Nature Reserve; *23* W Sayan Mountains, Tuva; *24* E Sayan Mountains; *25* Irkutsk oblast; *26* Buryatia, Khamar-Daban Range; *27* Olekma River, Yakutia; *28* Chita oblast; *29* Amur oblast; *30* Khabarovsk kray, Amur River; *31* Sikhote-Alin Range, Sikhote-Alinsky Nature Reserve

4.2 Measurements (Table 4.1 and Fig. 4.3)

Weight of 1-day-old fawns in the Kavkazsky Nature Reserve: males 9.1 kg (ranging from 7.0–11.1 kg), females 8.7 kg (7.2–9.7 kg) (Alexandrov 1968). Figure 4.3 shows weight and head and body length versus age.

4.3 Habitat (Tables 4.2 and 4.3)

During the 18th century, deer were numerous on the plains of W Siberia, Kazakhstan and E Ural. Today, they occur mostly in mountainous areas. Prefer forested areas where shrubs or tall grass thickets give shelter. In the past, *C. e. bactrianus* inhabited only riparian forests, shrubs and reed thickets along the large rivers in M Asia, but since the 1960s it has also acclimatized to living in the mountains (Tikhomolov 1982; Flint et al. 1990a).

Terrain Form. Freshwater shores, flood plains, terraces of mountains (Kazakhstan, Tajikistan, Yakutia, Sikhote-Alin), plains, hills and bogs (Lithuania), foothills, valleys, slopes (Caucasus, Altai, Yakutia) (Alexandrov 1968; Zyryanov 1975; Baratashvili 1980; Bromley and Kucherenko 1983; Sokov 1987; Smirnov 1989a; Kozhechkin et al. 1990; Sobansky 1992).

Altitude. 800–1,900 m a.s.l. (Caucasus); 300–450 m a.s.l. and 1,400–2,300 m a.s.l. (Tajikistan); 450–1,900 m a.s.l. (Altai); 300–700 m a.s.l. (Stolby, Sikhote-Alin); 800–1,900 a.s.l. (Tuva) (Alexandrov 1968; Zyryanov 1975; Baratashvili 1980; Bromley and Kucherenko 1983; Sokov 1987, 1993; Smirnov 1989a; Kozhechkin et al. 1990; Sobansky 1992; Table 4.2).

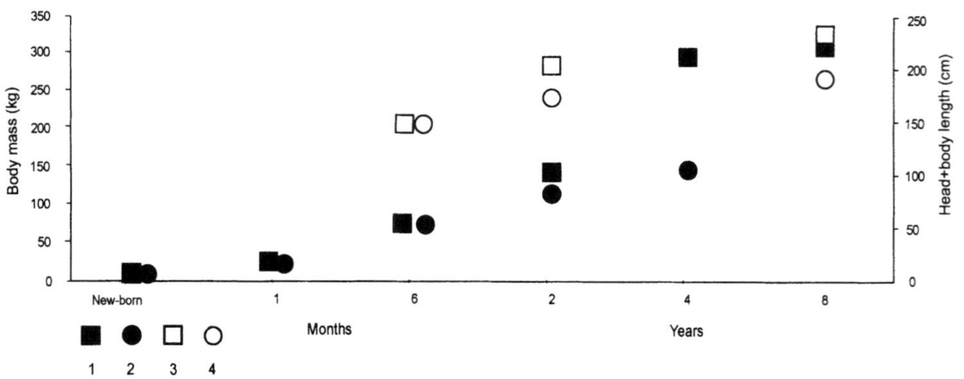

Fig. 4.3. Body mass and head + body length versus age of red deer in Kavkazsky Nature Reserve (Alexandrov 1968). *1* Male mass; *2* female mass; *3* male head + body length; *4* female head + body length

Table 4.1. *Cervus elaphus.* Measurements

Locality[a]	Longitude (°E)	Latitude (°N)	Gender	*n*	Body mass (kg)	Head and body length (cm)	Tail length (cm)
Belovezhskaya Forest[1]	24	53	Male	19	203 (186–234)	208 (173–228)	
			Female	7	139 (131–160)	182 (161–202)	
Voronezh Nature Reserve[2]	38	52	Male	48	203.6 (159–244)	226.4 (196–239)	14.2 (12–19)
			Female	43	131.7 (118–171)	188.5 (173–212)	13.3 (11–16)
Kavkazsky[3]	40	44	Male	6	310.0 (260.0–342.0)	234.0 (225.0–249.0)	13.0 (10.0–17.0)
Tajikistan[4]	68–70	69	Male	12		203 (196–214)	10 (8–11)
Tien Shan[5]	71–80	41–43	Male	5	323.4 (265–416)	232.2 (210–239)	11
			Female	11	155.3 (125–169)	208 (173–227)	13.7 (11–16)
Altai[6]	84–89	49–53	Male	7		227.0 (212–246)	15.0 (13–16)
			Female	7		210 (203–217)	14 (11–17)
E Sayan[7]	93–101	52–55	Male	6	263.3 (220–305)	282.6 (243–338.5)	
			Female	2	197.5 (150–245)	275.5 (274–277)	
Pri-Baikal[8]	102–109	52–56	Male	40		230 (204–246)	13 (11–16)
			Female	29		208 (190–226)	16 (13–18)
Sikhote-Alin[9]	133–140	43–51	Male	9	224 (184–254)	231 (210–252)	16 (14–18)
			Female	12	149 (120–182)	196 (173–216)	15 (12–17)

[a]The superscript numbers denote the following references: 1, Kozlo (1983a); 2, Prostakov (1996); 3, Alexandrov (1968); 4, Stroganov (1959); 5, Fedosenko (1984b); 6, Sobansky (1992); 7, Zyryanov (1975); 8, Sviridov (1983); 9, Bromley and Kucherenko (1983).

Table 4.2. Distribution of *Cervus elaphus* in different habitats in the Talas Alatau Range (% of observed animals) (Grachev and Smirnova 1980)

Habitat	Altitude (m a.s.l.)	January (%)	March (%)	May (%)	July (%)	September (%)	November (%)
Savanna	1,300–1,800	0	33	0	0	0	0
Sparse shrubs, *Juniperus sabina*	1,800–2,300	100	67	42	0	11	68
Subalpine	2,300–2,900	0	0	58	92	89	32
Alpine	2,900–3,400	0	0	0	8	0	0
Glaciers	3,400–4,200	0	0	0	0	0	0

Snow. Hoof pressure for one foot (when standing on 4 legs) was 490 g/cm^2 (ranging from 439–548) in Voronezh area, and 652 g/cm^2 (594–720) in the Crymsky Nature Reserve, age and gender were not determined (Nasimovich 1955). Deer can move easily in fresh snow of 20–30 cm depth. Deer concentrations are common where snow is <50–60 cm and where there is plenty of food (tree and shrub twigs). Males can use areas where the snow is 100 cm deep. The limit for females and calves is 60 cm (Nasimovich 1955). In Sayan, the critical snow depth was 70–75 cm (Dulkeit 1964; Zyryanov 1975). In the N Sikhote-Alin and Amur region, snow >80 cm leads to deer mortality of at least 20–30% (Bromley and Kucherenko 1983).

Water. From time to time deer drink water, usually from streams; in winter, snow is used. In hot summers deer bathe in water holes.

4.3.1 Biotope (Table 4.3)

4.4 Diet and Feeding Habits

4.4.1 Diet

Red deer have a much broader feeding niche than other Cervidae. For a list of plant genera eaten see Appendix 1, Table 4. Within the distribution area of red deer in the former USSR there are 12 genera of trees and shrubs that exist everywhere, but only *Salix*, *Populus* and *Sorbus* are used by red deer. The genera *Quercus* and *Fraxinus* are used in all places, except in Altai where these genera are absent. In many areas, feeding on *Tilia*, *Sambucus*,

Table 4.3. *Cervus elaphus*. Habitat[a] (*W* winter, *S* summer)

Habitats		Locality[b]									
		Belovezhskaya Forest (1)[1], Voronezh (2)[2]		Kavkazsky (1)[3], Tajikistan (2)[4], Kazakhstan (3)[5]			Altai (1)[6], Stolby (2)[7]		Pri-Baikal (1)[8], Yakutia (2)[9], Sikhote-Alin (3)[10]		
		1	2	1	2	3	1	2	1	2	3
Ecoregions		212	212	M252	M321	M341	M251	M251	M134	M134	M212
Mature forest											
Coniferous, abundant overgrowth	W	+	+	+++		x				++	++
	S	+	+	+		x	+			+++	+++
Coniferous, good grass cover	W								++		
	S								++		
Coniferous in mountains, sparse overgrowth, mosses	W			+	r			+++			+
	S			+++			+++		+	++	+
Mixed deciduous-coniferous	W	+++	+				+++	r			++
	S	+++	+				+++				++
Broadleaf, abundant undergrowth	W	+	+	+++	+++	x		r			++
	S	+	+	+	+++	x					++
Young forest											
Coniferous	W	+++						r			
	S	+++	R								
Deciduous	W	r									
	S	r									
Fresh water shore											
Meadows	W				+					++	
	S	r			+				+	++	
Shrubs	W		r							++	
	S		r		++					+++	
Forest	W				+				+++	++	
	S				+		++		+	+++	
Tugai[c], *Phragmites* tickets	W				++	x					
	S				++						

Table 4.3. *Continued*

Habitats		Locality[b]									
		Belovezhskaya Forest (1)[1], Voronezh (2)[2]		Kavkazsky (1)[3], Tajikistan (2)[4], Kazakhstan (3)[5]			Altai (1)[6], Stolby (2)[7]		Pri-Baikal (1)[8], Yakutia (2)[9], Sikhote-Alin (3)[10]		
		1	2	1	2	3	1	2	1	2	3
Flood plain	W		+							+++	
	S		+							+++	
Bog	W	r	r								r
	S	r	r						x		r
Burnt forest, recovered	W							r		++	+++
Grassland, previously burnt	S										+++
Cut forest with regrowth	W		+						+++		
	S		+					+++	+++		
Shrubland[d]	W				++	x		+++	+++		
	S				++	x					
Forest clearings, edges	W		r								
	S		r								
Shrubby tundra, mountainous	W										
	S						+++				
Grassland	W							+++			
	S								x		
Meadow, wet		r									
Grassland, alpine	W			+							
	S			+++			+	+	x		
Grassland, subalpine	W			+							
	S			+++			+++	+++	x		

[a]Symbols indicate % of habitat use was >30% (+++); 20–30% (++): 10–20% (+); <10% (r); x indicates habitat use noted, no numerical data reported.
[b]The superscript numbers denote the following references: 1, Sablina (1955); 2, Mertz (1953); Prostakov (1996); 3, Alexandrov (1968); 4, Flerov (1952); Stroganov (1959); Sokov (1987, 1993); 5, Fedosenko (1984b); 6, Dmitriev (1938); Sobansky (1992); 7, Zyryanov (1975); 8, Shvetsov et al. (1984); 9, Egorov (1971); 10, Bromley and Kucherenko (1983).
[c]*Populus pruinosa, Eleagnus angustifolia.*
[d]*Caragana, Juglans, Prunus, Malus, Cerasus, Crataegus* on slopes, terraces.

Pinus, *Rubus*, *Euonymus*, *Acer*, *Carpinus*, *Ulmus* and *Viscum* was found (Baskin 1965). The importance of grass and herbs in the diet of deer is under discussion. *C. e. sibiricus* have a diet mainly composed of grass and herbs (Sablina 1970). Also, there is a predominance of grass in the summer food of *C. e. xanthopigus* (Kaplanov 1948). However, many authors stress that the capacity of red deer to live on grasses and herbs should not be overestimated (Bromley and Kucherenko 1983). Salt licks are used when the ground is free of snow (Nasimovich 1938).

4.4.2 Regional Patterns of Diet (Table 4.4)

In Kazakhstan, the significant role of spruce (*Picea schrenkiana*) in feeding is remarkable (Fedosenko 1984b). In the Tajikistan tugai, in summer and autumn, a significant part of the diet is seeds and fruits (Stroganov 1959). In Altai, alder and pine are rarely eaten; spruce and fir are not eaten at all. Deer rarely strip bark from frozen trees (Kaplanov 1948; Zyryanov 1975). In the Stolby Nature Reserve, *Caragana* and *Populus tremula* comprise 57% of the total consumption of woody plants. Pine is rarely eaten (only 2% of diet), only tips of young pines and seedlings are taken (Zyryanov 1975). In the Far East, the heavy use of mushrooms, acorns, berries and bark (in autumn and spring) is remarkable (Bromley and Kucherenko 1983).

4.4.3 Life Forms/Plant Parts Eaten (Table 4.5)

4.4.4 Seasonal Aspects of Diet (Table 4.6)

In early winter, when snow is thin and food is available, deer stay in the upper parts of mountains. Thereafter they move downwards gradually and spend the second part of winter where snow cover is less and better for feeding. In early spring, the animals concentrate where the first green vegetation appears and then move upwards to summer pastures. Some deer stay in mountain valleys all summer (Mamukelashvili 1987).

Consumption of Food/Day (Tables 4.7 and 4.8). In deer farms, adult animals are fed 8–10 kg hay and 2–2.5 kg oats (fresh weight) (Pyatkov and Pryadko 1971).

Feeding Niche. In field and shrub layers, rarely below ground.

Table 4.4. *Cervus elaphus*. The main food consumed (%)

Plants	Regions[a]					
	Voronezh[b,1]	Kavkazsky[c,2]	Kazakhstan[c,3]	Altai[c,4]	Yakutia[d,6] Winter	Yakutia[d,6] Summer
Abies		3				
Acer	36	2				
Acorus						10
Allium						20
Alnus					11	20
Betula	44	10		80	33	20
Betula nana						10
Caragana				80		
Carex					45	70
Carpinus		10				
Cerasus		1				
Cereals					45	70
Chosenia					22	
Corylus	61	6	18			
Cotoneaster						
Crataegus		0.1				
Daphne		0.1				
Epilobium						30
Equisetum					100	80
Euonymus	90	1				
Fagus		10				
Fraxinus						
Iris					22	40
Larix (needles)					22	10
Lichens					22	30
Lilium						30

Lonicera			14			
Malus	63					
Mosses						10
Mushrooms					33	40
Padus	68		5	100		
Picea			32			
Pinus (needles)					11	
Pinus (twigs)						
Pirus		1				
Populus tremula	64	23		95	11	10
Prunus		1				
Pulsatilla						20
Quercus	30	26				
Rhamnus	32					
Rhododendron		3		22		
Ribes		1		25–90		
Rosa		1	9			
Rubus		3	9		33	30
Salix	69	1	45		88	80
Sambucus		1				
Sanguisorba						20
Sorbus	51		59	100		
Tilia						
Ulmus	29	1				

[a]The superscript numbers denote the following references: 1, Prostakov (1996); 2, Alexandrov (1968); 3, Fedosenko (1980); 4, Sobansky (1992); 5, Egorov (1971).
[b]Percent of eaten plants.
[c]Percent of eaten twigs.
[d]Percent of studied stomach content.

Table 4.5. *Cervus elaphus.* Life forms/plant parts eaten (x indicates the plant type used)

Life forms	Plant parts eaten							
	Whole plant	Leaves	Stems/ twigs	Bark	Roots/ bulbs	Flowers	Fruits	Seeds
Trees		x	x	x			x	x
Shrubs		x	x			x		
Dwarf shrubs		x	x					
Herbs		x	x		x	x		
Grasses/sedges		x	x					
Horsetails	x							
Mosses	x							
Lichens	x							
Fungi	x							

Table 4.6. *Cervus elaphus.* Seasonal aspects of diet[a]

Locality[b]	Food	Winter	Spring	Summer	Autumn
Voronezh[1]	Trees	++		r	
	Shrubs	++		+	
	Dwarf shrubs	r		r	
	Herbs, grasses				
	Green			+++	
	Dry	+++			
	Tree lichens	r			
	Mushrooms	r		r	
Stolby Nature Reserve[2]	Trees	++	r	++	+++
	Shrubs	++	+	+	++
	Herbs, grasses				
	Green		+++	+++	
	Dry	++	+		+++
	Tree lichens	++	R		+

[a]Symbols indicate % of food use was >30% (+++); 20–30% (++): 10–20% (+); <10% (r).
[b]The superscript numbers denote the following references: 1, Prostakov (1996); 2, Zyryanov (1975).

4.5 Social Organization and Behavior

4.5.1 Social Life (Tables 4.9 and 4.10)

There probably are bonds between animals in family and mixed groups (Table 4.9).

Table 4.7. *Cervus elaphus*. Consumption of food (fresh weight) per day (kg)

Locality[a]	Season	Gender, age	No. of 24-h experiments	Food
Belovezhskaya Forest[b,1]	Summer		4	12.5 (12.0–12.8) grass
Kavkazsky[2]	Winter	Male	2	1.87 (1.4–2.3)
		Female	11	2.83 (2.1–3.5)
Altai[3]	Winter	Male[b]	1	5.8 (including 4.0 kg aspen twigs, 1.2 kg lichens (*Usnea*)
		Female	1	3.4
		Female, 2 years	1	2.2
Stolby Nature Reserve[4]	Winter			12 kg twigs, i.e. that is 40–60% of daily intake

[a]The superscript numbers denote the following references: 1, Malinovskaya (1975); 2, Alexandrov (1968); 3, Sobansky (1992); 4, Zyryanov (1975).
[b]In captivity.

Table 4.8. *Cervus elaphus*. Utilization of different foods in winter (Kavkazsky Nature Reserve. Males and females were tracked and food biomass eaten was estimated by weighing an equal amount to that which was grazed/browsed). (Alexandrov 1968)

Gender	Food eaten during 24 h (kg)							
	Snow depth (cm)	Twigs	Bark	Dwarf shrub (*Rubus caesius*)	Herbs + grasses	Tree lichens	Other food	Total weight
Male		0.3–0.6	0.2–0.4	0–0.15	0.24–0.5	0.13–0.6	0.2–0.4	1.4–2.3
Female		0.13–0.14	0–0.34	0.08–2.0	0–0.25	0.1–0.34	0–1.5	2.1–3.5
				%				
Male, Female	0–20	9		81	8	1	1	
	60–70	53		0	0	45	2	

Table 4.9. *Cervus elaphus*. Group composition (x indicates when the type of group was observed)

Group types (Seasonal events)	Winter Wintering	Spring Calving	Summer Summering	Autumn Rut
Singles (male; female + fawn)	x	x	x	
Families (female + fawn + yearling)	x	x	x	
Mixed groups (some females + their fawns + their yearlings, lone females, young males)	x			
Groups of yearlings		x	x	
Harems (male + some females + their fawns)				x
Bachelor groups of males	x	x	x	x
Aggregations	x			

Red deer herds scatter when disturbed (Zyryanov 1975; Baskin 1976; Fedosenko 1980; Prostakov 1996). In Altai, in winter the density of deer (in shrubby valleys and on southern slopes in sparse coniferous forest) is 4–9 animals/10 km^2. In summer, the density of males in the upper reaches of streams on slopes near and above the forest line is 2 animals/10 km^2. In summer, the density of females and young deer (in diverse habitats, on all parts of slopes, including river valleys) is 7 animals/10 km^2 (Dmitriev 1938). The leader of female and mixed groups is usually an adult female; male groups are led by the largest male. The leaders determine rhythm of feeding, time of salt licking and direction of movement. Males usually bring up the rear. During rut, a dominant male can take females from a competitor (Baskin 1976; Fedosenko 1980; Table 4.10).

4.5.2 Home Range

On a yearly basis deer move over a range of 6–8 km within an area of 20–30 km^2 (Bromley and Kucherenko 1983). Wintering area <200–300 ha in Sikhote-Alin; in Stolby 400–800 ha, at the end of winter only 50–100 ha (Zyryanov 1975). In the Voronezhsky Nature Reserve, female groups occupy about 200 ha in winter (Mertz 1953, 1957); in the Kavkazsky Reserve, 40–50 ha; in Bashkortostan, 23–63 ha (Gordiyuk 1985). In the Voronezh oblast, daily movement during feeding is 1–6 km (Prostakov 1996). Daily movement of deer in the Kavkazsky Nature Reserve is about 1.6 km (0.8–2.6 km), with 2–4 resting places (Alexandrov 1968); in the Sikhote-Alin 2–3 km (up to 4 km). In Bashkortostan, daily routes in winter are 1–2 km (1.2–2.3 km) with 5–6 rests, 10–13 defecations and about 5–7 urinations (Gordiyuk 1985). The length of the daily route strongly depends on snow conditions,

Table 4.10. *Cervus elaphus.* Group size in different seasons

Locality[a]	Winter	Spring	Summer	Autumn (rut)	Year average
Belovezhskaya Forest[1]	3.5–5.0	1.7–4.4	1.4–2.6	1.6–3.2	2.9
Voronezh[2]	4.4–6.4	2.0–2.2	1.77–2.03	3.4–3.9	3.4
Kavkazsky[3]	4.9–5.1	4.5–5.1	4.4–4.5	3.4–3.8	4.2
Stolby Nature Reserve[4]	2.6–2.8	1.9–2.5	1.3–1.6	1.9–2.0	2.0
Tuva[5]					2.1

[a]The superscript numbers denote the following references: 1, Shostak (1975a); 2, Prostakov (1996); 3, Alexandrov (1968); 4, Zyryanov (1975); 5, Smirnov (1989a).

being half as long when snow depth is >50 cm. In Stolby, a daily route, when the snow depth is 45–80 cm, is 3.4 km (1.2–6.5) (Zyryanov 1975). Within the home range there are trails, feeding places, fresh water in summer and "otstoi" refuges on rocks (Kaplanov 1948; Alexandrov 1968; Pereladova 1981; Balĉauskas 1990).

4.5.3 Mode of Movement

Usually walking or trotting. Disturbed animals move by galloping or jumping. Jumps can be 3–4 m in length on average, maximum 6.5 and 1–2 m in height (Flerov 1935). Disturbed deer sometimes move by an alert trot – a smoothed trot with head highly elevated. When playing they move by stotting where fixed front and back legs are put down alternately and with clatter. Digs snow well when depth is <40 cm (Zyryanov 1975). Good swimmer: capable of swimming across large and turbulent rivers and crossing open waters up to 4 km wide. When feeding on aquatic plants they ensure that the ears are not submerged (Bromley and Kucherenko 1983; Sokov 1987; Sobansky 1992).

4.5.4 Dispersal/Migration

In winter, home ranges are reduced to 20–25% (Stolby Nature Reserves), 60–70% (Kavkazsky and Voronezhsky Nature Reserves) of summer habitats (Dmitriev 1938; Zharkov 1940; Mertz 1953; Zyryanov 1975). Deer will travel 4–15 km to visit salt licks. Seasonal migrations are usually short from some tens of kilometers in Kazakhstan and Sikhote-Alin, to 50–140 km in the Altai and W Sayan regions. The altitudinal shifts involved are from hundreds to 1,000–1,500 m in the Caucasus (Nasimovich 1955; Alexandrov 1968; Fedosenko 1980; Bromley and Kucherenko 1983; Sobansky 1992). Snow conditions strongly affect the distribution of deer.

4.5.5
Diurnal Activity (Table 4.11)

There are 7–9 feeding periods every 24 h (Prostakov 1996).

4.5.6
Shelters

Places with good visibility (on slopes, at fringes of forest) are used for resting. Before lying down, deer scratch away snow (Alexandrov 1968; Fedosenko 1980; Bromley and Kucherenko 1983). In summer, deer often use the same plot for rest. On very rare occasions females have been observed to make nests for newborn foals when snow came unexpectedly (Kaplanov 1948; Bromley and Kucherenko 1983).

4.5.7
Escape Behavior

In the Far East and Kazakhstan, red deer are very vigilant, especially when approaching salt licks and when antlers are in velvet. Stags look around up to 40 times per hour. On being disturbed, flight distance is about 400–

Table 4.11. *Cervus elaphus.* Diurnal activity (shaded parts denote activity; *W* winter, *S* summer)

Locality		Time of day (h)							
		0–3	3–6	6–9	9–12	12–15	15–18	18–21	21–24
Kazakhstan[1]	W								
	S								
Tadijkistan[2]	S								
Tadijkistan[3]	S								
Altai[4]	W								
	S								
Sikhote-Alin[5]	W								
	S								

The superscript numbers denote the following references: 1, Fedosenko (1984b); Grachev (1988); 2, Flerov (1935); 3, Sinelnikov (1964); 4, Sokov (1987); Sobansky (1992); 5, Bromley and Kucherenko (1983).

500 m. After the initial burst of movement, animals usually stop again and try to identify the danger. A disturbed group will subsequently move up to 3–4 km away. In mountains, as a defense against predators (wolf, wolverine), deer use places where they are protected from behind by a rock wall and can protect themselves with antlers (stags) or front legs. Deer often use water to escape predators, by crossing large and turbulent rivers, or staying in water. Escaping deer prefer to run up slopes. Wolves usually try to direct deer down slopes and then take them if they fall (Formozov 1946; Fedosenko 1980; Kudaktin 1998).

4.6 Reproduction (Tables 4.12 and 4.13)

The start of rut is indicated by: the appearance of patches (about 3 × 4 m) with disturbed soil and vegetation; bathing holes in streams or wet soil dug by males with hooves and antlers; roaring of males. Male territories are established and protected. Dominance depends on mass of animal and size of antlers (Tsalkin 1944; Fedosenko 1980). At the start of rut, the stronger males start roaring and occupying territories and their activity continues during the entire rutting period. Between neighboring territorial males, battles rarely take place, but they are very aggressive to single invaders (Kanshina and Shibanov 1976; Pereladova 1981). In the Kavkazsky and Voronezhsky Nature Reserves and the Sikhote-Alin there are activities, which can be characterized as "leks", i.e., many males are gathered for a lengthy period, roaring and demonstrating aggressive behavior. In the "leks", 50 or more males can be found within 1–1.5 km^2 (Pereladova 1981). In Tajikistan (*C. e. bactrianus*) males and females remain in their home ranges during the day and gather in the "leks" only at night (Pereladova 1988). In the Kavkazsky Nature Reserve deer roar from 16:00–17:00 h in the afternoon to 9:00–10:00 h in the morning. Frequency of roaring differs between "leks" (Pereladova 1981). The harem size is normally between 2–4, with a maximum of 11, but there are up to 18 females per male in Altai, 1–4 in Kavkazsky, 1–5 more rarely in Kazakhstan (Dmitriev 1938; Alexandrov 1968; Fedosenko 1984b; Sobansky 1992). Harem size depends on overall density of population, and increases during the rutting season. A very characteristic feature of deer rut is the constant attention that males show females. The behavior of the male stimulates maturing follicles in the female ovary (Mashkovtsev 1949). Most females have mated by the end of the rut (Nasimovich 1936; Alexandrov 1968). Before calving, the female isolates herself in a quiet place. The calves stay in the same place for 7 days, but will start following the mother after 2–3 weeks (Dmitriev 1938). Females start to leave the calves when they are 3 months old. They feed them twice a day, in the morning and evening. Lactation continues up to rut. Calves start to eat green plants when 3–4 weeks old (Bromley and Kucherenko 1983; Fedosenko 1984b; Tables 4.12 and 4.13).

Table 4.12. *Cervus elaphus.* Breeding parameters

Locality[a]	Age at sexual maturity (years)		Participation in breeding (years)		Pregnancy (days)
	Male	Female	Male	Female	
Belovezhskaya Forest[1]				2.5	210 (177–231)
Voronezh[2]	1.5	1.5			240–250
Kavkazsky[3]					230–235
Kazakhstan[4]	1.5	1.5–2.5	5.5	1.5–2.5	240–245
Far East[5]	1.5	1.5	3–4	1.5	

[a]The superscript numbers denote the following references: 1, Sablina (1955); Shostak (1976); 2, Merts (1953); Prostakov (1996); 3, Nasimovich (1936); Alexandrov (1968); 4, Fedosenko (1980); 5, Bromley and Kucherenko (1983).

Table 4.13. *Cervus elaphus.* Parameters of fertility

Locality[a]	Barrenness (%)	Proportion of females with *n* embryos (%)			Litter size
		1	2	3	
Belovezhskaya Forest[1]	8–33				
Voronezh[2]		87	13	0	1, Rarely 2
Kazakhstan[3]	7–8				1, Very rarely 2
Sikhote-Alin[4]	17	75	21	4	

[a]The superscript numbers denote the following references: 1, Shostak (1975a); 2, Prostakov (1996); 3, Fedosenko (1980); 4, Bromley and Kucherenko (1983).

4.6.1 Seasonal Breeding Events (Table 4.14)

Shedding antlers: Belovezhskaya Forest (Belarus) 10 March–15 April, Kazakhstan 1–20 April, and Stolby Nature Reserve 25 March–14 April. Removing velvet from antlers: Belovezhskaya Forest 1–11 July, Caucasus 1–11 July, Kazakhstan 15–21 August, and Altai 22–25 August (Dmitriev 1938; Tsalkin 1945; Sablina 1955; Zyryanov 1975; Fedosenko 1980).

Maximum Length of Life. In deer farms males live on average for 16–18 years, maximum 25 years (Zalessky 1930). Females can breed up to 15–18 years of age (Belonogov 1977).

Table 4.14. *Cervus elaphus.* Seasonal breeding events

Locality[a]	Rutting season			Calving season		
	Preparation to rut	Basic rut	Last demonstrations	First calving	Basic calving	Last calves
Belovezhskya Pushcha[1]	15–27 Aug.	8 Sept.–1 Oct.	16–23 Oct.	28 Apr.–7 May	Second part of May–mid-June	
Voronezh[2]		Second part of Sept.–end Oct.	Early Nov.	Early May	10 May–30 June	18 July
Moldova[3]	28 Aug.–3 Sept.	10 Sept.–early Oct.	25–28 Oct.		May	
Crimean Peninsula[4]	6 September	20 Sept.–1 Oct.	20 October		5–15 June–late June	
Kavkazsky[5]	15 Aug.	20 Sept.–10 Oct.	5–10 Nov.	7 Apr.–5 May	15 May–10 June	18 July
Kazakhstan[6]		25 Sept.–5 Oct.	Late Oct.	28 May		22 June
Tajikistan[7]	21 Aug.	Middle Sept.–10 Oct.	12 Oct.	5–10 May	20–27 May	2–10 June
Altai[8]	26 Aug.	15–20 Sept.	Mid of Oct.	20 May	Mid of June	Early July
Stolby Nature Reserve[9]	28–30 Aug.	15 Sept.–4 Oct.	Early Nov.		End of May–10 June	
Yakutia[10]	6–15 Sept.		Middle of Oct.		The first half of June	
Pri-Baikal[11]	10–21 Sept.	17 Sept.–4 Oct.	7–25 Oct.			
Amur[12]		20–25 Sept.			Mid–end May–early June	
Sikhote-Alin[12]	End Aug.–early Sept.	15–20 Sept.			Second part of Apr.–early May	

[a]The superscript numbers denote the following references: 1, Sablina (1955); Shostak (1976); 2, Prostakov (1996); 3, Chegorka (1984); 4, Kormilitsin (1970); Dulitskii and Pereladova (1980); 5, Zharkov (1940); Alexandrov (1968); 6, Fedosenko (1984b); 7, Sokov (1987); 8, Dmitriev (1938); Sobansky (1992); 9, Zyryanov (1975); 10, Egorov (1971); 11, Shvetsov et al. (1984); 12, Bromley and Kucherenko (1983).

4.7 Mortality

In some years, infections cause heavy deer mortality. Usually hard and deep snow cover, together with a shortage of food and poaching, are the main mortality factors. For example, during the dramatic winter of 1965/1966 in Altai the snow depth was 1.4–3.5 m and 50% of the deer population died (Sobansky 1992).

4.7.1 Predators

In the Far East, large predators kill 10–15% of the deer population annually. When snow cover is deep (>70–80 cm) and crusted, every meeting between deer and wolves leads to the death of the deer (Bromley and Kucherenko 1983). In Altai, some bears and wolves have become specialized to prey on deer calves. Females do not protect calves against wolf and bear, but do protect them against lynx and wolverine (Sobansky 1992; Table 4.15). There are differences in predation between wolf families: 54–62% of wolf scats from a family, specialized in deer hunting, contained deer remnants, while 28–51% of wolf scats from other families contained deer remnants (Kudaktin 1986). Among deer preyed on by wolf, 72% of males and 63% of females were older than 6 years in Kavkazsky Nature Reserve, 92% males and 50% females, respectively in Sayan, (Zavatsky 1986; Kudaktin 1998; Table 4.16).

4.7.2 Diseases, Helminths and Ectoparasites (Appendix 2, Tables 4, 5)

Anthrax, cattle plague, pasteurellosis, necrobiosis, necrotic stomatitis, pyroplasmosis, tuberculosis, and foot-and-mouth disease (Shostak and Vasilyuk 1976; Nasimovich 1989).

4.8 Population Structure and Densities (Tables 4.15, 4.16 and 4.17)

The ratio of adult males:females in the Belovezhskaya Forest Nature Reserve in 1945–1973 was 1:1.2–1:2, in the Kavkazsky Nature Reserve in 1941–1966, 1:0.9–1:2.6; in the Voronezhsky Nature Reserve 1:2.4; in the Altaisky Nature Reserve, in 1984–1988, 1:1.9. In all nature reserves, a decrease in the proportion of males was observed as the deer populations increased (Alexandrov 1968; Shostak 1975a; Sobansky 1992; Prostakov 1996; Table 4.17).

Table 4.15. *Cervus elaphus*. Survival of calves during the first year of life

Locality[a]	Year	Calves/100 females				Calves in population (%)			
		June	September	December	March	June	September	December	March
Kavkazsky Nature Reserve[1]	1963–1964	61	30	24	16	27	15	13	9
Altai Mountains[2]	1978					20	10		
Kazakhstan[3]		90	70						
Tuva republic[4]						27	22		14

[a]The superscript numbers denote the following references: 1, Alexandrov (1968); 2, Sobansky (1992); 3, Fedosenko (1980); 4, Smirnov (1990).

Table 4.16. *Cervus elaphus*. Causes of mortality, determined by examination of carcasses found by game wardens (%)

Locality[a]	Causes of mortality												
	Disease (1), starvation (2)		Wolf (1), bear (2), lynx (3), wolverine (4), tiger (5), snow leopard (6), dog (7)							Avalanches, trauma, floating of ice (1), wounded during rut, too old (2), poachers (3), road accident (4)			
	1	2	1	2	3	4	5	6	7	1	2	3	4
Belovezhskaya Forest[1]	22		58		7				1			9	3
Voronezh[2]	6		8						11	2	2	16–21	14
Kavkazsky[3]	17	7	31	(9–40)	4							12	
Bashkortostan[4]	4	23	21	11	14							8	
Tajikistan[5]			x		x		x[b]		x				
Kazakhstan[6]	x	x	x	x	x	x		x	x	x		x	
Altai[7]	1	9	43–54	2–4	2–7	6			3	12–25		4	
Stolby Nature Reserve[8]	1	11–40	5–8	2–4	2–9	6–10			16–29	1.8	2	8–16	
Far East[9]			x	x	x	x	x						

[a]The superscript numbers denote the following references: 1, Sablina (1955); Okarma et al. (1995); 2, Ryabov and Ovchinnikova (1964); Prostakov (1996); 3, Alexandrov (1968); Kudaktin (1998); 4, Gordiyuk (1980); 5, Flerov (1935); 6, Fedosenko (1980); 7, Dulkeit (1964); 8, Zyryanov (1975); Suvorov (1989); 9, Bromley and Kucherenko (1983); Matyushkin (1992).
[b]Before 1960s.

Table 4.17. *Cervus elaphus.* Gender and age composition of populations (%)

Locality[a]	Year	Males	Females	Yearlings	Calves
Belovezhskaya Forest[1]	1945–1973	25–40	39–50	19–26	12
Voronezh oblast[2]	1983–1990	18–25	52–66	5–10	10–15
Carpathians[3]	1977	29	54		17
Kavkazsky Nature Reserve[4]	1941–1966	26–46	43–64		8–14
Dzungarian Alatau Range[5]	1980	31	50		19
Altai Mountains[6]	1984–1988	25–35	53–64	2–12	3–6
Stolby Nature Reserve[7]	1955–1971	29	57		14
Yakutia[8]	1953–1964	31	34	16	19
Cis-Baikal[9]	1958–1972	26	38	12	24
Tuva republic[10]	1971–1987	37	49		14
Sikhote-Alin Mountains[11]	(1965–1969)	25	45–47	5–9	21–24
Sikhote-Alinsky Nature Reserve[12]	(1998)	18	53	9	20

[a]The superscript numbers denote the following references: 1, Sablina (1955); Shostak (1975b); 2, Prostakov (1996); 3, Kerechun (1980); 4, Alexandrov (1968); 5, Fedosenko (1980); 6, Sobansksy (1992); 7, Zyryanov (1975); 8, Egorov (1965); 9, Sviridov (1978); 10, Smirnov (1989b); 11, Bromley and Kucherenko (1983); 12, Myslenkov (pers. comm.).

Table 4.18. *Cervus elaphus.* Total number (official data) of animals in Russian Federation. (Borisov et al. 1992; Mirutenko 1996, 2000)

Locality	No. of animals (×1,000)			
	1981	1990	1996	1999
Total for all areas	149.2	188.6	168.8	178.3
Kaliningrad oblast	1.4	1.4	1.1	1.3
NW European Russia	0.04	0.1	0.04	0.04
C European Russia	4.7	4.6	3.8	4.5
S European Russia	1.9	2.2	3.4	2.1
Volga region	1.3	2.2	1.9	2
Ural region	0.9	0.7	0.7	0.8
N Caucasus	5.1	8.5	4.4	5
W Siberia and Altai	10.5	10.1	8.3	9.4
E Siberia	74.5	115.4	83.6	94
Far East	48.9	43.4	61.6	59.2

4.8.1 Population Density and Size (Tables 4.18, 4.19 and 4.20)

The numbers of some red deer subspecies were estimated in the 1980s: *C. e. hippelaphus* 9,000 (Fadeev 1990); *C. e. montanus* 8,700–9,700 (Muntyanu et al. 1992); *C . e . bactrianus* 700 (Sokov 1987); *C. e. sibiricus* 32,000 (Kuzyakin 1985); *C. e. xanthopygos* 51,000 (Bannikov and Pivovarova 1983).

Table 4.19. *Cervus elaphus*. Population density and size in nature reserves

Species	Locality[a]	Year	Population density (animals/10 km²)	Total number of animals (×1,000)
C. e. maral	Kavkazsky Nature Reserve[1]	1934	2–3	0.9–1.0
		1937	6	1.9
		1939	8	2.3
		1945	6	1.7
		1951	7	2.2
		1955	14	1.4
		1960	16	4.2
		1963	25	6.5
		1966	30	8.0
		1972		8.1
		1982		3.3
		1992		1.8
		1999		1.3–1.5
	Borzhomi Nature Reserve[2]	1934		0.08
		1940		0.8
		1944		1.4
		1964		2.0
		1970		1.4
		1999		0.15
C. e. bactrianus	Total[3]	1990		0.9
		1997		0.6
	Tajikistan (total)[3]	1965		0.26
		1970		0.34
		1975		0.4
		1980		0.43
		1985		0.41
		1993		0.65
	Tigrovaya Balka Nature Reserve[4]	1946–1949		0.5–0.6
		1990		0.35
		1997		0.1
	Ramit Nature Reserve[4]	1990		0.15–0.2
		1997		8–12 animals
	Uzbekistan Kyzyl-Kumsky Nature Reserve[4]	1997		0.15–0.22
	Kazakhstan Karachingil (Ili River) Nature Reserve[4]	1990		0.08
		1997		0.2
C. e. sibiricus	Stolby Nature Reserve[5]	1971	6–7	0.34–0.5
		1998		0.1–0.12
C. e. xanthopygos	Sikhote-Alinsky Nature Reserve[6]	1948	6	
		1996	Max 2.1	

[a]The superscript numbers denote the following references: 1, Alexandrov (1968); Kudaktin (pers. comm); Trepet (1997); 2, Arabuli (1975); Kudaktin (pers. comm.); 3, Chernyshev (1958); Stroganov (1959); Bannikov (1972); Sokov (1987); Sapozhnikov (1967); Sokov (1993); 4, Sinelnikov (1964); Pereladova et al. (1997, 1999a); 5, Zyryanov (1975); Knorre et al. (1999); 6, Kaplanov (1948); Myslenkov (pers. comm.).

Table 4.20. *Cervus elaphus.* Population density and size in hunting grounds

Locality[a]	Year	Population density in forest (animals/10 km²)	Population size (×1,000)
Lithuania[1]	1988		2.15
	2001	2.4	4.73
Latvia[1]	1988		8.75
Belarus[1]	1988		5.0
	1990–1993		5–7
Ukraine (all)[1]	1988		16.0
C S European Russia[b2,13]	1956		1.1
	1965	0.95	0.7
	1970	1.8	1.55
	1975	2.3	1.7
	1980	1.6	1.5
	1985	2.0	1.9
	1990	1.5	1.7
	1995	2.8	2.4
Carpathians[1,3]	1960		2.0
	1970		6.3
	1974		8.6
	1988		8.7
Crimean Peninsula[1]	1988		1.9
N Caucasus[4,13]	1979	1–9	6.3
	1988		8.2
	1991		6.4
	1995		4.4
Georgia[1]	1988		1.4
Azerbaijan[1]	1988		1.3
Kazakhstan (all)[5]	1980	2–30	10.5
	1992–1999		10–15
Dzungarian Alatau Range[5]	1969	19.2	7.5
Kyrgyzstan[6]	1972	10.8	0.13
Altai Mountains[7,13]	1975	1–20	8.5
	1995		7.6
W Sayan Mountains[8]	1972	2–11	
W and E Sayan Mountains[8]	1980–1992	1.5	8.5
Tuva republic[8]	1975	0.2–1.2	
	1992	0.6–4	10.4
Buryatia republic[11]	1980–1992	0.8	36
Transbaikal area[11]		0.5	
Khamar-Daban Mountains[11]	1980 (Summer)	0.5–1.7	
	1980 (Winter)	2–9	
Chita oblast[11,13]	1980–1992	0.9	20
	1995		19
	1998		26.5
Krasnoyarsk kray[9,13]	1965	1–1.3	8.0
	1974	1–11	7.0
	1980		11.5
	1985		8.7
	1990		9.5
	1995		7.3

Table 4.20. *Continued*

Locality[a]	Year	Population density in forest (animals/10 km^2)	Population size (×1,000)
Khakassiya republic[10,13]	1973	0.9	1.3
	1991		2.9
	1995		6.9
Irkutsk oblast[10,13]	1965	1.2–1.3	5
	1991		25.5
	1995		28.8
Khabarovsk kray[12,13]	1983		14–18
	1991		26.8
	1995		18.4
Amur oblast[12,13]	1983	0.4–0.5	6–7
	1995		7.9
Sikhote-Alin Mountains[12]	1983	1.5–3.3	20–24
	1995		29.8
Sakhalin[13]	2000		0.35

[a]The superscript numbers denote the following references: 1, Drew et al. (1989); Kozlo (1995); Baleišis et al. (2002); 2, Solomatin (1975); Bannikov et al. (1983); Prostakov (1996); 3, Kerechun (1975, 1980); 4, Kotov and Ryabov (1963); 5, Fedosenko (1984b); Bekenov et al. (1992); Bekenov (1999); 6, Yanushevich et al. (1972); 7, Sobansky (1975, 1992); 8, Smirnov (1990, 1994); 9, Zyryanov (1975); 10, Lavov (1974); 11, Shvetsov et al. (1984); 12, Bromley and Kucherenko (1983); 13, Mirutenko (1996, 2000); Puzansky (1999).
[b]Voronezh + Lipetsk + Belgorod + Kursk + Tambov oblast.

Table 4.21. *Cervus elaphus.* Damage to the undergrowth in the Belovezhskaya Forest Nature Reserve (Korochkina and Bogdanovich 1975)

Year	1960		1967		1973	
Density of *Cervus elaphus* (animals/10 km^2)	14.5		36.7		28.2	
	No. of stems	Stems damaged (%)	No. of stems	Stems damaged (%)	No. of stems	Stems damaged (%)
Pine	2,104	27	71	34	114	8
Spruce	1,035	0	3,726	1	2,406	20
Birch	684	1.5	96	25	171	21
Aspen	495	47	267	47	203	89
Oak	243	47	202	38	173	32
Carpinus	8	0	0	0	97	76
Other	4	0	0	0	6	0
Total	4,573	20	4,362	7	3,170	26

4.8.2
Impact on Vegetation (Table 4.21)

In the Belovezhskaya Forest a human-induced peak in deer numbers occurred in the early 1900s and deer (together with European bison and large numbers of domestic cattle pastured in the forest) eliminated forest undergrowth (Table 4.21). As a result, there is a 20-year gap in the forest succession (Severtsov and Sablina 1953; Sablina 1955; Bannikov and Lebedeva 1956; Jêdrzejewska et al. 1997). In the Stolby Nature Reserve, plant mortality exceeded 10% for *Caragana*, *Salix*, *Sorbus*, and *Populus* (Zyryanov 1975); in Kazakhstan 75% of undergrowth of *Picea schrenkiana* was damaged, *Salix* 11%, aspen 6%, *Lonicera*, *Rosa*, and *Juniperus* were also damaged (Fedosenko 1980).

4.9
Relations to Man

Meat, velvet, antlers and tail glands (employed in Chinese medicine) are used. In Altai, there are deer farms specializing in velvet antler production. The annual production of dried antler from maral (*C. e. maral*) is about 28 tons (Drew et al. 1989). Hunters killed 1,000–3,500 red deer in the Far East over the period from 1966–1978 (Bromley and Kucherenko 1983). Approximately 5,000 animals were officially hunted annually in the USSR (Bannikov and Pivovarova 1983). However, in the 1980s, it was estimated that the annual harvest (together with poaching) was 50,000 in the former USSR territories, 40,000 of these were harvested in the Russian Federation (Danilkin 1999).

5 Sika Deer Manchurian – *Cervus nippon*

Cervus nippon Temminck, 1838.
Artiodactyla, Cervidae, Cervinae

5.1 Status and Distribution (Fig. 5.1)

5.1.1 Taxonomic Remarks

Only *C. n. hortulorum* Swinhoe, 1864 (Far East) was found (Heptner 1989). It inhabits a natural range and has been introduced into the European part of the former USSR.

5.1.2 Status

In Russia, in its natural range, sika deer is decreasing in number and its hunting is banned. In the European part of the former USSR, it is bred for game and used for sport hunting (Bannikov and Pivovarova 1983; Makovkin 1999; Rozhnov 2001).

5.1.3 Distribution (Figs. 5.2 and 5.3)

Distribution Area. Ca. 102,000 km^2 (original range).

Occurrence in Ecoregions. Warm Continental Regime Mountains (M212) (natural); Warm Continental (212), Prairie (252), Prairie Regime Mountains (M252) (acclimatized).

Fig. 5.1. Sika deer

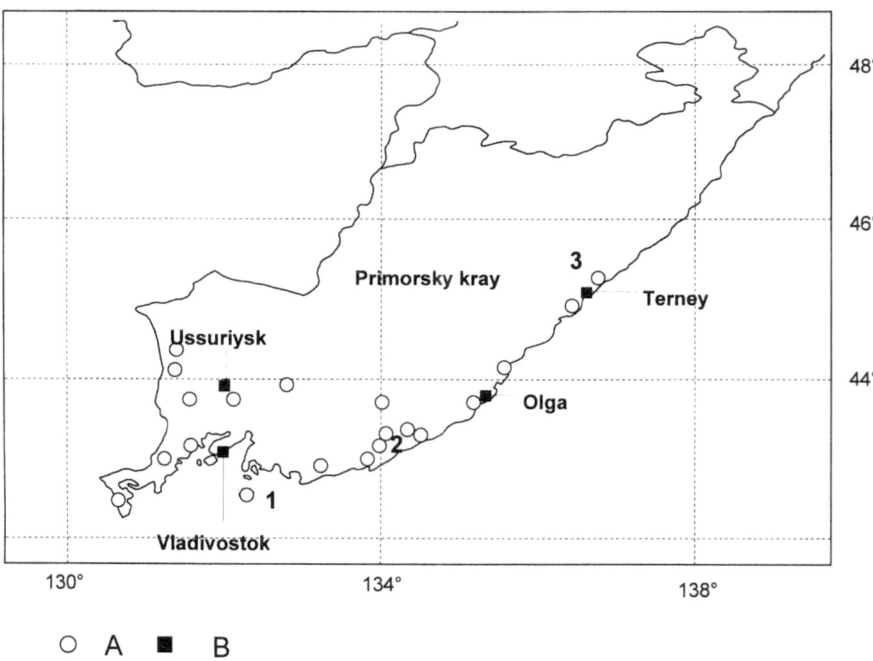

Fig. 5.2. *Cervus nippon*. Distribution map in the Far East (Makovkin 1999). *A* local populations; *B* cities; *1* Askold Island; *2* Lazovsky Nature Reserve; *3* Sikhote-Alinsky Nature Reserve

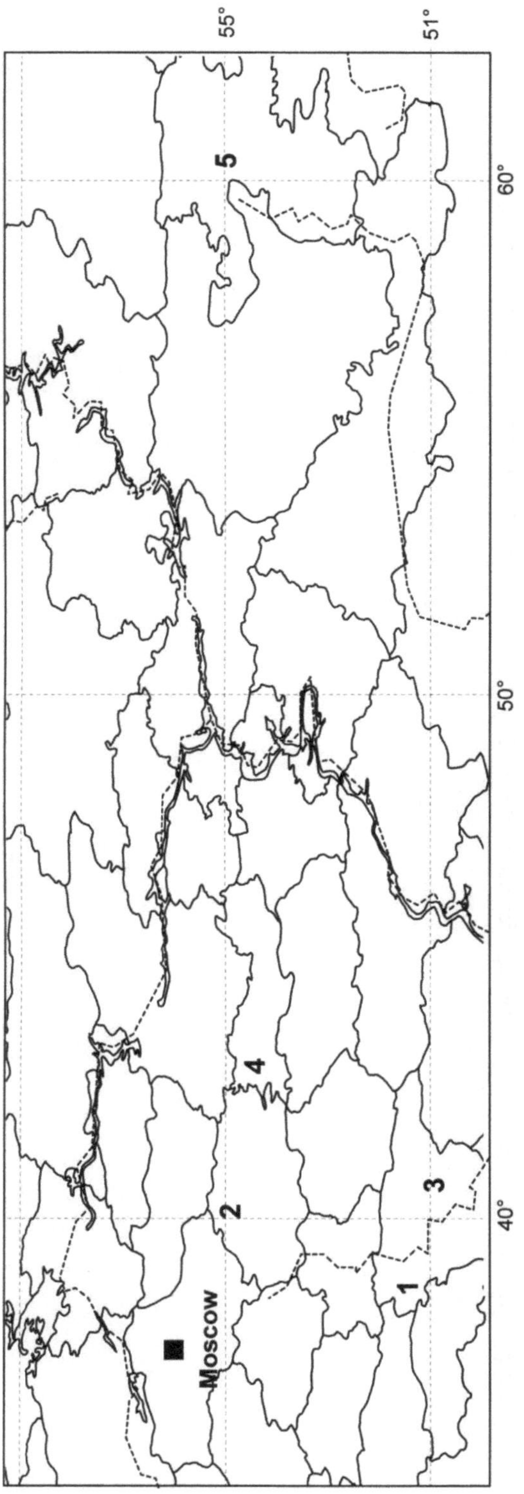

Fig. 5.3. *Cervus nippon.* Populations in the area of introduction (Bannikov et al. 1983; Prostakov 1996). *1* Voronezh oblast; *2* Oksky Nature Reserve; *3* Khopersky Nature Reserve; *4* Mordovsky Nature Reserve; *5* Ilmensky Nature Reserve

Table 5.1. *Cervus nippon*. Measurements

Locality[a]	Longitude (°E)	Latitude (°N)	Gender	*n*	Body mass (kg)	*n*	Head and body length (cm)	Tail length (cm)
Lazovsky Nature Reserve[1]	134	43	Male	5	117.4 (104–139)	5	173.2 (168–186.7)	18 (17–19)
			Female	12	73.1 (60–92.6)	12	162.4 (149–176.1)	16.3 (14–17)
Khopyorsky Nature Reserve[2]	40	52	Male	42	138.3	66	190.6	
			Female	34	89.6	60	174.7	
Voronezh oblast[3]	38–43	49–52	Male	54	114.5 (77–146)	59	170.1 (150–196)	18.4 (17–21)
			Female	60	68.3 (61–102)	66	158.2 (143–178)	16.6 (14–19)

[a]The superscript numbers denote the following references: 1, Prisyazhnyuk (1981); 2, Petrashov (1977); 3, Prostakov (1996).

5.2 Measurements (Table 5.1)

5.3 Habitat (Table 5.2)

Broadleaf forest, on seashores and on nearby slopes of mountains (Dormidontov 1977).

Terrain Form. Seashores; flood plains; foothills, valleys and slopes of mountains.

Altitude. 0–500 m a.s.l. (Far East) (Bromley 1956).

Snow. Snow depth of more than 40–50 cm is limiting (Bromley and Kucherenko 1983). Pressure of hoof is on average 940–980 g/cm^2, maximum 1,110 g/cm^2 (Bromley 1956). Fawns and young animals have the highest pressure, while animals of 5 years or older have the lowest. Prefer sites where snow cover lasts for less than 140 days.

5.3.1 Biotope (Table 5.2)

5.4 Diet and Feeding Habits

5.4.1 Diet

Thin twigs from trees and shrubs, buds, bark, acorns (in winter), herbs and tree leaves, flowers and grass in summer and also mushrooms in autumn. In S European Russia, acorns comprise up to 40% of the diet in autumn (Protoclitova 1961; Prisyazhnyuk and Prisyazhnyuk 1974). For a list of plant genera eaten see Appendix 1, Table 5. Sika deer need salt (obtained from sea algae and salt water).

Feeding Behavior. Picks small pieces of food from each plant (Bromley 1956).

Life Forms/Plant Parts Eaten (Table 5.3)

Feeding Niche. In field and shrub layers.

Table 5.2. *Cervus nippon.* Habitat[a]

Habitat type		Locality[b]	
		Far East[1]	Voronezh oblast[2]
Mature forests	Pine plantations		r
	Spruce-fir	r	
	Mixed deciduous-coniferous		+
	Aspen		r
	Oak on elevated places		+++
	Oak restored after burns	+++	
	Broadleaf in valley with undergrowth	++	+
Young forests	Young thickets of *Quercus, Tilia, Fraxinus, Corylus, Lespedeza, Juglans*	+++	
Shores of rivers and lakes			+
Flood plain			r
Alder forest			r

[a]Symbols indicate percent of habitat use was >30% (+++); 20–30% (++): 10–20% (+); <10% (r); x indicates habitat use noted, no numerical data reported.
[b]The superscript numbers denote the following references: 1, Bromley (1956); 2, Prostakov (1996).

Table 5.3. *Cervus nippon.* Life forms/plant parts eaten (x indicates use)

Life forms	Plant parts eaten							
	Whole plant	Leaves	Stems/ twigs	Bark	Roots/ bulbs	Flowers	Fruits	Seeds
Trees		x	x	x			x	x
Shrubs		x	x	x			x	x
Herbs		x	x			x		
Grasses/sedges		x	x					
Fungi	x							

5.5 Social Organization and Behavior

5.5.1 Social Life

In the 19th century, 20–60 animals were found in each group in the Far East, today the group consists of 7–9 animals, except during spring and summer, when females with fawns live separately (Bromley 1956). In introduced populations, the average size of groups is 1.6–1.9 in summer and 3–4.1 in winter (Prostakov 1996). During rut, the dominant male keeps some females within its territory. Later, young males join these groups (Table 5.4).

5.5.2 Home Range

Move less than 4–7 km from the center of the home range. In summer, the home ranges for single animals are 1–2 km^2 and for groups, 4–6 km^2; in winter, the range is less, about 20–30 ha. During rut, a male's territory is about 4 km^2, with 6–7 "rutting points", if the harem consists of 4–5 animals. The territory can be up to 9 km^2 if the harem includes 14–16 animals (Bromley 1956).

5.5.3 Mode of Movement

Steps are 75–110 cm, normal jumps are 3–3.6 m, maximum jumps are 6–8 m (Bromley and Kucherenko 1983). Good swimmer, capable of crossing a 10-km-broad seawater stretch.

5.5.4 Escape Behavior

Vigilant, detects man in the forest by listening at 300–350 m, they run away at 200–250 m (Bromley 1956).

Table 5.4. *Cervus nippon.* Size and composition of groups (%)

Locality[a]	Composition	No. of animals per group				
	Sexual composition	1	2–4	5–10	11–20	>20
Askold Island (Far East)[1]	Males	35	45	20	0	0
	Females with fawns	9	42	27	17	5
	Mixed herds	0	23	32	19	26
	Seasonal composition					
	Summer	14	46	23	15	2
	Autumn	12	40	32	4	12
	Winter	0	28	28	31	13
Voronezh oblast[2]	Seasonal composition	14	73	13	0	0

[a]The superscript numbers denote the following references: 1, Prisyazhnyuk and Prisyazhnyuk (1974); 2, Prostakov (1996).

5.6 Reproduction (Table 5.5)

At the start of the rutting season males arrange "rutting points", about 4 m^2 where they scrape away soil and trample down vegetation, fray trees with their antlers, urinate, lie in the mud and roar. Battles among males are rare and do not lead to the death of competitors. In the middle of the rut males have harems of 3–5 females. In Voronezh oblast the average number of fawns per female was 1.08; in 92% of cases, females were observed with 1 fawn, in 8% with 2 fawns (Prostakov 1996).

5.6.1 Seasonal Breeding Events (Table 5.6)

Maximum Length of Life. 11–14 years, maximum 20 years (Dormidontov 1969).

Table 5.5. *Cervus nippon*. Breeding parameters. (Bromley 1956)

Age at sexual maturity (years)		Participation in breeding (years)		Length of Pregnancy (days)	Barrenness	Litter size
Male	Female	Male	Female			
1.5–2.5	1.5	3–4	2.5, Rare 1.5	233–241	Rare	1, Very rarely 2

Table 5.6. *Cervus nippon*. Seasonal breeding events

Locality[a]	Rutting season			Calving season		
	Preparation to rut	Basic rut	Last appearances of rut	First calves	Main calving	Last calves
Far East[1]	29 September–11 October	15–25 October	31 October–4 November	Last April	Mid May	End May
Khopyorsky, Voronezh[2]	6 September	Second part of September–End November	2 December	6 May	End May–June	6 July

[a]The superscript numbers denote the following references: 1, Bromley (1956); 2, Prostakov (1996).

5.7 Mortality

Heavy snow causes mortality (Far East). In years of heavy snow, numbers of sika deer were reduced to 50% of the original population size (Bromley and Kucherenko 1983). Among the dead animals, 32% were fawns, 40% were adult females and 28% were adult males (Bromley 1956).

5.7.1 Predators

Grey wolf, dogs, yellow-throated marten, lynx, tiger and leopard (Bromley 1956). Wolf predation is highest during March–April, when there is crust

Table 5.7. *Cervus nippon.* Gender and age composition of populations (%)

Locality[a]	Year	Total population (%)		
		Males	Females	Fawns
Far East[1]	1938	21	64	15
Askold Island[2]	1898	31	46	23
	1930	35	40	25
	1935	24	43	33
	1960	17	62	21
Voronezh oblast[3]	1988–1991	23 (20–28)	57 (49–60)	20 (16–23)

[a]The superscript numbers denote the following references: 1, Vendland (1938); 2, Prisyazhnyuk (1978); 3, Prostakov (1996).

Table 5.8. *Cervus nippon.* Total number of animals in each year of study (official data) in the Russian Federation. (Borisov et al. 1992; Mirutenko 1996, 2000; Danilkin 1999)

Locality	Total number of animals (×1,000)		
	1981	1990	1995
NW of Russian Federation	0.19	0.3	–
Central regions of European Russia	2.92	6.5	3.1
Volga-Vyatka region	0.2	0.6	–
Volga region	0.42	1.13	1.44
N Caucasus region	0.52	1.15	0.3
Ural region	0.69	0.6	0.08
Far East region	2.0	7.0	5.9
Total	7.2	18.5	10

on top of the snow and slippery ice on rivers. Of the deer found dead, 77% were killed by wolves, 10% by dogs and 3% by lynx (Bromley 1956). In European Russia, the main predators are wolf, dog and hybrids of wolf and dog (Ryabov et al. 1994).

5.7.2 Diseases, Helminths and Ectoparasites (Appendix 2, Tables 4 and 5)

5.8 Population Structure and Densities (Tables 5.7, 5.8 and 5.9)

Between 1945–1950, in the Lazovsky Nature Reserve, the population of deer included 13% aged ≥5 years, 18% aged 3–5 years, 13% aged 2–3 years, 24% aged 1–2 years and 32% were fawns (Bromley 1956).

5.8.1 Population Density and Size (Tables 5.8 and 5.9)

Table 5.9. *Cervus nippon*. Population density and size in different localities

Locality[a]	Year	Population density (animals/10 km^2)	Total number of animals (×1,000)
Oksky Nature Reserve[b,1]	1938–1992		0.05
Mordovsky Nature Reserve[c,1]	1952–1982		0.19
Khopyorsky Nature Reserve[c,1]	1961		0.5
	1966	42	0.7
	1972		1.5
	1985		2.55
Voronezh (hunting husbandries)[1]	1990	2.6–8.8	0.25
Ilmen Nature Reserve[b,1]	1940–1979		0.03
Far East[2]	1900		10
	1950		1.2
	1960		1
	1973		0.7
	1983		4.1
	1990		7
	1998		5.12
Askold Island	1998		0.1
Lazovsky Nature Reserve[3]	1993	12–20	0.73
	1996		0.39
Sikhote-Alinsky Nature Reserve and surrounding[4]	1997	1.3	0.13

[a]The superscript numbers denote the following references: 1, Shtarev (1966); Dormidontov (1969); Bannikov and Prisyazhnyuk (1977); Filonov (1993); Prostakov (1996); 2, Bromley (1956); Pikunov et al. (1973); Ryashchenko (1976); Danilkin (1999); Makovkin (1999); 3, Makovkin (1997, 1999); 4, Voloshina (1997).
[b]Introduced population.
[c]Introduced population with intensive human involvement in population regulation.

5.9 Relations to Man

Protected species in the Far East. Farmed in the Far East for velvet antlers (Mirolyubov and Ryashchenko 1948) and in many game husbandries in European Russia for hunting as a form of sport. In the 1980s, in the Far East, there were 60,000 farmed sika deer, in Altai 15,000 (Khakhin and Prisyazhnyuk 1985). Strong impact on vegetation was found when the population density was ≥50 animals/10 km^2 (Zobov 1995; Makovkin 1999).

6 Moose – *Alces alces*

Alces alces Linnaeus, 1758.
Artiodactyla, Cervidae, Odocoileinae

6.1 Status and Distribution (Fig. 6.1)

6.1.1 Taxonomic Remarks

Three subspecies occur: *A. a. alces* Linnaeus, 1758 (European Russia, Ural Mountains, W Siberia, Altai Mountains); *A. a. pfizenmayeri* Zukowski, 1910 (E of Yenisey River to Chersky Range); *A. a. buturlini* Chernyavsky et Zhelesnov, 1982 (NE Siberia); *A. a. cameloides* Milne-Edwards, 1867 (Amur region, Sikhote-Alin Mountains) (Chernyavsky and Zheleznov 1982; Heptner 1989).

6.1.2 Status

Numerous, hunted (Filonov 1983b; Sokolov and Lebedeva 1989; Smirnov 1994; Lomanov and Lomanova 2000).

6.1.3 Distribution (Fig. 6.2)

Distribution Area. Ca. 14,418,000 km^2.

Occurrence in Ecoregions. Tundra (124), Subarctic (131, 132, 133, 134, 135, 136), Subarctic Regime Mountains (M131, M132, M133, M134, M135), Warm Continental (212), Warm Continental Regime Mountains (M212), Prairie (252, 253), Prairie Regime Mountains (M251), Temperate Steppe (331, 332).

Fig. 6.1. Moose

6.2
Measurements (Table 6.1)

6.3
Habitat (Table 6.2)

Mostly forests. Also found in forest steppe and forest tundra, as well as in forests on river flood plains, and shrub-covered ravines of tundra and steppe. Inhabits lowlands as well as mountains up to the tree line (Ustinov 1960; Petrovsky et al. 1967; Pavlov 1976; Makarova 1981).

Terrain Form. Seashores; shores of freshwater; flood plains; lowlands and bogs; plains; mountain plateaus, foothills, valleys, and gentle slopes (Kaplanov 1935, 1948; Dmitriev 1938; Semenov-Tien-Shansky 1948a; Laptev 1958; Egorov 1971; Yazan 1972; Timofeeva 1974; Bromley and Kucherenko 1983; Domnich 1986; Baidavletov 1988; Sobansky 1992; Baskin 1994).

Altitude. 200–1,700 m a.s.l. in Altai (Baidavletov 1988).

Temperature. Inhabits areas with a temperature range from –60 to +30°C.

Snow. Snow depth >70–80 cm is limiting. However, moose can successfully inhabit areas with a snow depth up to 200 cm. During such conditions, they keep to 2–3 magisterial trails. When threatened, moose prefer to use these

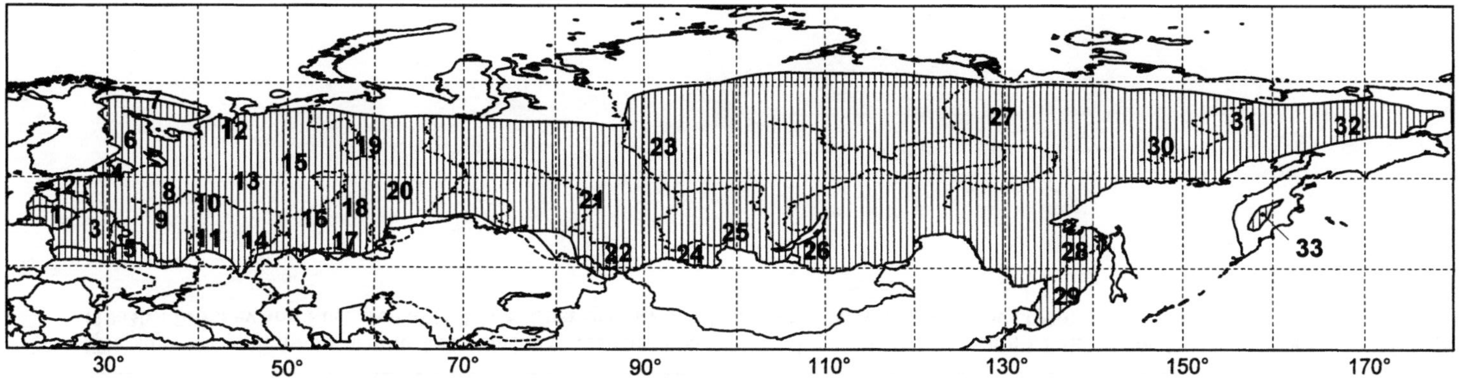

Fig. 6.2. *Alces alces.* Distribution map (Heptner 1989). *1* Lithuania; *2* Estonia; *3* Belarus; *4* Leningrad oblast; *5* Chernigov oblast; *6* Karelia; *7* Kola Peninsula; Laplandsky Nature Reserve; *8* Darwinsky Nature Reserve; *9* Moscow oblast; Tula oblast; Tulskie Zaseki Nature Reserve; *10* Oksky Nature Reserve; *11* Voronezh oblast – Belgorod oblast; Centralno-Chernozemnyi Nature Reserve; *12* Arkhangelsk oblast; *13* Kostroma oblast; *14* Saratov oblast; Tambov oblast; *15* Kirov oblast; Udmurtiya; *16* Tatarstan; Mordovsky Nature Reserve; *17* Orenburg oblast; Buzuluk area; *18* Ilmensky Nature Reserve; *19* Pechora River; Pechoro-Ilychsky Nature Reserve; *20* Sverdlovsk oblast; *21* Tomsk oblast; *22* Altai Mountains; *23* Enisey River; *24* Tuva; *25* E Sayan Mountains; *26* Buryatia; *27* Yakutia; *28* Amur River; *29* Sikhote-Alin Range; *30* Kolyma River; *31* Omolon River; *32* Anadyr River; Chukotka; *33* Kamchatka

Table 6.1. *Alces alces.* Measurements

Locality[a]	Longitude (°E)	Latitude (°N)	Gender	*n*	Body mass (kg)	Head and body length (cm)	Tail length (cm)
Belovezhskaya Forest[1]	24	53	Male	81	(336–418)	(246–260)	
			Female	60	(304–354)	(236–246)	
Voronezh oblast[2]	38–43	49–52	Male	56	319.5 (285–345)	251.2 (220–278)	17.5 (15.1–19.4)
			Female	58	296.6 (271–330)	243.1 (216–271)	16.5 (15.2–18.6)
Kazakhstan (Altai)[3]	84–87	49–50	Male	3	373.6 (301–448)	279.6 (268–292)	9.0 (8 –10)
			Female	4	295.0 (240–362)	257.7 (238–271)	
Yakutia[4]	106–163	55–68	Male	3	396.3 (364–440)	265 (256–278)	13
			Female	2	369 (340–398)		
Amur oblast[5]	137	50	Male	17	299.0 (240–423)	249.5 (118–286)	
			Female	18	280.8 (231–323)	243.2 (215–280)	
Omolon River[6]	160	63	Male	7	(300–744)	(270–346)	
			Female	7	(350–546)	(274–320)	
Chukchi okrug[7]	158–180	64–68	Male	5	(550–645)	(292–315)	
			Female	2	(456–472)		

[a]The superscript numbers denote the following references: 1, Kozlo (1983b); 2, Prostakov (1996); 3, Baidavletov (1984); 4, Egorov (1971); 5, Abramov (1963b); Bromley and Kucherenko (1983); 6, Chernyavsky and Domnich (1989); 7, Zheleznov (1990).

Table 6.2. *Alces alces*. Habitat[a]

Habitat type	Season[b]	Locality[c]												
		Karelia (1)[1], Kostroma (2)[2], Pechora Basin (3)[3]			Leningrad oblast (1)[4], Moscow, Tver (2)[5], Bryansk (3)[6]			Sverdlovsk (1)[7], Tomsk (2)[8], Altai (3)[9]			Yenisey, Yakutia (1)[10], Sayan, Amur (2)[11]		Kolyma (1)[12], Sikhote-Alin (2)[13]	
		1	2	3	1	2	3	1	2	3	1	2	1	2
(Ecoregion)		134			212			M131	253	M251	132, 136	M133, M134	M132	M212
Mature forest														
Coniferous	W	+	++	x	++	r		+		x	x	x	+	x
	S	r	r		r			r			x	x	++	
Mixed deciduous-coniferous[d]	W		+++	x	r	r	++	++		+++				
	S		r		r			+++						
Broadleaf	W	+	+	x	+	r	++	r			x	x		
	S		r	x	r			r			x			
Young forest														
Coniferous	W						+++	r		++				
	S							r		+++				
Mixed deciduous-coniferous	W	+						r						
	S							r						
Forest stand														
Coniferous	W								x					
	S								x					
Broadleaf	W			x					x					
	S			x					x					
Freshwater shore														
Shrubs	W									+++			+	
	S			x						+++			+++	x
Forest	W	+	r		r	++			x	+	++	x	+++	
	S	++	+	x	+				x	+	r	x	+++	x

Table 6.2. *Continued*

Habitat type	Season[b]	Locality[c]												
		Karelia (1)[1], Kostroma (2)[2], Pechora Basin (3)[3]			Leningrad oblast (1)[4], Moscow, Tver (2)[5], Bryansk (3)[6]			Sverdlovsk (1)[7], Tomsk (2)[8], Altai (3)[9]			Yenisey, Yakutia (1)[10], Sayan, Amur (2)[11]		Kolyma (1)[12], Sikhote-Alin (2)[13]	
		1	2	3	1	2	3	1	2	3	1	2	1	2
Bog	W		r	x	r		+	++			r	x		
	S	+++	++	x	++			++			x	x		
Burnt forest	W	+								+++	+++	x		x
	S	+						r			x	x		x
Cut forest	W	+	+++		+++	++				+++		x		
	S	+	r		++									
Shrubland	W			xx									r	
	S			x										
Forest clearings,	W	r			+	+++					r			
edges	S				++					++				
Tundra, forest-	W												+	
tundra	S			x									++	
Grassland	W							r					r	
	S			x				+	x					

[a]Habitat used in >30% (+++); 20–30% (++): 10–20% (+); <10% (r); x indicates habitat use noted, no numerical data reported.
[b]W–winter, S-summer.
[c]The superscript numbers denote the following references: 1, Ivanter (1968); Troitsky (1972); Markovsky (1978, 1981); 2, Bogatyrev (1982); Baskin (1994); 3, Teplov (1960); Yazan (1972); 4, Timofeeva (1974); 5, Yurgenson (1964); Danilov (1950, 1951); Alexandrova and Krasovsky (1957); Kaletskaya (1961); Zablotskaya (1964); 6, Vatolin (1980); 7, Dvornikov and Dvornikova (1985); 8, Kaplanov (1935); Laptev (1958); 9, Dmitriev (1938); Baidavletov (1988); Sobansky (1992); 10, Egorov (1971); Tsarev (1975); Zyryanov (1975); 11, Abramov (1963b); Dulkeit (1964); Bromley and Kucherenko (1983); 12, Sleptsov (1975); Chernyavsky and Domnich (1989); Zheleznov (1990); 13, Kaplanov (1948); Bromley and Kucherenko (1983).
[d]All types of mixed coniferous-broadleaf forest.

trails to escape. A hoof pressure of 420–500 g/cm^2 (Nasimovich 1955) means that moose sink through snow of almost any thickness.

6.3.1
Biotope (Table 6.2)

6.4
Diet and Feeding Habits

6.4.1
Diet

Specialized to live on woody species and herbs. Twigs and bark of trees and shrubs are eaten in winter; leaves of trees and shrubs, aquatic plants, herbs and grass during seasons without snow. Coniferous trees and shrubs are also eaten. Toxic plants comprise 13–33% of the moose diet, some of them being the main food species: *Pinus*, *Juniperus*, *Equisetum*, *Caltha*, *Ranunculus*, *Padus*, *Solidago*, *Convallaria* and *Calla*. However, *Aconitum* and *Cicuta* are deadly for calves (Alexandrova and Krasovsky 1957; Knorre 1959; Egorov and Krivosheev 1965; Baskin 1994). In most areas, salt licks are not used, but in Yakutya, Amur and the Sikhote-Alin they are very important (Bgatov et al. 1988; Zaumyslova 1989). Plants, which are rich in water, are used to satisfy thirst (Ditsevich 1988; Gusev 1989). For a list of plant genera eaten see Appendix 1, Table 6.

6.4.2
Life Forms/Plant Parts Eaten (Table 6.3)

6.4.3
Seasonal Aspects of Diet (Table 6.4)

6.4.4
Consumption of Food /Day (kg) (Tables 6.5 and 6.6)

Feeding Niche. In field and shrub layers.

6.5
Social Organization and Behavior

6.5.1
Social Life

Live as solitary animals or in small groups. Mother and calves live together. Strong personal connections between a pair of tame females were found,

Table 6.3. *Alces alces.* Life forms/plant parts eaten (x indicates the type used)

Life forms	Plant parts eaten							
	Whole plant	Leaves	Stems/twigs	Bark	Roots/bulbs	Flowers	Fruits	Seeds
Trees		x	x	x		x		
Shrubs		x	x					
Dwarf shrubs		x	x					
Herbs		x	x		x			
Grasses/sedges		x	x					
Horsetails		x	x					
Lichens[a]	x							
Fungi	x							

[a]Tree lichens.

Table 6.4. *Alces alces.* Preferred seasonal food[a]

Locality[b]	Food	Winter	Spring	Summer	Autumn
Leningrad oblast[1]	Coniferous trees and shrubs	+++			+
	Deciduous trees and shrubs	+++	++	++	+++
	Dwarf shrubs	r	+	r	+
	Terrestrial herbs and grass	r	+	+++	
	Aquatic plants	r	+	+++	r
Altai Mountains[2]	Coniferous trees and shrubs	r			
	Deciduous trees and shrubs	+++		+++	
	Dwarf shrubs				
	Terrestrial herbs and grass			+	
	Aquatic plants			++	

[a]Symbols used indicate food type consumed: heavily and is one of main foods (+++); often (++); occasionally (+); rarely (r).
[b]The superscript numbers denote the following references: 1, Timofeeva (1974); 2, Baidavletov (1988).

indicating a capacity for personal recognition (Baskin 1976). In large aggregations (in Altai, maximum 270 animals; Kostroma, maximum 75), social connections are absent, and animals scatter in different directions when disturbed. There are no leaders (Baskin 1994).

6.5.2 Home Range

Settled way of life is usual all year-round. However, in summer in Yakutia and Sikhote-Alin, moose cover long distances to salt licks. Clearly visible

Table 6.5. *Alces alces.* Consumption of food (fresh weight/day) (kg)

Locality[a]	Season	Consumption of food by age group		
		Calf	1–3 years old	>3 years old
Leningrad oblast[1]	Winter			10.2 (5–21)
Moscow oblast[2]	Autumn	7.7	12	20
	Winter	7.7	11.2	13
	Summer	2.4	15	35
Pechora Basin[3]	Winter	7.5		
	Spring			33

[a]The superscript numbers denote the following references: 1, Timofeeva (1974); 2, Kaletsky (1967); 3, Lanina (1959).

trails 10–20 km in length were observed in wintering areas and also between summer feeding grounds and the rich salt grounds in Yakutia (Egorov 1965). In summer, there are daily movements upwards to feed in alpine meadows, then downwards to rest in the coniferous forest (Baidavletov 1988). In Pechora, where there are autumn migrations, the moose were observed to be more mobile in the first half of winter. In mountains moose select shaded slopes with loose snow. When snow becomes deeper, mobility declines. Under these conditions, animals concentrate in a few places with good feeding conditions and many trails develop (Yazan 1972; Baidavletov 1988; Baskin 1994). In late winter, the home range is less than 2.5–10 ha. Daily roaming is <700–900 m in Lapland (Semenov-Tien-Shansky 1948a), and about 500 m in the Darwinsky Nature Reserve (Nasimovich 1955). Moose do not defend territory, but females with very small calves protect small areas against invaders. During their second summer, most yearlings (tame) moved up to 120 km from their mother, but some individuals stayed where they were born. Females are strongly attached to their places of calving and mating and return year after year to almost the same few meters (Knorre 1959; Kozhukhov 1973; Mikhailov 1973; Bogomolova and Kurochkin 1984; Minaev 1992).

6.5.3 Mode of Movement

Usual movements and walk are quite sluggish. Speed of gallop is 30 km/h, a fast trot 15 km/h, a normal trot is 7–10 km/h (Yazan 1972). They can jump <2 m in height. Good swimmers (can swim 2–3 km in lakes) and divers (Semenov-Tien-Shansky 1948a; Sobansky 1992). In swamps, moose crawl on their belly with the forelegs stretched out in front (Kaplanov 1935). Digging food up from below the snow occurs, but is not typical.

Table 6.6. *Alces alces*. Movement, feeding and defecation by adult moose (both sexes)/24 h, during winter, in Leningrad oblast (Timofeeva 1974)

	The first half of winter				The second half of winter				Average/day
	November	December	January	Average	February	March	April	Average	
Number of animals studied	3	8	6		7	10	5		
Travel distance (m)	1,060	1,526	1,526	1,540	1,772	1,083	910	1,287	1,369
Number of browsed twigs	2,082	1,683	1,668	1,740	1,820	1,490	1,498	1,347	1,480
Number of browsed trees	434	413	397	412	294	241	321	263	318
Weight of food (g)	11,466	8,680	8,654	9,110	11,125	10,130	7,458	10,428	10,167
Weight of dried food (g)	6,086	4,643	4,630	4,866	5,505	5,350	3,873	5,386	5,306
Number of pellet groups	18	9	14	12	18	23	28	22	19
Weight of fresh pellets (g)	3,887	3,213	4,316	3,667	5,889	6,441	5,586	6,483	5,572
Weight of dried pellets (g)	1,887	1,592	1,667	1,660	2,244	2,691	2,844	2,637	2,389

6.5.4 Dispersal/Migration

In Lithuania, 20–30 km autumn and spring migrations occur; animals migrate on a broad front (Baleisis 1990). The moose migrate from areas where snow depth is up to 70 cm (Pechora), or where snow, about 50 cm deep (Amur), makes willow shrubs inaccessible (Yazan 1972; Bromley and Kucherenko 1983). The length of migration reaches 300 km in W Siberia and 200–500 km in C Siberia (Laptev 1958; Tsarev 1975). The extent of migrations in the mountains reaches 100–150 km in Altai. During summer many bulls and barren females live in the subalpine zone (Baidavletov 1988; Sobansky 1992). The front of migrations is generally wide (50–150 km); however, in some places, moose move on a fairly narrow front, permitting mass catches by fence traps (Yazan 1961). Speed of migration is 1.5–2 km/h (Yazan 1972). In the same areas (Pechora, Amur), both migratory and settled populations occur.

6.5.5 Diurnal Activity (Table 6.7)

In hard frost, below –20 °C, moose are active for only 1–2 h/day and spend the rest of the time lying down. Only migratory animals are more active in very cold weather (Yazan 1972). In summer, after the appearance of blood-sucking insects, moose shift to a nocturnal mode of life. They spend the daytime in water, and only emerge onto windy open slopes and large clearings.

Table 6.7. *Alces alces.* Diurnal activity

Locality	0–3	3–6	6–9	9–12	12–15	15–18	18–21	21–24
Pechora Basin[1]								
Autumn								
Winter								
Transbailkal[2]								
Summer								

Black shading, most animals are active; gray shading, a few animals are active.
[a]The superscript numbers denote the following references: 1, Yazan (1972); 2, Skalon (1961).

6.5.6 Shelters

Harassment by blood-sucking insects is reduced by standing in water up to the neck, or staying in open, windy places. Otherwise, moose rest in places protected by thickets from one side and open on the other sides, with a preferred visibility of 70–90 m (Baskin 1994). Moose do not paw the snow or the ground before lying down.

6.6 Reproduction (Table 6.8)

Age of maturity is 1.5 years. Males participate in breeding from 2.5 years of age, females from 1.5 years of age. Knorre (1959) found that the females need to attain >300 kg to become pregnant, but Yazan (1972) found pregnant yearlings weighing <300 kg. Polycyclicity of females occurs at 18–21 day intervals over 2.5 months, and therefore, the age of embryos can vary by 2–2.5 months (Timofeeva 1974; Kozhukhov 1989). Time of heat extends for only 3–4 h. Pregnancy lasts for 215–243 days (Knorre 1959; Kozhukhov 1989). Female fertility (Table 6.8) declines if the autumn is early and cold.

Table 6.8. *Alces alces.* Fertility in different localities

Locality[a]	Barrenness (%)	Proportion of females with *n* embryos (%)			Average number of embryos/female
		1	2	3	
Estonia[1]	51	65.9	34.1	0	1.3
Leningrad oblast[2]	46[b]	79.6	20.2	0.2	1.2
Moscow oblast[3]	51[b]	72.7	27.2	0.03	1.3
Tambov oblast[4]	31[b]	61.5	38.5	0	1.4
Arkhangelsk oblast[3]	53[b]	65.2	34.5	0.3	1.35
Kirov oblast[5]	10[c]	47.8	52.2	0	1.5
Pechora Basin[6]	12[c]	72	28	0	1.3
Bashkortostan[7]	45	65.5	34.5	0	1.3
W Altai[8]	12.8	41.5	58.5	0	1.6
Novosibirsk oblast[9]	10.5	54.7	45	0.3	1.5
Tomsk oblast[10]	35.7	70.9	29.1	0	1.3
Yakutia[11]		69	31	0	1.3
NE Siberia[12]	44	78	22	0	1.2

[a]The superscript numbers denote the following references: 1, Kiyli (1985); 2, Timofeeva (1974); 3, Filonov (1983b); 4, Kheruvimov (1967); 5, Glushkov (1980); 6, Knorre (1959); 7, Gordiyuk (1982); 8, Baidavletov (1988); 9, Kiryukhin (1990); 10, Lyalin and Aushev (1989); 11, Egorov (1971); 12, Zheleznov (1990).
[b]Females ≥2.5 years.
[c]Females ≥1.5 year.

Barrenness occurs more often in lean females (Yazan 1972). Moose females give birth to 1–2 calves. The number of newborn in populations correlates with snow depth in the previous winter (Teplov 1960; Ling 1973; Kudryashova 1980; Filonov 1983a, 1993). The question of monogamy or polygamy is discussed (Likhachev 1955; Knorre 1959). Concentration of bulls showing rutting behavior (bellowing, physical marks and scent left on ground and trees) on small plots, indicates the existence of "leks". In these "leks", many bulls throw their antlers (Baskin 1994). A bull can probably mate with several females, but it will do so in different areas and at different times (Yazan 1972; Minaev 1992). The strongest bull monopolizes the females.

6.6.1
Seasonal Breeding Events (Table 6.9)

Shedding of antlers in Pechora starts in November and peaks during January (Yazan 1972); in the St. Petersburg region, November–January (Rusakov 1979).

Maximum Length of Life. Tame moose live 18–20 years and can reproduce until up to 17 years of age (Kozhukhov 1989).

6.7
Mortality

In the Darwinsky Nature Reserve, in 1950–1985, the survival of calves was negatively correlated to moose density, and to snow depth (Filonov and Kaletskaya 1988). In the Pechora, most calves die during spring migration. Calf mortality becomes less if females complete the migration before calving (Yazan 1972). Higher mortality occurs among twin calves (Timofeeva 1974). For the Darwinsky, Mordovsky and Oksky Nature Reserves, the average level of mortality from natural causes in 1948–1981 was relatively constant at 1.5–12.0% (Filonov 1983a, 1993).

6.7.1
Predators

Wolves killed 10–82% of moose found dead by game wardens, with the maximum in small nature reserves, where carcasses were found more often; bears killed 7–64% (maximum in Sikhote-Alin), wolverines 5–18% (Pechora and Yenisey areas), tigers 9% (in Sikhote-Alin Mountains). Percentage of animals killed by drowning during ice break was 22–50%, although drowned moose are probably overestimated (Semenov-Tien-Shansky 1948a; Filonov 1983a, 1989; Danilov 1987; Pavlov 1990;

Table 6.9. *Alces alces.* Seasonal breeding events

Locality[a]	Rutting season			Calving season		
	First demonstrations	Basic rut	Last demonstrations	First calving	Main calving	Last calves
Karelia[1]	August	September–first half of October			May–June	
Leningrad oblast[2]	Early August	August–September	Early October	April	First half of May	Early June
Kostroma[3]	End of August	September	Early October		First half of May	
Oksky[4]	End of August	September	26 September	Mid-April	20 April–15 May	
C S European Russia[5]	End of July	September	End of October	April	May	June
W Altai[6]	Second half of August	September–10 October		Early May	Second half of May	
Yakutia[7]	5–10 September	20 September–10 October	Middle of October		End of May–10 June	
Amur oblast[8]	Middle September	Mid-September–end of October			May–first half of June	
Far East[8]	End of August–early September	September–early October			May	
Kolyma[9]	End of August–early September	20 September–end October	November	12 May	20 May–10 June	

[a]The superscript numbers denote the following references: 1, Ivanter (1968); 2, Timofeeva (1974); Rusakov (1979); 3, Baskin (1994); 4, Zykova (1964); 5, Kheruvimov (1969); 6, Baidavletov (1984); 7, Egorov (1965); 8, Bromley and Kucherenko (1983); 9, Chernyavsky and Domnich (1989).

Table 6.10. *Alces alces.* Gender and age composition of populations (%)

Locality[a]	Reserve (R), hunting grounds (H)	Years	Male/female ratio		Animal gender and age (%)			
			Observed	Hunted	Males	Females	Yearlings	Calves
Lapland Nature Reserve[1]	R	1930–1940			39	40		21
Lapland Nature Reserve	R	1963–1975			33	45	8	14
Karelia republic[2]	H	1964–1967	0.96:1	1.38–1.77:1				
Leningrad oblast[3]	H	1962–1970	1.08:1	1.24:1	70–78		7–12	15–18
Darwinsky Nature Reserve[b,4]	R	1950–1974	0.56:1		27	44	9	20
Oksky Nature Reserve[5]	R	1947–1971	0.49:1		21	35	13	31
Tulskie Zaseki Nature Reserve[6]	R	1940–1941			21	13	33	33
Centralno-Chernozemnyi Nature Reserve[b,7]	R	1963–1974			30	20	20	30
Mordovsky Nature Reserve[8]	R	1949–1971			25	44	9	22
Tatarstan republic[9]	H			1.32:1				
Tambov oblast[10]	H			0.61:1				
Saratov oblast[11]	H		1.37:1					
Pechora Basin[12]	H		0.86:1	1.08:1		59	27	14
Pechoro-Ilychsky Nature Reserve[13]	R	1938–1948			45	40		15
Pechoro-Ilychsky Nature Reserve [13]	R	1955–1959			74		11	15
Bashkir republic[14]	H	1934–1937		1.17:1	42	35		23
Ilmensky Nature Reserve[b,15]	R	1948–1960			39	43	2	16
Yakutia republic[16]	H				26	26	18	30
Tuva republic[17]	H	1976–1985			45	43	12	
Tuva republic[17]	H	1989				82		18
Omolon Basin[18]	H		0.87:1			69	11	20
Chukchi okrug[19]	H	1972–1979			35	35	13	17
Anadyr Basin[19]	H		1.32:1			71	17	12
Sikhote-Alinsky Nature Reserve[20]	R	1937–1939		0.40:1	30	56	6	8

[a]The superscript numbers denote the following references: 1, Semenov-Tien-Shansky (1948a); 2, Troitsky (1972); 3, Timofeeva (1974); 4, Kaletskaya (1961); 5, Kiseleva et al. (1965); 6, Zablotskaya (1964); 7, Filonov (1983b); 8, Borodina et al. (1970); 9, Zaripov and Znamensky (1964); 10, Kheruvimov (1969); 11, Filonov (1983b); 12, Yazan (1972); 13, Knorre (1959); Teplov (1960); 14, Filonov (1983b); 15, Filonov (1983b); 16, Egorov (1965); 17, Smirnov (1990); 18, Chernyavsky and Domnich (1989); 19, Chernyavsky et al. (1980); 20, Kaplanov (1948).
[b]Average data for May–August for newborn, May–November for adults, April for yearlings.

Table 6.11. *Alces alces.* Fluctuations of population density (animals/10 km^2 of forest area), in ecoregions and localities, within the European part of the former USSR territory (1700–1995). Approximate data based on historical evidence of witnesses[a] (before 1960) and reviews (since 1960). (Aspisov 1930, 1956; Danilov 1949, 1951; Serzhanin 1961; Bannikov and Teplov 1964; Baskin 1998)

Ecoregion	134		212			252		332
Locality	Karelia republic	Kostroma oblast	Leningrad oblast	Moscow oblast	Belarus	Chernigov oblast	Tatarstan	Orenburg oblast
Year								
1700			0.5	1	1	1	7	
1750	0.5		0.5	1		1	7	
1800	1	0.5	0.5	0	2	1	0	0
1850		1	1	0.5		0.5	0.5	0
1860	0.5	3	3	1	3	3	1	0
1870	1	3	3	1	1		1	
1880	0.5	3	3	3		0.5	3	
1890	0.5	3	5	3		0.5	3	
1900	3	0.5	7	1			3	0.5
1910	3	0.5	3	1			7	0
1920	0.5	3	0.5	0.5	0.5	0	0.5	0
1930	0.5	0.5	0.5	0.5		0.5	0.5	0
1940	1	1	1	0.6			0.2	0.1
1950	1.3	0.3	1.3	1.0			1.1	0.5
1960	1.3	0.4	4.5	5.8			5	1.7
1970	1.7	1.2		5.0	2.0		7	0.5
1980	2.9	2.4	5.9	5.6	4.2	3.3	7.4	4.9
1990	3.2	2.2	3.5	6.2			5.8	8.7
1995	1.7	1.0	1.6	2.2			4.2	3.9

[a]Before 1940, a point scale is used: 0–no moose; 0.5 moose/10 km^2 of forested area – very few ; 1 – scattered but permanent populations with limited hunting; 3 – moderate densities, moose are usually hunted, with each hunter killing one or two moose; 5 – high density, moose hunting grounds are evaluated from an economical perspective, hunting success reaches 5–6 moose/hunter/winter; 7 moose/10 km^2 of forested area – very high density, with some hunters killing more than a dozen moose/winter.

Neifeld 1992; Myslenkov and Voloshina 1992). In Karelia 80% of wolf scats contained moose remnants (Danilov 1987). Wolf predation is correlated to the ratio of moose to wolf, which is 121 : 1 in the taiga zone (ecoregion 134); 372 : 1 in the mixed forest zone (ecoregion 212); 131 : 1 in forest-steppe zone (ecoregion 252) (Chervonnyi 1975; Filonov 1989). Brown bear attack moose most often during rut, when bulls are less vigilant, and in spring when bulls and yearlings are weak (Yazan 1972; Zavatsky 1975).

6.7.2 Diseases, Helminths and Ectoparasites (Appendix 2, Tables 4, 5)

6.8 Population Structure and Densities (Table 6.10)

6.8.1 Population Density and Size (Tables 6.11 and 6.12)

Two great declines of moose numbers in eastern Europe were observed: the first in 1800–1850, and the second in 1920–1940 (Aspisov 1930, Kulagin 1932; Yurgenson 1935; Kirikov 1959, 1960, 1966; Baskin 1998).

Table 6.12. *Alces alces.* Population size from official data

Locality[a] Years	Population size (×1,000)			
	1981	1990	1995	1999
Estonia[1]	10	12	6.1	8.6
Latvia[2]		17		
Belarus[3]			25–30	
Kazakhstan[4]		5.3		
Russian Federation[5]	796	826	673	631
Kaliningrad oblast	1.3	1.7	0.6	0.4
N European Russia	135	143	130	92
NW European Russia	50	45	31	23
C European Russia	104	107	65	57
Volgo-Vyatka region	34	39	28	25
S European Russia	5	0.8	2.3	2.5
Volga region	23	23	11	13
Ural region	125	125	95	85
N Caucasus	0.9	1	0.4	0.2
W Siberia and Altai	68	93	90	102
E Siberia	105	101	106	113
Far East	145	146	114	117

[a]The superscript numbers denote the following references: 1, Tönisson (1997); Tönisson et al. (2000); 2, Gaross and Mikheeva (1992); 3, Kozlo (1995); 4, Bekenov et al. (1992); 5, Glushkov (1985); Borisov et al. (1992); Lomanov and Lomanova (1996, 2000).

Table 6.13. *Alces alces.* Harvest (Official data) (Glushkov 1985; Borisov et al. 1992; Lomanov and Lomanova 1996, 2000)

Locality	Years	Harvested moose (×1,000)
Russian Federation	1946–1955	75.4
	1956–1965	167.6
	1966–1975	258.1
	1976–1985	573.4
	1986–1988	203.8
	1991–1995	224.0
	1998	16.1
Kazakhstan	1990–1991	0.25

In the 1950s–1970s, the virgin moose population of the European part of the USSR recovered to its former population density and known historical borders. Since 1990, a decline in moose populations over all their range has been observed (Lomanov 1988, 1995; Danilkin 1998). The modern moose decline has coincided with diminishing carrying capacity of their habitats. Between 1960 and 1990, the virgin forests of European Russia, the Urals and S Siberia were clear-cut. These areas are now covered by regenerated forest which have a limited carrying capacity (Mamaev et al. 1991).

6.8.2 Impact on Vegetation

Browsing by moose can damage young pine and spruce forests, especially plantations (Timofeeva 1974; Kuznetsov 1987; Abaturov and Smirnov 1992; Dunin and Kozlo 1992; Galako et al. 1994; Randveer 1995).

6.9 Relations to Man

Moose are hunted for meat, skin, and antlers as trophies (Table 6.13; Bannikov et al. 1983). Also domesticated for milk production (Knorre 1961; Kozhukhov 1973; Mikhailov 1976). For tame moose the average period of lactation is 113 days (maximum 160 days); yield per lactation 190 kg (max. 575 kg), milk/day is 1 kg (max. 7.4 kg). Moose milk is used to treat human ulcers (Kozlov et al. 1992; Treus et al. 1992). Moose damage plantations of pine and spruce (Smirnov 1987; Lyubchenko et al. 1997).

7 Roe Deer – *Capreolus capreolus, C. pygargus*

Capreolus capreolus Linnaeus, 1758,
Capreolus pygargus, Pallas, 1771.
Artiodactyla, Cervidae, Odocoileinae

7.1 Status and Distribution (Fig. 7.1)

7.1.1 Taxonomic Remarks

Sokolov et al. (1992b) gave convincing arguments that two separate species exist: *Capreolus capreolus* Linnaeus 1758 and *Capreolus pygargus* Pallas 1771. The first species has no subspecies division, the second one includes: *C. p. pygargus* Pallas 1771 (Volga area, Ural Mountains, N Kazakhstan, Siberia to E Baikal lake, Altai Mountains, Yakutia) and *C. p. tianschanicus* Satunin 1906 (S Kazakhstan, Tien Shan Mountains, Transbaikal, Far East).

7.1.2 Status

Numerous, hunting species (Stubbe and Danilkin 1992; Smirnov 1994; Danilkin 1999).

7.1.3 Distribution (Fig. 7.2)

Distribution Area. Ca. 9,027,000 km^2.

Occurrence in Ecoregions. Subarctic (133, 134, 136), Subarctic Regime Mountains (M131, M134), Warm Continental (212), Warm Continental Regime Mountains (M212), Prairie (251, 252, 253), Prairie Regime Mountains (M251, M252), Mediterranean Regime Mountains (M262), Tropical/Subtropical Steppe (315), Tropical/Subtropical Desert Regime Mountains (M321), Temperate Steppe (331, 332), Temperate Desert (342).

Fig. 7.1. Siberian roe deer

7.2 Measurements (Table 7.1)

Body size increases from west to east.

7.3 Habitat

Elevated or mountain areas (<3500 m), dry plains with a mosaic landscape, where there are small (100–500 ha) forest islands and neighboring areas of steppe or meadow. Shrub or tall grass thickets are important for shelter. Avoids vast territories covered by closed forest. Snow depth >50 cm is limiting (Sablina 1955; Pasternak 1955; Rusakov 1978; Balĉauskas 1990; Danilkin 1999).

Terrain Form. Shores of fresh water, flood plains, bogs, plains, hills, foothills, slopes and mountain valleys (Ushkov 1947; Savinov 1961; Arabuli 1963, 1967; Ayzin 1969; Zyryanov 1975; Bromley and Kucherenko 1983; Sludsky and Baidavletov 1984; Prostakov and Obtemperansky 1986; Sobansky 1992).

Altitude. 50–250 m a.s.l. (Belarus, Amur), 200–1000 m a.s.l. (Ural, Transbaikal, Caucasus), 200–1,800 m a.s.l. (Trancaucasia, Kyrgyzstan); 900–

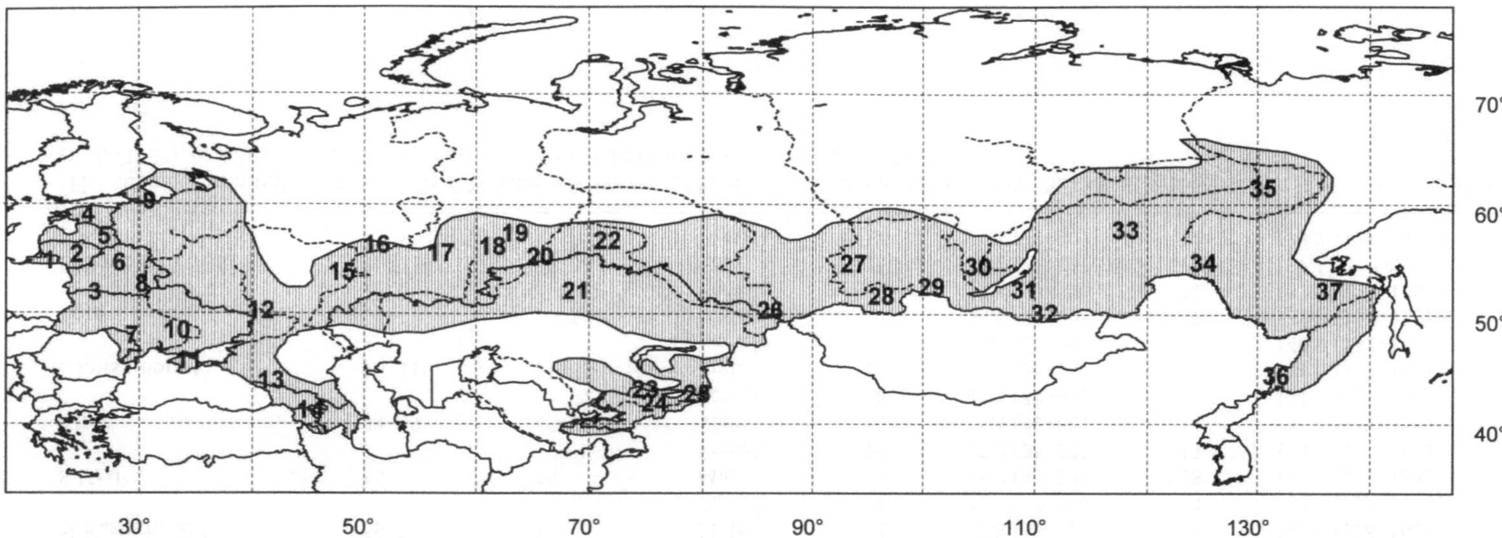

Fig. 7.2. *Capreolus* spp. Distribution map (Danilkin 1999). *1* Kaliningrad oblast; *2* Lithuania; *3* Volyn oblast; *4* Estonia; *5* Latvia; *6* Belarus; *7* Moldova; *8* Polesye; *9* Leningrad oblast; *10* Ukraine; *11* Crimean Peninsula; *12* Voronezh oblast; Belgorod oblast; *13* N Caucasus, Kavkazsky Nature Reserve; *14* Transcaucasia; *15* Volga area; *16* Volga – Vyatka area; *17* Bashkortostan; *18* Ilmensky Nature Reserve; *19* Sverdlovsk oblast; *20* Tyumen oblast; Kurgan oblast; *21* N Kazakhstan; *22* W Siberia; *23* S Mountains of Kazakhstan; *24* Kyrgyzstan; *25* Tien Shan Mountains; *26* Altai Mountains; *27* Krasnoyarsk kray; Stolby Nature Reserve; *28* W Sayan Mountains, Tuva; *29* E Sayan Mountains; *30* Angara River; Irkutsk oblast; *31* Buryatia; *32* Sokhondinsky Nature Reserve, Chita oblast; *33* E Siberia; *34* Amur oblast; *35* Yakutia; *36* Kedrovaya Pad Nature Reserve; Primorsky kray; *37* Amur River, Khabarovsk kray

Table 7.1. *Capreolus* spp. Measurements

Locality[a]	Longitude (°E)	Latitude (°N)	Gender	*n*	Body mass (kg)	*n*	Head and body length (cm)
Lithuania[1]	21–27	54–56	Male	131	29.2	112	119.6 ± 0.4
			Female	213	26.1	176	118.9 ± 0.3
Voronezh, Belgorod[2]	35–43	49–52	Male	59	27.7 (21–34)	59	119.1 (100–137)
			Female	62	26.0 (19–32)	62	116.3 (94–136)
Georgia[3]	40–47	41–44	Male	15	25.2	15	116 (110–129)
			Female	16	24.3 (18–30)	16	117.6
SE Kazakhstan[4]	67–83	42–46	Male	32	35.0 (30–43)	39	128.4 (120–136)
			Female	24	32.0 (28–39)	30	126.7 (116–137)
N Kazakhstan[4]	59–79	51–55	Male	8	43.8 (33–53)	8	141.8 (128–162)
			Female	4	41.5 (38–47)	4	140.5 (139–143)
S Ural[5]	60	56	Male	25	48.0 (43–54)	28	143.8 (137–153)
			Female	16	42.2 (35–50)	13	143.1 (135–151)
Altai[6]	84–89	49–53	Male	14	48.6 (41–60)	14	141.4 (132–156)
			Female	6	44.5 (38–47)	10	141.0 (135–148)
Irkutsk oblast[7]	95–118	51–60	Male	16	45.3 (36–54)	17	141.0 (137–145)
			Female	17	43.0 (36–55)	17	140.9 (132–150)
Transbaikal[8]	104–110	50–54	Male	7	36.6 (32–40)	7	135.1 (131–139)
			Female	11	33.4 (30–39)	11	132.3 (126–138)
Primorsky kray[9]	131–139	42–48	Male	15	39.2 (32–49)	15	137 (127–147)
			Female	12	34.3 (30–40)	12	134.7 (128–142)

[a]The superscript numbers denote the following references: 1, Bluzma (1975); 2, Prostakov (1996); 3, Arabuli (1966); 4, Pole (1974); 5,7, Danilkin et al. (1992); 6, Sobansky (1992); 8, Smirnov (1978); 9, Bromley and Kucherenko (1983).

2,100 m a.s.l. (S Transbaikal) (Dmitriev 1938; Nasimovich 1939; Ushkov 1947; Pasternak 1955; Sablina 1955; Arabuli 1963, 1967; Timofeeva 1985; Ayzin 1969; Darman 1986, 1990; Orlova 1986).

Snow. European roe deer avoid snow >30–40 cm; Siberian roe deer avoid >40–50 cm. In the steppes, inhabited areas have a snow depth <25 cm. Formozov (1946) and Heptner (1989) suggested that 50 cm snow is the factor which determines the northern distribution. Hoof pressure: 320 (300–360) g/cm^2 (Dulkeit cited in Nasimovich 1955).

Water. Can survive on moisture available in plants. Use snow in winter (Prostakov 1990; Danilkin 1996).

7.3.1 Biotope (Table 7.2)

In various regions of Lithuania, 3–100% of the roe deer population inhabit arable land (Balĉauskas 1988).

7.4 Diet and Feeding Habits

7.4.1 Diet

Herbs (55%), trees and shrubs (25%), cereals and sedges (16%).

In northern and Siberian parts of their range, in winter, roe deer prefer to feed on twigs, dry herbs and grasses, fallen leaves, mosses, and tree lichens (Rusakov 1979; Egorov 1965). Herbs are preferred during winter, grasses and sedges during summer (Holisova et al. 1992). Roe deer eat mushrooms throughout the year. For a list of plant genera eaten, see Appendix 1, Table 7. In seasons without snow cover, especially in summer, roe deer visit artificial salt licks (Nasimovich 1938; Prostakov 1990; Panichev 1998).

7.4.2 Life Forms/Plant Parts Eaten (Table 7.3)

Consumption of Food/day. In captivity, consumption of fresh food: calves eat 1.1 kg, 1–3-year-old animals eat 1.7 kg, >3-year-old animals eat 2.3 kg. Siberian roe deer in captivity eat 2.5–3.5 kg (Padaiga 1975; Danilkin 1996). Free-ranging roe deer, in the Khingan range in winter, eat 1.5 kg of twigs plus 0.5–0.6 kg wilted grass (fresh weight) (Darman 1986). Free-ranging roe deer in Belarus, in winter: males eat 1.4–1.9 kg of twigs, females 1–1.2 kg (fresh weight) (Dunin et al. 1999).

Table 7.2. *Capreolus* spp. Habitat[a]

Habitat type	Locality[b]											
	Leningrad oblast (1)[1], Belarus, Amur (2)[2], south-central European Russia (3)[3]			W Siberia (4)[4], Kazakhstan (5)[5], Ural (6)[6]			E Siberia, Transbaikal, Yakutia (7)[7], Primorsky (8)[8], Altai (9)[9]			Crimean Peninsula, Caucasus (10)[10], Transcaucasia (11)[11], Kyrgyzstan (12)[12]		
(Ecoregions)	212 1	212 2	252 3	253 4	342 5	M131 6	M134 7	M212 8	M251 9	M252 10	M262 11	M341 12
Mature forest												
Coniferous, light	r	r	++		r		r		+++	x		x
Coniferous, dark	r								r			
Coniferous, mixed	+	r					+++			+++	x	
Mixed deciduous-coniferous	+	r			r	x	+++		x	+++	x	x
Broadleaf		++	++					x		+++	x	x
Forest stand												
Coniferous				x	x	x	x					
Mixed deciduous-coniferous				x	r	x	+++	x	r			
Plantation strips					+++							
Broadleaf			x									
Fresh water shore												
Meadows		r										
Shrub	+	++			x		++					x
Forest		+	+	x			++				x	x
Burnt forest						x	+++	x				
Cut forest	r		r			x	+++					
Shrubland, edges	+	+++	+	x		x	x		x	++		x
Forest-steppe		+++					+++		++			
Grassland	++		x			x	x	x			x	x
Swamp	r		+							+		
Arable lands	+	r	+									

[a]Symbols indicate % of habitat use was >30% (+++); 20–30% (++): 10–20% (+); <10% (r); x indicates habitat use noted, no numerical data reported.
[b]The superscript numbers denote the following references: 1, Rusakov (1979); 2, Sablina (1955); Darman (1982, 1986); Tyshkevich (1999b); 3, Novikov and Timofeeva (1964); Prostakov and Obtemperansky (1986); Prostakov (1996); 4, Metelsky (1976); 5, Pole (1966); Baidavletov and Murzov (1982); Sludsky and Baidavletov (1984); 6, Ushkov (1947); Sokolov and Danilkin (1981); 7, Fetisov (1953); Zyryanov (1975); Sokolov and Danilkin (1981); 8, Bromley and Kucherenko (1983); 9, Dmitriev (1938); Savinov (1961); Tsybulin (1990); Sobansky (1992); 10, Dinnik (1910); Pasternak (1955); 11, Nasimovich (1939); Arabuli (1963, 1967); 12, Ayzin (1969).

Table 7.3. *Capreolus* spp. Life forms/plant parts eaten (x indicates the type used). (Rusakov 1979; Holisova et al. 1992)

Life forms	Plant parts eaten							
	Whole plant	Leaves	Stems/ twigs	Bark	Roots/ bulbs	Flowers	Fruits	Seeds
Trees		x				x	x	x
Shrubs		x				x	x	x
Dwarf shrubs		x	x			x	x	
Herbs		x	x			x	x	x
Grasses/sedges		x	x					
Horsetails		x						
Mosses	x							
Lichens	x							
Fungi	x							

Feeding Niche. In field and shrub layers.

7.5 Social Organization and Behavior

7.5.1 Social Life (Tables 7.4 and 7.5)

Disturbed groups scatter, confirming that the personal connections between individuals are weak or absent (Baskin 1976; Prostakov 1990; Danilkin 1996). However, in 86% of cases, females with calves are leaders of groups (Danilkin et al. 1995). Males are strongly territorial. Territories are marked by scarring trees with antlers, as well as leaving secretions of scent glands, located on the head and neck (Danilkin 1996). Females also protect their territories against other females (Bromley and Kucherenko 1983).

7.5.2 Home Range

Fidelity to their home range is an important feature of roe deer. A network of trails, resting places and piles of pellets, tracks with scent from hoof scent glands, are found within the home range. The central area of the home range is mostly used for rest (Danilkin 1992; Sokolov and Gromov 1993; Danilkin et al. 1995; Prostakov 1996). Male home ranges usually overlap with some home ranges of females. In the Stolby Nature Reserve, the daily home range was about 40 ha; males and females occupied about 500 ha each during winter (Zyryanov 1975). In Belgorod oblast: adult males (whole

Table 7.4. *Capreolus* spp. Group composition[a]. (Danilkin 1999)

Group types	Seasons and seasonal events			
	Winter Wintering	Spring Calving	Summer Summering	Autumn Rut
Singles (male, female + fawn)	++	+++	+++	
Families (female + fawn + yearling)	++	+	r	++
Mating couples (male + female)				++
Groups of yearlings	+	r	r	+
Aggregations	+++			++

[a]Symbols used indicate group type observed: very often (+++); often (++); occasionally (+); rarely (r).

Table 7.5. *Capreolus* spp. Number of animals per group(% of observed groups)

Locality[a]	Season	No. of animals per group				
		1	2	3	4	≥4
Leningrad oblast[1]		39	16	10	35	0
CS European Russia[2]	Winter	13–27	25–65	20–56	0	0
	Summer	48–59	20–27	20–27	0	0
Stolby Nature Reserve[3]	Winter	23	25	25	23	4
Sokhondinsky Nature Reserve[4]	Winter, January	21	32	0	47	0
	Winter, February	5	20	30	45	0
	Summer	77	22	1	0	0

[a]The superscript numbers denote the following references: 1, Rusakov (1979); 2, Prostakov (1990); 3, Zyryanov (1975); 4, Orlova (1986).

year) occupy 385 ha (189–845), adult females (whole year) occupy 459 ha (237–701) (Danilkin 1999). In Amur area: the winter home range is 900–14,400 ha (Danilkin et al. 1995). In the Far East: adult male home range is 50–75 ha, adult female home range is 5–15 ha. Daily route: in the St. Petersburg area is 0.5–10.5 km (Rusakov 1979), in the Stolby Nature Reserve the route is 1.9 km in winter, in neighboring hunting grounds it is 3.8 km (Zyryanov 1975). Daily winter route in Transbaikal is 5.2 km (Smirnov 1978) and in the Khingan mountains it is 0.7–2.1 km (Darman 1990).

7.5.3 Mode of Movement

Walk, saltatorial-cursorial running, jumps, and swims (Gambaryan 1974; Hildebrand 1988). Escape by jumping when chased. In many areas

(Leningrad oblast, C S European Russia, Ural) roe deer regularly dig in snow to reach food, but in general this behavior is not typical. Stand on back legs to reach branches of trees.

7.5.4
Dispersal/Migration

Past seasonal migrations have occurred relating to changes in feeding conditions and snow. In the last part of the 19th century there were mass migrations in Ural (Turkin and Satunin 1902). Upward migrations in Caucasus between 500–1,000 m a.s.l., the migrations were 5–30 km, maximum 40–50 km (Nasimovich 1955). In Amur region and the Far East, annual autumn (from the first half of October) migrations occur to avoid deep snow (>30 cm) in the northern part of the range and to reach wintering places with plenty of birch shrubs (food biomass about 160 kg/ha, fresh weight), extent 100–300 km, maximum 400 km. The migrations continue for 10–40 days in autumn and 21–100 days in spring. The speed of travel is 4–11 km/day in autumn and 1.5–6 km/day in spring (Formozov 1946; Bromley and Kucherenko 1983; Danilkin et al. 1993). There are no migrations in Yakutia where winter and summer grounds overlap (Egorov 1971).

7.5.5
Diurnal Activity (Table 7.6)

Daily activity is divided into 6–12 periods. Main activity periods occur during sunrise and sunset (Sokolov and Danilkin 1981).

7.5.6
Shelters

Rake snow clear to lie on grass and leaves.

7.5.7
Escape Behavior

Very vigilant. Keeping a look out for 50% of the active period; often staying in a vigilant pose for up to 15 min. Signals for disturbance: special vocal signal ("bark"), strike the ground with the front leg hoof; "signal jump" and demonstration during flight of white "mirror" around tail. Flight distance in an open place is 300–400 m, in dense forest 75–100 m. Greatly disturbed roe deer can run up to 14 km, but usually stay within 300–400 m, then start to make circles. Mother keeps her fawns at some distance from each other to decrease the risk of all the fawns being taken at the same time (Zyryanov 1975; Bromley and Kucherenko 1983).

Table 7.6. *Capreolus* spp. Diurnal activity

Locality	Season	Time of day							
		0–3	3–6	6–9	9–12	12–15	15–18	18–21	21–24
Leningrad oblast[1]	Winter								
Transbaikal[2]	Winter								
	Spring								
	Summer								
	Autumn								

Black shading, most animals are active; gray shading, some animals are active.
[a]The superscript numbers denote the following references: 1, Rusakov (1979); 2, Smirnov (1978).

7.6 Reproduction (Tables 7.7 and 7.8)

Fierce fights among males during rut are common. Males mark their territories by disturbing the ground and use their antlers to scar and break shrubs and small trees in the surrounding area. Harems are not present in roe deer populations. Males mate within their territory. Male follows (or pursues) a female while the couple is moving in circles, resulting in the creation of round trails, or plots, from trampled vegetation. After mating the male searches for another female. Female heat lasts for 4–5 days. Embryos do not develop until after about 170 days (Tsaplyuk 1977). In Ukraine, the latency period is 60–75 days (Steklenev 1980). 1–2 year old females usually mate during November–December and there is no latency period (Lavov 1978). Calving occurs in thickets of grass and shrubs. Calves hide for the first 6–8 days, after which they follow the mother. At 1 month age they start to eat grass and leaves, after 2 months they feed by themselves. Lactation continues for 1.5 year, interrupted by the rut (Bromley and Kucherenko 1983).

7.6.1 Seasonal Breeding Events (Table 7.9)

Maximum Length of Life. 11–12 years (Stubbe and Danilkin 1992), or 10–15 years (Bromley and Kucherenko 1983). In Buryatia, where hunting pressure is high, roe deer have an average age of 2.5 years, maximum 7.5 years (Shvetsov et al. 1984).

Table 7.7. *Capreolus* spp. Breeding parameters

Locality[a]	Age at sexual maturity (years)		Participation in breeding (years)		Length of pregnancy (days)	Litter size
	Male	Female	Male	Female		
Kazakhstan[1]	1.5	0.5	2.5	1.5	264–318	2 (1–3)
Kyrgyzstan[2]	1.5	1.5	2.5	1.5	270	
Ukraine[3]					235–240	1–2

[a]The superscript numbers denote the following references: 1, Tsapluk (1977); Sludsky and Baidavletov (1984); 2, Yanushevich et al. (1972); 3, Steklenev (1980).

Table 7.8. *Capreolus* spp. Fecundity in different localities (% of observed females)

Locality[a]	No. of calves/female				
	1	2	3	4	Average
NW Russia[1]	39	50	11	0	1.7
S C Russia[2]	43	49	6	2	1.7
Far East[3]	22	63	15	0	1.9

[a]The superscript numbers denote the following references: 1, Prusaite et al. (1977); Rusakov (1979); 2, Prostakov (1996); 3, Bromley and Kucherenko (1983).

7.7 Mortality

Calf mortality during the first year can be between 20–90% (Stubbe and Danilkin 1992), in the Cis-Baikal area the figure is 38%. Causes of mortality in Russia, as a percentage of dead animals found: predators 36%, hunting and other human activity 57%, catastrophes (e.g. floods) 4%, and other (e.g. road accidents, poisoning by agricultural fertilizers) 3% (Perovsky 1988). In very snow rich winters in the Far East (snow depth 1–1.5 m) mass migration occurred and mortality reached extremely high levels because of starvation, drowning during the crossing of thin, snow-covered ice, and high predation and poaching pressure. During such situations, a three- to fourfold decline in numbers was observed (Barancheev 1962; Rakov 1965; Kucherenko and Shvets 1977).

Table 7.9. *Capreolus* spp. Seasonal breeding events

Locality[a]	Rutting season			Calving season		
	First demonstrations	Main rut	Last demonstrations	First calving	Main calving	Last calves
Leningrad oblast[1]	Early August	August–September	Early Oct.	Early April	May	Early June
Belarus[2]		Mid July–mid August			May–July	
CS European Russia[3]	10 July	20 July–20 September		Late April	May–20 June	End of June
Ukraine[4]	May–June	July–August			June	
Ural[5]	4 August	10 August–10 September	18 September	10 May	20 May–20 June	23 June
Kazakhstan[6]		August–September			20 May–July	
Kyrgyzstan[7]		Mid August	End of September		May	
Buryatia[8]		First half of August	End of September			
Yakutia[9]	July	August-early September		End of April–early May	End of May–mid June	
Amur[10]					End of May–early June	
Primorsky kray[11]	End of June	August-early September	Very end Oct.	Mid April	End of May–early June	25 June

[a]The superscript numbers denote the following references: 1, Rusakov (1979); 2, Sablina (1955); 3, Prostakov (1990); 4, Steklenev (1980); 5, Kiselev (1979); Danilkin (1992); 6, Pole (1973); Sludsky and Baidavletov (1984); 7, Yanushevich et al. (1972); 8, Fetisov (1953); Shvetsov et al. (1984); 9, Egorov (1971); 10, Darman (1990); 11, Bromley and Kucherenko (1983).

7.7.1 Predators

In central Russia wolf, fox, raccoon dog, marten and dogs are predators of roe deer. In Siberia and adjacent areas, and also the Far East, wolf, lynx, brown bear, tiger, fox, yellow-throated marten and raptors are predators (Bromley and Kucherenko 1983). Wolf predation is estimated to account for 25% of roe deer found dead in NW Russia, 12% in central-southern European Russia, 66–79% in Ural (Rusakov 1979; Filonov 1989), 3% in W Sayan (Zavatsky 1986), and 9% in Transbaikal (Smirnov 1978). Lynx predation accounts for 11% of dead roe deer in Ural (Filonov 1983a), up to 89% in the Ilmensky Nature Reserve (Dvornikov 1984), 37% in Transbaikal (Smirnov 1978). In Sayan and Altai roe deer constitute 46% of all lynx prey. Roe deer remnants are found in 8–20% wolf scats in the Kazakhstan mountains (Fedosenko and Sludsky 1981).

7.7.2 Diseases, Helminths and Ectoparasites (Appendix 2, Tables 4 and 5)

Brucellosis, tuberculosis, rabies, necrobiosis, pasteurellosis (Rukhlyadev 1964; Sedov et al. 1992).

7.8 Population Structure and Densities (Table 7.10)

Male:female ratio of adult roe deer in Lithuania was 1 : 1.31–1 : 1.56, in Ural 1 : 1.38, in Altai 1 : 1.33, in Transbaikal 1 : 1.19 and in Amur 1 : 1.12 (Bluzma 1975; Smirnov 1978; Danilkin 1992; Sobansky 1992).

Table 7.10. *Capreolus* spp. Gender and age composition of populations (%)

Locality[a]	Adults		Yearlings		Calves
	Male	Female	Male	Female	
Lithuania[1]	16–22	25–29	15–20		35–37
SC European Russia[2]	31	44			25
Carpathians[3]	27	50			23
Altai[4]	28	31	12	16	13
Transbaikal[5]	30	37	7	6	20

[a]The superscript numbers denote the following references: 1, Bluzma (1975); 2, Pobedinsky (1997); 3, Lunkashu et al. (1986); 4, Sobansky (1992); 5, Smirnov (1978).

7.8.1
Population Density and Size (Tables 7.11 and 7.12)

During 1985–1993, Siberian roe deer increased 2.6 times in number, while the European roe deer increased 1.6 times (Danilkin 1996). In Lithuania, population density was 2–33 animals/10 km^2 on arable lands and 95–385 animals/10 km^2 in forests (Bluzma and Baleišis 1992).

7.8.2
Impact on Vegetation

In the Baikal area 50% of larch seedlings are damaged or killed where roe deer are numerous (Lyamkin and Sokolova 1992). In Amur area and south-central Russia roe deer depress the growth of lime trees (Darman 1982; Prostakov 1996).

7.9
Relations to Man (Table 7.13)

Hunted species. Products: meat, antlers (trophy), and skin. In the USSR, during the 1980s, 10,000–40,000 European roe deer and 5,000–10,000 Siberian roe deer were harvested annually (Sokolov and Lebedeva 1989).

Table 7.11. *Capreolus* spp. Recent numbers of animals obtained from official data. (Bekenov et al. 1992; Borisov et al. 1992; Mirutenko 1996, 2000; Bekenov 1999; Dunin et al. 1999)

Locality Years	Population size (×1,000)			
	1981	1990	1995	1999
Belarus			40	
Russian Federation	330	759	741	669
Kaliningrad oblast	7.6	10.5	9.3	6.5
NW European Russia	6.5	7.2	7.3	10
C European Russia	13.9	29.9	21.4	25.3
Volga–Vyatka area	0.02	0.6	0.4	0.4
Central-southern European Russia	7.1	11.4	11.6	14.2
Volga area	17.1	17.7	16	19
N Caucasus	16.1	20.4	14	13.5
Ural	28.0	85.6	145.7	140.7
W Siberia	23.8	85.6	110.5	112
E Siberia	142.2	233.6	223	165.5
Far East	68.5	288.3	181.6	161.9
Kazakhstan		30	15–20	15–20
N and NW Kazakhstan		10		
S mountains of Kazakhstan		20		

Table 7.12. *Capreolus* spp. Population size and density in different areas

Locality[a]	Year	Population density in hunting grounds (animals/10 km^2)	Population size (×1,000)
Belarus[1]	1985	1.1	
	1990–1993		23–28
Belarus (Polesye)[2]	1998	4.4 (1–7)	
Estonia[3]	1945		5
	1971		43
	1999		29
CS European Russia[b4]	1970	2.2	3.2
	1975	4.1	5.8
	1980	6.1	7.5
	1985	6.9	7.9
	1990	6.7	8.7
	1995	8.6	12.2
Kavkazsky Nature Reserve[5]	1999		0.4–0.6
Bashkortostan[6]	1961		0.4
	1971		4
	1980		5.4
	1998	0.5–1	9.9
Ilmen Nature Reserve[7]	1984	18–33	
Kyrgyzstan[8]	1966	10–130	6.3
	1985	0.9	
Kazakhstan[9]	1985	0.3	
Altai[10]	1960		8
	1985		15
	1998	1.5–2	42.3
Angara forest-steppe[11]	1970	1.2	
E Sayan[12]	1970	0.6–1.2	
Baikal surroundings[13]	1970	0.5–2.5, max. 18	
E Siberia[14]	1985	0.4–1.7	
	1995	1–2	223
Amur[15]	1975	3.3–4.2	100
	1979		40–45
	1981		35
	1998	3–5	97.8
Khabarovsk[16]	1960		25–30
	1970		18–20
	1974		8–10
	1980		8–10
	1995	1–1.5	26
Kedrovaya Pad[17]	1930		0.7
	1989		0.7
Krasnoyarsky kray[18]	1975		17.5
	1980		22.7
	1985		13
	1995	0.5–1	45
Kurgan oblast[19]	1981		7.7
	1985	Max. 100	11
	1990		32
	1994	9.6	70

Table 7.12. *Continued*

Locality[a]	Year	Population density in hunting grounds (animals/10 km²)	Population size (×1,000)
Latvia[20]	1971	40	57
Lithuania[21]	1968	40 (8–110)	68.8
	1985	6.1	
	1990	33–66	
	2001		69.0
Moldova[22]	1980	5–50	8
N Caucasus[23]	1961		12.4
	1971		9.4
	1980		12.6
	1995	1–1.5	14
Primorsky kray[24]	1968		40–60
	1974		13–15
	1979	1.0–1.5	10–12
	1998	1.5	26.7
Sverdlovsk oblast[25]	1930		20
	1941		Decline
	1945		5
	1946		1
	1950		2.5
	1965		10
	1968		2–5
	1971		4
	1998	0.5–1	14.6
Transbaikal[26]	1953		62
	1970	0.3–4.4	9–16
Ukraine[27]	1961		43.2
	1965		80
	1969	8–16	120
	1985	2.4	150
W and N Kazakhstan[28]	1962	1–53	8.5
	1979	1–16	6.3
Volyn oblast[29]	1978		13.5
	1986	35–37	12.3

[a]The superscript numbers denote the following references: 1, Sablina (1955); Kozlo (1995); 2, Dunin et al. (1999); 3, Tönisson et al. (2000); 4, Prostakov (1996); 5, Kudaktin (pers. comm); 6, Fadeev (1994); Mirutenko (1996, 2000); Danilkin (1999); 7, Dvornikov and Dvornikova (1985); 8, Yanushevich et al. (1972); 9, Bekenov et al. (1992); 10, Sobansky (1990b); Mirutenko (1996, 2000); Danilkin (1999); 11, Lavov (1974); 12, Lavov (1974); 13, Smirnov (1990); 14, Smirnov (1990); Mirutenko (1996, 2000); Danilkin (1999); 15, Bromley and Kucherenko (1983); Mirutenko (1996, 2000); Danilkin (1999); 16, Bromley and Kucherenko (1983); Mirutenko (1996, 2000); Danilkin (1999); 17, Korkishko (1992); 18, Smirnov (1990); Mirutenko (1996, 2000); Danilkin (1999); 19, Ostanin (1994); Mirutenko (1996, 2000); Danilkin (1999); 20, Danilkin (1992); 21, Padaiga (1969); Bluzma (1975, 1990); Bluzma and Baleišis (1990); Baleišis et al. (2002); 22, Muntyanu et al. (1980); 23, Stubbe and Danilkin (1992); Mirutenko (1996, 2000); Danilkin (1999); 24, Bromley and Kucherenko (1983); Mirutenko (1996, 2000); Danilkin (1999); 25, Kiselev (1992); 26, Smirnov (1980, 1990); 27, Krainev (1975); 28, Baidavletov and Murzov (1982); Sludsky and Baidavletov (1984); 29, Kolisnik (1990).

[b]Voronezh + Lipetsk + Belgorod + Kursk + Tambov oblast. Density was calculated on forested area.

Table 7.13. *Capreolus* spp. Harvest data. Obtained from official data. (Bromley and Kucherenko 1983; Smirnov 1994; Danilkin 1999; Mirutenko 2000)

Locality	Year	Harvested roe deer (×1,000)
Russian Federation	1960	18.7
	1965	14.7
	1970	10.7
	1975	8.6
	1980	10.1
	1985	10.6
	1990	25.8
	1995	24.5
	1997	18.9
	1999	17.4
Ural	1970	1.5–5
	1999	4.8
Tuva	1975	2
	1999	0.1
Khabarovsk kray	1968	0.7
	1978	0.05
	1999	0.3
Amur oblast	1968	4.8
	1978	1.4
	1999	1.5
Primorsky kray	1968	1
	1978	0.8
	1999	1.2

Stubbe and Danilkin (1992) and Danilkin (1999) suggest that between 1980–1990, of the total 350,000–400,000 European roe deer in the USSR, 80,000–114,000 were harvested annually, and of the 300,000–350,000 Siberian roe deer 50,000–100,000 were killed. Most of them were killed illegally. In 1997, of the 613,000–651,000 roe deer, of both species in Russia, only 19,000 were officially harvested (Danilkin 1999) (Table 7.13).

8 Reindeer – *Rangifer tarandus*

Rangifer tarandus Linnaeus, 1758.
Artiodactyla, Cervidae, Odocoileinae

8.1 Status and Distribution (Fig. 8.1)

8.1.1 Taxonomic Remarks

Five subspecies occur: *R. t. tarandus* Linnaeus, 1758 (European part of Russia excluding Ural Mountains); *R. t. pearsoni* Lydekker, 1902 (Novaya Zemlya Archipelago); *R. t. sibiricus* Murray, 1886 (Siberian tundra and Arctic Ocean islands); *R. t. valentinae* Flerov, 1933 (forest zone of Siberia, in the east up to Stanovoy Range, S Siberian Mountains, Altai); *R. t. phylarchus* Hollister, 1912 (Kamchatka Peninsula, regions adjoining the Sea of Okhotsk, Amur region) (Heptner 1989). Despite the fact that wild and tame reindeer populations live in the same area and there is an exchange of some individuals, Shubin and Ionova (1984) and Shubin and Ephimtseva (1988) found rather different frequencies of alleles in two loci. They concluded that the two forms evolved separately. In Novaya Zemlya, among 6,000 reindeer living on these islands, in 1986, 10% demonstrated color features of tame reindeer, as a result of hybridization since 1928–1933, when 604 tame reindeer were acclimatized there (Zubkov 1935; Novikov 1983a; Kupriyanov and Belikov 1986; Khakhin 1998). Feral populations of reindeer inhabit the Wrangel Island, domestic reindeer from Chukchi Peninsula were introduced in 1948, and have become feral since 1972 (Baskin and Skogland 1997).

8.1.2 Status

According to the Red Data book of the Russian Federation (Rozhnov 2001) *R. t. pearsoni* is rehabilitating. A population of reindeer of the Altai and Sayan Mounatains is rare (named *R. t. angustifrons* Flerov, 1932). Other populations are hunted (Syroechkovsky 1986; Sokolov and Lebedeva 1989; Zheleznov 1990; Paponov 2000).

Fig. 8.1. Reindeer

8.1.3
Distribution (Fig. 8.2)

Distribution Area. Ca. 12,589,000 km² (estimated historically); ca. 11,572,000 km² (modern range).

Occurrence in Ecoregions. Tundra (121, 122, 123, 124, 125), Subarctic (131, 132, 133, 134, 135, 136), Tundra Regime Mountains (M121, M122), Subarctic Regime Mountains (M131, M132, M133, M134, M135), Warm Continental (M212, M251).

8.2
Measurements (Table 8.1 and Fig. 8.3)

8.3
Habitat (Table 8.2)

Terrain Form. Sea shores, flood plains, lowlands, bogs, plateau, uplands, foothills, troughs, slopes, and tops (Vodopyanov 1970; Azarov 1976; Dvoryadkin and Shvets 1976; Shargaev 1976; Kostyuk et al. 1979; Bromley and Kucherenko 1983; Shvetsov et al. 1984; Mosolov 1985; Zyryanov and Rozhkov 1988; Kuksov et al. 1989; Zheleznov 1990; Smirnov 1994).

Altitude. 0–1,400 m a.s.l. (Kola Peninsula, N Ural Mountains, Putoran Mountains, Chukchi Peninsula, Kolyma Upland), 400–2,700 m a.s.l. (preferably 1,400–1,800 m a.s.l.) (Altai Mountains, Sayan Mountains, Transbaikal).

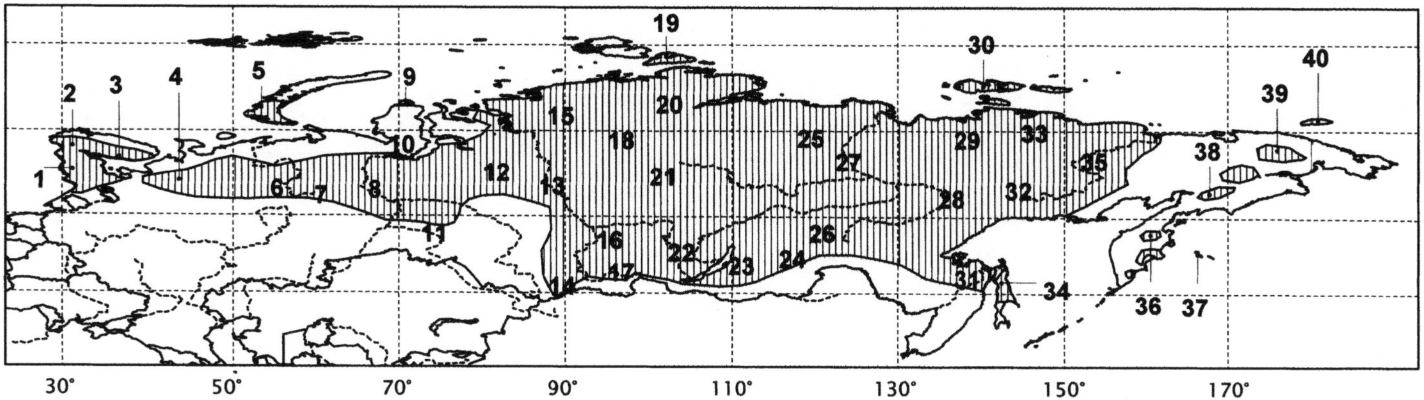

Fig. 8.2. *Rangifer tarandus.* Distribution map (Perovsky 1984; Vershinin et al. 1984; Heptner 1989; Zheleznov 1990; Zhukov 2000). *1* Karelia; *2* Laplandsky Nature Reserve; 3 Tersky Coast; *4* Arkhangelsk oblast; *5* Novaya Zemlya archipelago; *6* Komi, Pechora River; *7* Ural Mountains, Konda River; *8* Ob River; *9* Belyi Island; *10* Yamal Peninsula; *11* Khyanty-Mansy okrug; *12* W Siberian Plain; *13* Enisey River; *14* Altai Mountains; *15* Pyasina River; *16* Krasnoyarsk kray; *17* Sayan Mountains, Tuva; *18* Putoran Mountains; *19* Severnaya Zemlya archipelago; *20* Taymyr Peninsula; *21* Evenky okrug; *22* Irkutsk oblast; *23* Transbaikal, Buryatia; *24* Chita oblast; *25* Olenek River; *26* Olekma River; *27* Lena River; *28* Aldan River; 29 Yana River; *30* Novosibirskie archipelago; *31* Amur River; *32* Okhotsk-Kolyma Upland; *33* Indigirka River; *34* Sakhalin Island; *35* Kolyma River; *36* Kamchatka Peninsula; *37* Komandorskiye archipelago; *38* Parapolsky Lowland; *39* Chukchi Peninsula; *40* Wrangel Islands

Table 8.1. *Rangifer tarandus*. Measurements

Locality[a]	Longitude (°E)	Latitude (°N)	Gender	n	Body mass (kg)	Head and body length (cm)	Tail length (cm)
Murmansk oblast[1]	29–41	67–69	Male	25	91.8 ± 1.8	186.3 ± 1.6	15.3 ± 0.37
			Female	31	79.1 ± 1.0	173.3 ± 1.3	14.2 ± 0.2
Altai[2]	88–90	49–53	Male	5	172 (147–202)	203.0 (200.0–205.0)	16.4 (14.0–20.0)
			Female	3	98.0 (82–116)	167.0 (162.0–170.0)	18.8 (18.0–19.0)
Yakutia[3]	135–147	67–70	Male	21	179.8 (147–210)	203.8 (189–214)	14.8 (15–17)
Chukchi okrug[4]	169–173	64–66	Male	14	148.8 (80–180)	204.0 (180–210)	14.2 (12–15.5)
			Female	10	111.6 (72–112)	199.5 (194–205)	14.1 (13–15)

[a]The superscript numbers denote the following references: 1, Makarova (1976); 2, Sobansky (1992); 3, Popov et al. (1980); 4, Zheleznov (1990).

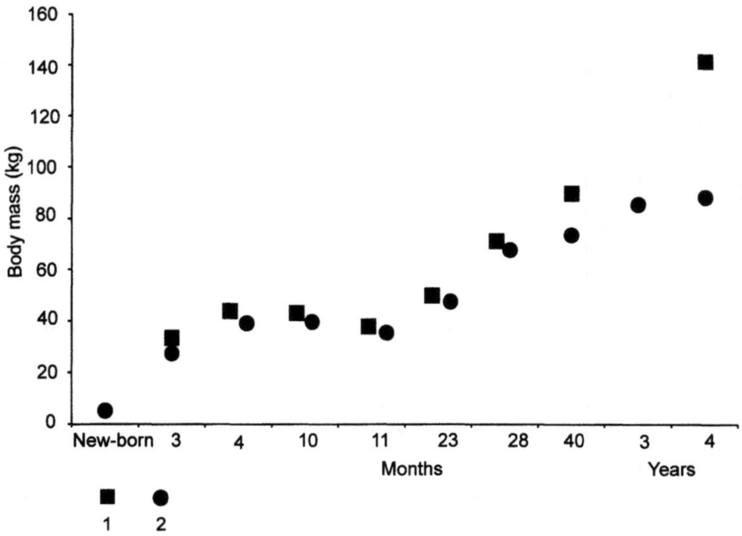

Fig. 8.3. *Rangifer tarandus.* Body mass relations versus age (Pavlov et al. 1989; Zyryanov and Vinokurov 1990). *1* Males; *2* females

In summer, reindeer prefer to move upwards to escape insects (Averin 1948; Fetisov 1955; Moskvitin et al. 1976; Lineytsev 1983; Zyryanov 1990; Sobansky 1992; Smirnov 1994).

Temperature. Segal (1959) and Davydov (1960a,b) found increased metabolism when the air temperature dropped below –25 °C, but Sokolov and Kushnir (1986) found that adult reindeer can keep temperature homeostasis, without additional energetic expenses, down to –61 °C. In cold and windy winters, reindeer lose weight more rapidly (Baskin 1970). They suffer when the air temperature exceeds +15 °C. Frequency of breathing reaches 70–80 times/min when the temperature is +23–25 °C, but only 12 times/min when temperature at +6 °C, and 5 times/min when the temperature at –25 °C (Slonim 1962; Chermnykh et al. 1989).

Snow. Snow >50–60 cm impedes movement of single animals, 80–90 cm of snow impedes the movement of herds. In European taiga, reindeer were mostly found where snow depth was <50–90 cm (Fadeev 1985). Reindeer can dig through snow 90–120 cm deep. However, digging in snow >70–80 cm requires more energy than the animal can obtain from food (Nasimovich 1955). In some taiga areas with snow depths of 150–200 cm, tree lichens are eaten instead of ground lichens (Sablina 1980). Hoof pressure is 140–180 g/cm^2 (Semenov-Tien-Shansky 1948b; Dulkeit 1959).

Table 8.2. *Rangifer tarandus*. Habitats

Habitat	Season[b]	Locality[c]								
		Kola Peninsula (1)[1], Arkhangelsk, Komi (2)[2]		W Siberia, Evenky okrug (1)[3], Taymyr (2)[4], Yakutia (3)[5]			Altai, Sayan (1)[6], Transbaikal (2)[7]		Okhotsk – Kolyma (1)[8], Kamchatka (2)[9]	
		1	2	1	2	3	1	2	1	2
(Ecoregion)										
Tundra										
Arctic	W					x				
	S				x	x				
Lichens–mosses	W					x				
	S				x	x				
Mosses–dwarf shrubs	W					x				
	S				x	x				
Seashore tundra	W									+++
	S				x	+++				+++
Forest tundra										
Mixed deciduous-coniferous with lichens	W		+++		x					
Bogs	W			x	x					
	S			x				r		
Pinus pumila shrubs with lichen cover on soil	W								+	+
Highlands										
Dwarf shrub tundra	W	x					x			
	S						x			
Shrubs (pine, birch, alder)	W						x			
	S			x	x					

Table 8.2. *Continued*

Lichen tundra	S						+++			
Grass tundra	W						+++			
	S						+++			
Subalpine coniferous sparse forest	W	x			x		x	x		
	S				x		x			
Alpine tundra	W	x			x		x		x	+++
	S		x		x		x	+++	x	++
Taiga										
Coniferous forest (sparse) with ground lichens	W	x		x	x	x	x	+++		
	S			x			x	r		
Coniferous forest with tree lichens	W		+++					+		
Birch forest	S							+	+	
Flood plain (boggy) with shrubs and grass	W					x		r	+	
	S			+++				r		
Shrub belt near alpine tundra	S									
Burns with grass	W		+					x		
Bogs with sparse pine	W	x	+	x						
	S	x		+++						
Bogs inside forest	W	x		x						
	S	x								

[a]Symbols indicate % of habitat use was >30% (+++); 20–30% (++): 10–20% (+); <10% (r); x indicates habitat use noted, no numerical data reported.
[b]W, Winter; S, summer.
[c]The superscript numbers denote the following references: 1, Semenov-Tien-Shansky (1948b); Makarova (1988); 2, Polezhaev (1976); Fadeev (1985); 3, Makridin (1962); Azarov (1976); Shargaev (1976); Yudin and Belov (1976); Semyanov and Starovoitov (1985); Zhukov (2000); 4, Sdobnikov (1958); Krechmar (1966); 5, Skalon (1961); Egorov (1965); Egorov and Egorova 1967; Zyryanov and Zyryanov (1976); Revin (1989); 6, Fetisov (1955); Shaposhnikov (1956); Sokolov (1983); Zyryanov (1990); Sobansky (1992); 7, Smirnov (1976); Zyryanov and Zyryanov (1976); 8, Zheleznov (1990) 9, Averin (1948); Mosolov (1986).

Water. Drink water in summer; reindeer like the brackish seawater in river estuaries. During winter, they eat snow and lick water from the top of ice formed on rivers.

8.3.1 Biotope (Table 8.2)

8.4 Diet and Feeding Habits

8.4.1 Diet (Table 8.3)

Feeding on lichens is closely linked to the morphology of teeth (Sokolov 1959), stomach and intestines and their physiology (Akaevsky 1939; Gulchak 1954). The hooves are adapted for digging in snow (Tarasov 1956). Lichens, as energy rich food, are important for survival (Table 8.3).

The role of winter green plants and other foods has been discussed (Alexandrova 1940; Andreev 1940, 1968; Syroechkovsky 1986). In Novaya Zemlya, Taymyr, and the Chukchi Peninsula plants other than lichens had a significant role in the winter diet (Alexandrova 1940; Michurin and Makhaeva 1962; Michurin and Vakhtina 1968; Shchelkunova 1982). Reindeer can adapt to low intake of lichens and use green shoots of *Eriophorum*, *Carex*, and *Equisetum*, as well as dry grass. Specially selected by the Chukchi, the "khargin" breed of tame reindeer (Rumyantsev 1976) use about 50% of green plants, 7–8% of dry grass, 7% mosses and only 19–22% lichens in winter. Reindeer on the Novosibirskie Islands exist almost without lichens (Kischinsky 1971).

Table 8.3. *Rangifer tarandus*. Lichen utilization in different areas as percent of winter diet

Locality[a]	Years	Proportion of lichens in diet (%)
Kola Peninsula[1]	1937	80–90
	1970s	27–57
Novaya Zemlya[b,2]		20
W Siberia[3]		50–90
Taymyr[4]		26–38
Yakutia[5]		40–90
Chukchi okrug[6]		25

[a]The superscript numbers denote the following references: 1, Semenov-Tien-Shansky (1977); Makarova (1988); 2, Alexandrova (1937); 3, Kupriyanov (1990); 4, Michurin (1965a); 5, Egorov (1965); 6, Zheleznov (1990).
[b]Tame reindeer.

For list of plant genera eaten see Appendix 1, Tables 8 and 9.

Green mosses make up 24–36% of the food intake on Taymyr (Michurin and Vakhtina 1968), 14.7% of the diet in Altai (Shaposhnikov 1955). Also, mosses contribute to 24–39% of the diet of tame reindeer on Yamal (Igoshina and Florovskaya 1939; Alexandrova 1940). An abundant water supply in summer is very important to restore body tissues after winter starvation and for fattening in autumn (Baskin 1970). Reindeer find additional mineral substances in old, shed antlers, bones, and by feeding on snow with urine and excrement. In the E Siberian mountains, reindeer use mineral water as well as the excrement of some rodents, especially *Pteromys volans*.

8.4.2 Life Forms/Plant Parts Eaten (Table 8.4)

8.4.3 Seasonal Aspects of Diet (Table 8.5)

To consume freshly growing plants in spring, reindeer follow melting snow northwards, to the sea, and to elevated areas where summer comes later (Avramchik 1939; Gulchak 1954; Vakhtina 1964; Baskin 1970, 1989). During winter, in addition to *Cladonia* and *Cetraria* lichens, reindeer of the Stanovoy Range consume significant quantities of *Equisetum*; in early spring they consume the shoots of *Carex* and *Eriophorum*. In late spring and early summer they take young leaves of *Salix*, *Betula*, *Alnus* and *Potentilla*. In autumn, mushrooms are very important in the diet (Vodopyanov 1971). In Sayan, leaves and young shoots of dwarf shrubs, especially *Betula nana* and *B. rotundifolia*, was found to form a significant part of the diet (Sokolov 1983). In Sakhalin, reindeer consume mostly *Cladonia* and *Cetraria* during winter and in summer herbs and shrubs.

Table 8.4. *Rangifer tarandus*. Life forms/plant parts eaten (x indicates plant part used)

Life forms	Plant parts eaten							
	Whole plant	Leaves	Stems/twigs	Bark	Roots/bulbs	Flowers	Fruits	Seeds
Trees		x						
Shrubs		x	x					
Dwarf shrubs		x	x					
Herbs		x	x			x		
Grasses/sedges		x	x			x		
Horsetails		x	x					
Mosses	x							
Lichens	x							
Fungi	x							

Table 8.5. *Rangifer tarandus*. Seasonal utilization of different plant life forms (percent of stomach content)

Locality[a]	Food	Winter	Spring	Summer	Autumn
N Taymyr[1]	Shrubs, dwarf shrubs			14–79	8–20
	Herbs			2–38	
	Grass/sedges			25–72	51–66
	Dry plants	5–17		0–5	66
	Mosses			2	3–8
	Algae			1–3	
	Lichens	26–38		1–7	4–5
	Mushrooms			1–5	
Putoran Mountains[2]	Larch needles				2–10
	Shrubs, small twigs				5–18
	Shrub leaves				1–4
	Vaccinium leaves				0–1
	Herbs				1
	Green grass/sedges				1–4
	Grass/sedges				43–64
	Dry plants				1–3
	Horsetail				0–2
	Lichens				9–38
Yakutia[3]	Shrubs, leaves	1–4		54	
	Grass	14–22		45	
	Mosses	1–2		2	
	Tree lichens	43–59			
	Foliose lichens	9–20			
Chukchi okrug[4]	Shrubs	5	25	20	16
	Green grass			78	
	Dry plants, winter-green plants	58	61		67
	Mosses, mushrooms	6	3		1
	Lichens	25	10		11
	Other	6	1	2	5

[a]The superscript numbers denote the following references: 1, Lysenko and Shelkovnikova (1989); 2, Mironenko (1984); Larin et al. (1990); 3, Safronov et al. (1990); 4, Zheleznov (1990).

Consumption of Food/Day. About 5 kg (fresh weight) lichens (Table 8.6). They eat 12–40% of the lichens available in a snow crater (Andreev 1968).

Feeding Niche. In bottom, field and shrub layers.

8.5 Social Organization and Behavior

8.5.1 Social Life (Tables 8.7 and 8.8)

Live in families (females with calves), herds and gatherings. In big herds there is probably no personal recognition between animals, but recogni-

Table 8.6. *Rangifer tarandus.* Consumption of food in summer (kg/day). (Zyryanov and Kolpashchikov 1981)

Food	Fresh weight (kg/day)	Dry weight (kg/day)
Birch	0.9	0.3
Willow	2.7	0.7
Grass	11.3	1.8
Total	14.9	2.8

Table 8.7. *Rangifer tarandus.* Group size in forest reindeer populations

Group size	Locality[a]								
	Arkhangelsk (1)[1], Ob–Irtysh (2)[2]			Altai (1)[3], S Evenky okrug (2)[4]		S Yakutia (1)[5], Sayan (2)[6]		Tuva (1)[7], Sakhalin (2)[8]	
	1	2		1	2	1	2	1	2
		Male groups	Female groups						
Mean	8		12	6	13	3	4		
Range	5–20	2–8	3–30	3–8	2–55	1–7	2–15	5–11	7–8

[a]The superscript numbers denote the following references: 1, Fadeev (1985); 2, Bakhmutov and Azarov (1981); Belov (1990); 3, Sobansky (1992); 4, Kuksov et al. (1989); 5, Egorov (1965); Revin (1989); 6, Sokolov (1983); 7, Smirnov (1994); 8, Dvoryadkin and Shvets (1976).

Table 8.8. *Rangifer tarandus.* Seasonal changes in group size on Taymyr (Yakushkin 1976)

Group size	Winter	Spring migration (before calving)	Early summer (after calving)	Middle summer	Autumn migration
Average	52	12–16	18–22	19–48	308
Maximum	300	700	650	100,000	20,000

tion was found in small groups (Zyryanov and Zyryanov 1976). During rut, bulls try to keep harems in the same migratory herds, but often fail. In summer, during insect harassment time, condensed groups protect the animals inside the group from insects, which mostly attack animals on the outside of the group. The least protection against insects appears in herds of about 30 reindeer, about three quarters of animals are protected in herds of 1,000 animals, and there is no significant increase in protection when herd size exceeds 2,000 reindeer (Baskin 1970, 1974). In practice, the smallest group, which can be managed by humans is 35, maximum size of tame reindeer herds is 1,000–1,500 in Taymyr, 3,000–3,500 in Kamchatka, and 5,000 in Chukchi Peninsula (Sdobnikov 1935; Baskin 1974, 1992).

Hierarchy depends on the size of antlers, which differs with age; both sexes have antlers. Reindeer is the only species in the Cervidae where both sexes have antlers. Bulls shed antlers in late autumn. In winter, they protect themselves with their front legs (boxing matches). Antlers give dominance to females and yearlings during competition for craters in snow, which animals dig to reach food. Calves use the protection of the mother. During rut, combats between bulls take up to 30 min, on some days, after the last combat, winners are not challenged by competitors. In summer, reindeer feed by trotting up to and biting flowers, leaves, and tips of shoots. This type of behavior does not depend on the size of the herd. Initially, animals will feed close together in large herds, however, the animals will start to compete to be first to untouched pasture. This results in a cessation of feeding as they race to get ahead of one another. This "competitive movement" disperses the herd throughout the pasture, at which stage, normal grazing is resumed (Baskin 1970; Vodopyanov 1971). Following the leaders and using their experience and alertness, assists the animals in a herd to escape danger; leaders are important during migrations, finding ways to cross rivers. In small groups, of up to 35 animals, permanent leaders were observed. In larger groups there are potential leaders, who become the actual leaders, depending on the situation. In tame reindeer herds, 7–12% of animals can serve as leaders (Naumov and Baskin 1969; Baskin 1974, 1989).

8.5.2 Home Range

During winter and summer, reindeer groups live mostly within a 100-km radius and do not have home ranges, in the sense of using the same pasture places, trails, resting spots, etc. In spring and autumn they migrate along more or less traditional routes. In Karelia, the summer home range of females was 6–7.8 km^2 of land area. In taiga uplands of S Yakutia, reindeer roamed within 3–6 km. In Sayan, groups of reindeer lived on the same slope of a river valley, and rarely made trips more than 25–30 km (Zyryanov 1990). As well as numerous migratory reindeer, there are probably settled reindeer populations on the tundra of NW Taymyr, along the coast of the

Kara Sea and the Yenisey River and also to the south, from the Taymyr Lake. They find appropriate winter conditions where snow cover is <50–60 cm and where strong winds blow away snow (Michurin and Vakhtina 1968; Uchitkin 1990).

8.5.3 Mode of Movement

Usual movements are walk, trot, sometimes amble. When trotting, the animal is supported on three legs, which is important when moving through snow or swamps. Good swimmers, able to cross broad Siberian rivers and sea straits, up to 120 km wide (Pinegin 1932). The speed of swimming is 5.5 km/h (Michurin 1965b). Reindeer are excellent at digging in snow to find food. Can climb very steep slopes covered with hard snow and can cross-stony fields. However, reindeer prefer smooth places.

8.5.4 Dispersal/Migration (Fig. 8.4 Tables 8.9 and 8.10)

The migratory herds lag about 15–20 days behind the line of melting snow. In the middle of summer there is a short migration back, and then

Table 8.9. *Rangifer tarandus.* Migratory characteristics. (x indicates behavior observed)

Character of terrain	Forest			Tundra	
	Mountain	Plain			
Way of life	Settled		Migratory	Settled	Migratory
Locality[a]					
Karelia[1]		x	x		
N Yamal[2]				x	
Taymyr[3]	x[b]			x	x[c]
Evenky okrug[4]		x	x		
Sayan[5]	x				
Yakutia[6]	x		x		x[c]
Novosibirskie Archipelago[7]				x	x
Transbaikal[8]	x	x			

[a]The superscript numbers denote the following references: 1, Bludnik et al. (1989); 2, Bakhmutov and Azarov (1981); 3, Krechmar (1966); Geller and Borzhonov (1984); Kolpaschikov (2000); 4, Kostyuk et al. (1979); Lavov (1976); Kuksov et al. (1989); Zhukov (2000); 5, Dulkeit (1964); Sokolov (1983); 6, Egorov (1971); Safronov et al. (1999); 7, Kishchinsky (1971); Kupriyanov and Belikov (1986); Safronov et al. (1990); 8, Vodopyanov (1970); Shargaev and Maturova (1976).
[b]There are short migrations up and down mountains.
[c]Summer in tundra, winter in forest.

Table 8.10. *Rangifer tarandus*. Times of migration

Locality[a]	Spring				Autumn	
	Males		Females		For whole population	
	Start	Finish	Start	Finish	Start	Finish
N Ural[1]			10 April–mid-May[c]			November[d]
Konda[2]			April–May			November–December
All Taymyr[3]					Mid-August	Mid-November
Pyasina	Late May		Early May	Mid-June	Early September	10 October
Putoran Mountains[b]	10 May	Late June	1 April	Late May	20 August–15 September	
Kheta	2 March		15 February		20 August.	
Moyero River	5 March–5 April		5 February–24 March			

[a]The superscript numbers denote the following references: 1, Polezhaev (1976); 2, Azarov (1976); Bakhmutov and Azarov (1981); 3, Krechmar (1966); Kolpashchikov (1979); Lineytsev (1983); Kolpashchikov et al. (1983).
[b]Intermediate point on the migration route.
[c]Arriving in these mountains.
[d]Departure.

movement again, to colder places, to use secondary vegetation. When snow conditions dramatically worsen, mass migrations happen in any season (Pinegin 1932; Baskin 1986; Zheleznov 1990). The extent of male migrations is usually much longer than that of females. In many areas migratory and settled populations co-exist during some seasons.

In all populations, females start spring migrations earlier than males in order to be on calving grounds early.

In Yakutia, three populations, inhabiting areas between the great rivers (Lena - Olenek, Yana - Indigirka, Indigirka - Kolyma), are described (Zyryanov and Zyryanov 1976; Safronov et al. 1990). They spend summer in northern lowlands along the seashores and migrate south, into taiga areas for wintering (Safronov et al. 1990). In Taymyr, the extent of the migrations reaches 1,000–1,200 km (Fig. 8.4), in Yakutia 600–700, maximum 1,000 km.

Populations of forest and mountain reindeer migrate for rather shorter distances. In Altai, there are short vertical and horizontal migrations over slopes to find more appropriate pastures, with favorable snow. However, some populations migrate over 80–100 km, maximum 150 km (Sobansky 1992). In Tuva, the snow factor evokes migrations of 50–70 km (Smirnov 1994). In Sayan and Altai, in early spring, females start migration from alpine pastures down to the upper part of the forest zone, where they spend the calving season. Males arrive there later. Males and females make the autumn migration together (Dmitriev 1938; Sokolov 1983). In the mountain area of the Kronotsky Nature Reserve (Kamchatka), reindeer carry out short migrations, from the tops of mountains, along river valleys to the sea in spring, and then back again in autumn (Mosolov 1986). There is contradictory information on the reindeer population inhabiting the great area between the Ob River and the Ural. Reindeer in this area spend the winter in pine forests, with lichen cover, on the Konda and Berezovo Rivers. In spring, some of the animals migrate to the Ural, while others disperse over swamps and the Ob River flood plain (Azarov 1976; Shargaev 1976). Reindeer of the Belyi Island migrate to Yamal. In the past, reindeer from the Novosibirskie Islands migrated south to the mainland, where they wintered in the Verkhoyansk Mountains and as a result, crossed the sea straits twice a year. Since the period of great decline, by 1930, these migrations ceased (Kishchinsky 1971). However, since 1980 reindeer have restored these migrations.

In the 1930s, Taymyr reindeer were more numerous east of the Taymyr lake, and mass migrations were observed through the Khatanga and Kheta Rivers. In the 1960s, most of the population inhabited the western part of Taymyr and the longest migration routes were observed through the Pyasina and Dudypta Rivers. By 1960, reindeer migrated in autumn from tundra of W Taymyr, past Norilsk and through the Putoran Mountains to N Evenky Okrug. A number of them crossed the Yenisey River on ice and a few thousand wintered in the tundra of W Taymyr. Following a growth

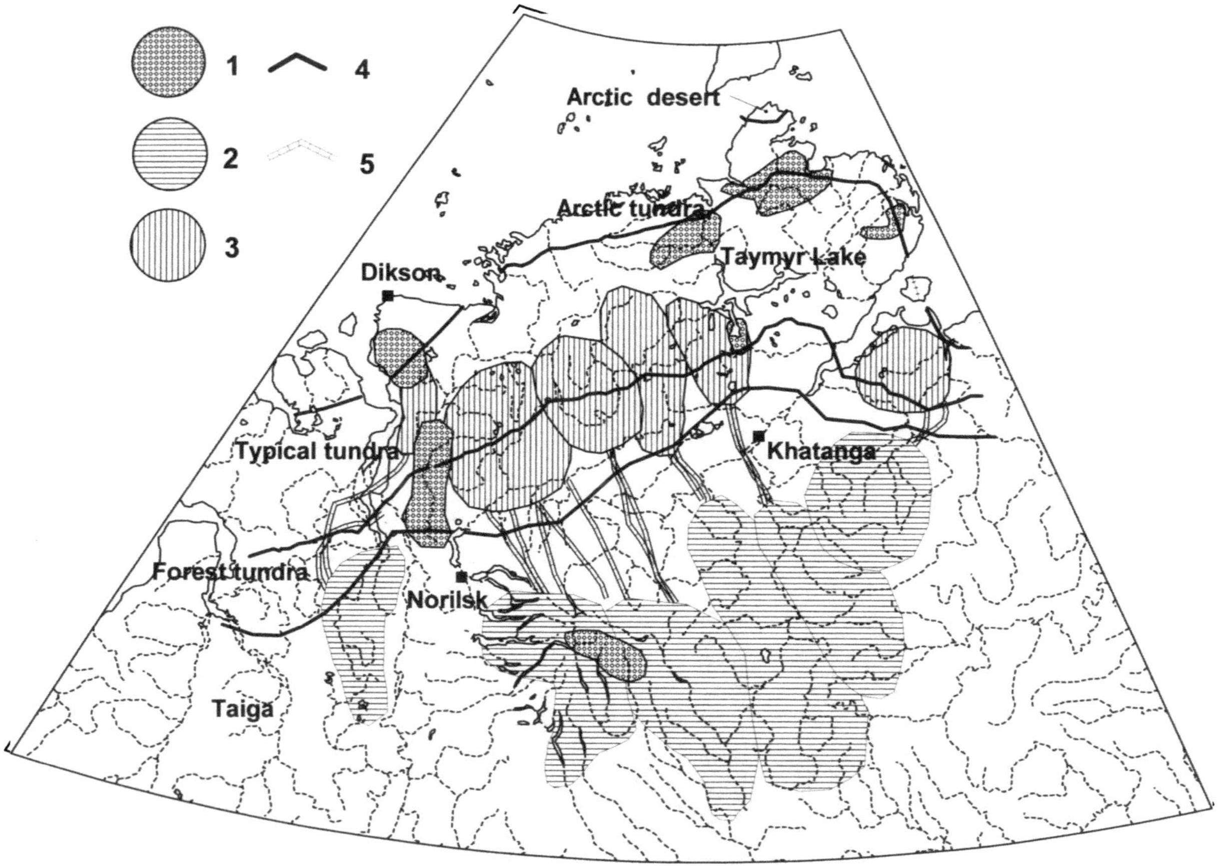

Fig. 8.4. Seasonal grounds and migrations on Taymyr in 1962–2000 (Krechmar 1966; Yakushkin et al. 1970; Malygin et al. 1978; Kuksov 1981; Geller and Borzhonov 1984; Uchitkin 1990; Kolpashchikov 2000). *1* Non-migratory populations, *2* winter grounds; *3* summer grounds; *4* borders of vegetation zones; *5* migratory ways

in the population number, in 1970 reindeer prolonged their migrations through the Putoran Mountains to N Evenky. Also, after crossing the Kheta River, they continued on to Yakutia, to the Olenek River. In spring, reindeer used the same routes to get to the calving grounds in the tundra near the rivers Agapa, Dudypta and Nizhnyaya Taymyra (Geller and Borzhonov 1984; Kolpashchikov and Pavlov 1995; Kolpashchikov 2000). Summer grounds were spread over most of the tundra of W and C Taymyr.

8.5.5
Diurnal Activity (Table 8.11)

During hot summers, in many areas, animals spend the day and early evenings on snow fields, which cool them and where there is reduced insect activity. Under especially strong insect harassment animals feed only 2 h/day (Vodopyanov 1984). In the forest areas of the Kola Peninsula, reindeer spend the summer days slowly running along the same trails to avoid insects (Semenov-Tien-Shansky 1977). Tame reindeer, at this time, often clump together into compact masses, protecting themselves from blood-sucking insects (Baskin 1970).

8.5.6
Shelters

To escape blood-sucking insect harassment, forest reindeer run back and forth along slopes, river and lake shores. However, when the density of the reindeer population increases and there are mountain tundra plots avail-

Table 8.11. *Rangifer tarandus*. Diurnal activity

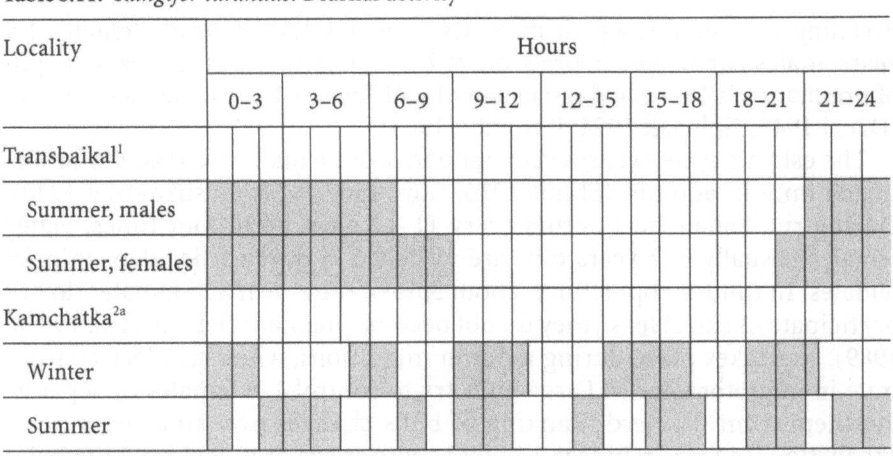

Shading indicates that animals are active. The superscript numbers denote the following references: 1, Vodopyanov (1970); 2, Baskin (1970).
[a]Tame reindeer when pasturing without human influence.

able, the animals prefer to go up mountains, use windy areas, snow fields, and gather in big herds. Using snowfields in summer and gathering in a big mass are also very common. In summer, gatherings can reach 100,000 (Baskin 1970; Geller and Pavlov 1972; Semenov-Tien-Shansky 1977; Yakushkin et al. 2001).

8.5.7 Escape Behavior

Large migratory herds are more alert than small herds; male herds are less alert than female herds. Reindeer like to cross a large river by following the track of other herds. If the herd in front is disturbed, the followers will stop and search for another crossing (Savelev 1977). Reindeer are not shy animals. During the flight from wolverines, they make the first stop after 300–450 m in an open place, and at 700–900 m in rough terrain. However, single individuals run away from wolverines for 3–4 km, and if the predator continues to attack, reindeer run away for 10–15 km (Lineytsev and Krashevsky 1988). At first, disturbed reindeer flee from man upwind and uphill, immediately after disturbance and there is the tendency to resume travel in the original direction during the final withdrawal (Baskin and Skogland 1997). Where reindeer rarely meet humans, they can be approached to within 80–100 m. Reindeer often approach man, tractors, and aircraft out of curiosity (Semenov-Tien-Shansky 1977; Sobansky 1992; Smirnov 1994).

8.6 Reproduction (Tables 8.12 and 8.13)

Breeding parameters: age at maturity of males 1.5–2.5 years, females 1.5 years; males participate in breeding at 2.5 years, females at 1.5 years; length of pregnancy is 192–246 days; usually 1 calf/female, 2 calves/female are rare (Druri 1949; Gulchak 1954; Borozdin 1963).

The estrous cycle occurs every 2 months, but females are ready to be fertilized only in autumn (Shmit 1936; Gorbunov 1939; Mashkovtsev 1940). During rut, female heat occurs every 11–22 days, up to four times. Fatter females (usually 4–7 years old) are in heat 5–6 days earlier than thinner females. In tundra populations, about 50% of the 2-year-old females do not participate in rut; at least, they do not become pregnant (Pavlov et al. 1985a, 1989). Rut takes place during autumn migrations, when reindeer concentrate in migratory herds. Large bulls try to control 4–8 females by separating them from the herd. Ranking of bulls changes over time, because of exhaustion. In tame reindeer, 1 bull (3 years of age or older) per 18 females yielded the maximum number of calves, using the minimum number of males, and this ratio of males to females permitted males to conserve enough fat to survive during winter. For example, a male to female ratio of

Table 8.12. *Rangifer tarandus*. Barrenness of reindeer (%)

Locality[a]	Years	Age (years)							All
		2	3	3–6	4–8	7–9	9	10	
Taymyr[1]	1975–1983		54		18		32		25 (12–40)
	1987–1990	81		11		13		20	12
Evenky okrug[2]	1975								16.5
Yakutia[3]	1990								20
Kamchatka[4]	1975								7

[a]The superscript numbers denote the following references: 1, Pavlov et al. (1985b); 2, Lavov (1976); 3, Safronov et al. (1990); 4, Fil (1976).

Table 8.13. *Rangifer tarandus*. Fertility of females versus age and body mass (Taymyr, 1981–1983) (Pavlov et al. 1985a,b, 1989)

Body mass (kg)	Pregnant females (%)	Age (years)	Pregnant females (%)
50–59	2.6–3.4	All ages ≥1.5	25.0
60–69	13.0–35.0	3	46.3
70–79	44.0–52.0	4–8	82.5
≥80	17.9–32.0	≥9	67.6

1:200 results in all the females being mated, but most of the exhausted males die during winter (Borozdin 1963; Baskin 1970).

Newborns can stand after 1 h; take the first suck within 5 h. During the first days, mothers are not able to recognize the calf by smell, and calves can follow any female, even dog or man. Mothers leave the birthplace after 5–7 h, when the calves are able to follow and run fairly quickly. After 7 days calves start to eat lichens. Up to 3-month-old calves follow only their mothers; waiting for them in the last suckling place and cannot survive in the herd without them. The socialization period ends during the third month, when calves start to follow a herd; thereafter orphans can survive. After 4 months (in late summer), when young, green food becomes rare, females drastically decrease the volume of milk, as well as maternal behavior. Ties between a mother and her calf usually exist up to the following spring, occasionally up to 3 years (Golovnev 1940; Baskin 1970).

8.6.1 Seasonal Breeding Events (Tables 8.14 and 8.15)

Adult bulls shed antlers in early winter, young and barren females in early spring, pregnant females 3 days after birth. Cleaning antlers from velvet

Table 8.14. *Rangifer tarandus*. Seasonal breeding events

Locality[a]	Rutting season			Calving season		
	Preparation to rut	Basic rut	Last appearances of rut	First calves	Basic calving	Last calves
Kola Peninsula[1]	August–September	5 October		18 May	1 June–end June	
Konda[2]		7–30 September			20–25 May	
Taymyr[3]	15–20 October		Mid November		Second half of June	
Yakutia						
Tundra[4]		Late October–early November		29 May	10–20 June	
Taiga[4]		Mid-October			Late May	
Transbaikal[5]		10 September–10 October			Late May–10 June	
Khabarovsk[6]		Late September–October				
Chukchi okrug[7]	Early October		Late October–early November	25 May		16–20 June
Kamchatka[8]	10 October	Mid-November	Late November			
Sakhalin[9]		Mid-September–mid-November				

[a]The superscript numbers denote the following references: 1, Semenov-Tien-Shansky (1977); 2, Shargaev (1976); 3, Krechmar (1966); 4, Safronov et al. (1990, 1999); 5, Vodopyanov (1970); Moskvitin et al. (1976); Revin (1989); 6, Shargaev and Maturova (1976); 7, Zheleznov (1990); 8, Fil (1973); Mosolov (1986); 9, Zagorodskikh and Reimers (1984).

Table 8.15. *Rangifer tarandus*. Changes in rut and calving in tame reindeer in relation to snowmelt and temperature. (Vinogradov 1936; Baskin 1970; Lydolph 1977)

	Average date of frost		Basic rut	Average date of temperature above 0°C		Start of snowmelt	Basic calving
	First drop to <0°C	Constantly <0°C		First rise to >0°C	Constantly >0°C		
Kola Peninsula	1 October	16 October	6–10 October	15 April	12 May	1 June	16–20 May
N Ural	4 October	6 October	6–20 October	30 April	23 May	1 June	26–30 May
Taymyr	21 September	29 September	16–20 October	17 May	31 May	10 June	26–30 May
Chukchi okrug	21 September	3 October	21–25 September	12 May	21 May	20 May	1–5 May
Wrangel	10 September	20 September	20–25 August	20 May	10 July	20 May	20–25 April
N Kamchatka	15 September	1 October	5–10 September	5 May	20 May	15 May	1–5 May

occurs in late summer (August). On average, calving takes place 15–20 days earlier. In most areas, wild reindeer calve 2–4 weeks later than tame reindeer.

Maximum Length of Life. In Taymyr, males live on average 4–5 years, maximum 14 years; females on average 6–7 years, maximum 19 years (Pavlov et al. 1985a).

8.7 Mortality

8.7.1 Causes of Mortality

High mortality, caused by overgrazing, has been observed only in Lapland during 1940–1967, when density of wild reindeer reached 1.2 animals/km^2

Table 8.16. *Rangifer tarandus.* Gender and age composition of populations

Locality[a]	Year	Animal gender and age (%)			
		Males	Females	Yearlings	Calves
Upper Pechora[1]		27	50		23
Ob–Irtysh[2]	1979, 1984	13	42	25	20
Taymyr[3]	1969	13	42	19	26
Evenky okrug[4]	1974	30	50		20
Olenek–Lena[5]	1965	17	36	20	27
	1975	12	43	17	28
	1981	25	37	10	28
	1985	36	49		15
	1990	27	33	17	23
	1994	25	33	22	20
Yana–Indigirka[5]	1980	29	37	14	20
	1987	24	32	25	19
	1993	22	32	24	22
Indigirka–Kolyma[5]	1987	15	37	23	25
Chukchi okrug[6]	1974	25	49		26
	1986	25	32	24	19
S Kamchatka[7]	1970–1974	20–23	53–54		22–27
Sakhalin[8]	1975		71	21–25	

[a]The superscript numbers denote the following references: 1, Sokolsky (1984); 2, Belov (1990); 3, Pavlov et al. (1971, 1985a); 4, Kuksov et al. (1989); 5, Tsybulsky (1995); Safronov et al. (1990, 1999); Vshivtsev et al. (1980); Egorov and Krechmar (1967); Safronov et al. (1990, 1999); 6, Kolpashchikov (1990); Zheleznov (1990); 7, Fil (1973, 1976); 8, Zagorodskikh and Reimers (1984).

(Semenov-Tien-Shansky 1977). Very cold weather in June, when migratory reindeer cross rivers, can lead to mass mortality among calves, due to inflammation of the lungs (Zabrodin and Pavlov 1983). Ice crust on snow caused great mortality on the Novaya Zemlya, especially in the winters of 1920–1921, 1931–1932, and 1980–1981 (4,500 reindeer died) (Novikov 1983a). The Novosibirskie Islands population became almost extinct in 1920–1930, following the death of many reindeer, when they crossed the Dmitriy Laptev Strait in autumn (Pinegin 1932). In Sayan, high mortality of calves occurred in winters with deep snow (Fetisov 1955). In Kamchatka, Fil (1976) found that 44% of the calves died during the first year, and 33.5% during the second year. According to Baskin (1983), the causes of newborn calf mortality in the tundra and taiga localities, respectively were as follows: 22 and 10% were feeble calves, 3 and 2% were deformed, 13 and 6% died during snow-storms, 4 and 12% were killed by predators, 2 and 4% died due to broken legs.

8.7.2 Predators

Wolf (in Putoran Mountains, 92% of scats had reindeer remnants), wolverine (26–76% in different seasons), and lynx (17%). In North Europe 23–82% of wolverine scats, up to 81% in W Siberia, up to 97% in S Siberian Mountains, up to 98% in the Far East, contained reindeer remnants (Novikov 1998). Brown bear, raven, golden eagle and sea eagle kill mostly calves. Wolf predation is equal among different gender and age groups in

Table 8.17. *Rangifer tarandus.* Gender and age composition of the Taymyr population in 1968–1989. The data is the result of tooth analyses of reindeer shot when crossed the Pyasina river during spring and autumn migration. (Pavlov et al. 1985a; Kolpashchikov 2000)

Age	Embryo	2–3 months	9–11 months	1–2 years	3–4 years	3–7 years	5–7 years	8–10 years	≥11 years
Male / Female ratio	1.04:1	1.08:1	1:1.2	1:1.3	1:1.03	1:2.0		1:1.78	1:3
Age composition (%)									
1968–1972				7-24	28	65–80	43	22	7
1981–1984					37		49	7–12	1–4
1987–1989					51		37	10	2

Table 8.18. *Rangifer tarandus*. Population density and size in different localities

Locality[a]	Year[a]	Population density (animals/10 km^2)	Population size (×1,000)
Altaisky kray	1980–1992[1]	0.04	0.3
	1997–1999[2]		0.35
Amur oblast	1979[3]	0.5	3.8–4
Bolshevik Island (Severnaya Zemlya Archipelago)[4]	1930		0.1
	1986		0.15–0.2
Buryatia	1975[5]	3–4	6–6.5
Chita oblast	1980–1992[5]	0.3	5.2
Chukchi okrug	1977	3–16	3–3.5
	1986	2–7	32
Evenky okrug	1974–1976[7]	9.4 (5–17)	70
	1979[7]		66
	1989[7]	3	7.5
	1999[2]		60
Islands of the Russian sector of Arctic[4]	1983		ca. 16
Kamchatka oblast[8]	1961		10
	1970		8
	1983		3.5
	1997		4.5
Khabarovsk kray	1979[9]	0.5	14–16
Khanty-Mansisky okrug[2]	1999		6.1
Kola Peninsula, Lapland Nature Reserve[10,b]	1929		0.1
	1932		0.21
	1936		0.4
	1940		0.9
	1949		0.4
	1957		2.0
	1961		4.4
	1965		10.5
	1966		12.6
	1967		12.6
	1969		12.5
	1970		22
	1981		0.1
	1988		0.2
	1990		0.2
	1993		0.2
	1999		1.3
Kola Peninsula, Tersky coast[10]	1969		8
	1985		2.5
Komi[11]	1950		10
	1999[2]		2.2
Krasnoyarsk kray, forest	1975[12]		70
	1980[12]		82
	1985[12]		83
Novaya Zemlya[13]	1979		10
	1981		6
	1996		10

Table 8.18. *Continued*

Locality[a]	Year[a]	Population density (animals/10 km^2)	Population size (×1,000)
Novosibirskie Archipelago[14]	1930		Very few
	1965		(17–18)
	1970		17–18
	1980		7.5
	1985		7.5–16
Sakhalin[15]	1968		0.84–2.1
	1975	0.5–8.6	4
	1999[2]		4.4
Sayan[16]	1980–1992	0.3	1.7
Sibiryakov Island[13]	1981		0.3
Taymyr, summer[17]	1959		110–120
	1965		120
	1966		250
	1969	7.4–9.7	333
Taymyr, winter	1969	2.8–3.0	
Taymyr, summer	1970[17]		340
	1972[17]		386
	1975[17]		450
	1978[17]		475
	1981[17]		510
	1985[17]		575
	1988[17]	0.2	570
	1990[17]		625
	1993[17]		560
	2000[18,c]		1,000
	2001[19]		350
Tomsk oblast[20]	1975		1.2–1.5
Transbaikal[16]	1969	0.47	
	1980–1992	0.1	5.0
Tuva[16]		0.4–0.6, max 9.2	
	1980–1992	0.2	2.0
Wrangel island[21]	1964		0.15
	1956		0.6
	1980		3.0
	1991		3.0
Yakutia, total[22]	1920		350
	1965		117–122
	1970		160
	1978		169
	1981		147
	1985		186–191
	1990		230
	1999[2]		250
Yakutia, forest[18]	1975–1976	2.8–6.6	
	1965		30–35
Yakutia, tundra[18]	1920		350
	1990		230

Table 8.18. *Continued*

Locality[a]	Year[a]	Population density (animals/10 km²)	Population size (×1,000)
Yakutia, Olenek–Lena[18]	1959		18–20
	1965		20
	1975		50
	1981		61
	1985		55–61
	1988		73
	1990		81
	1994		78
Yakutia, Yana–Indigirka[18]	1965		50
	1975		109
	1978		87
	1980	0.2–9	55
	1985		121
	1987		130
	1989		116
	1991		101
	1993		85
Yakutia, Indigirka - Kolyma[18]	1975		22
	1987		29
	1993		40
	1996		34
Yamal (northern tip) and Belyi Island[23]	1930		2.5
	1920		8
	1960		0.5
	1981		2

[a]The superscript numbers denote the following references: 1, Moskvitin et al. (1976); 2, Paponov (official data) (2000); 3, Bromley and Kucherenko (1983); 4, Novikov (1983a); Kupriyanov and Belikov (1986); 5, Smirnov (1994); 6, Zheleznov (1990); 7, Kostyuk et al. (1979); Kuksov et al. (1989); 8, Mosolov (1986, 1997); 9, Dvoryadkin and Shvets (1976); 10, Semenov-Tien-Shansky (1948b, 1989); Makarova (1988, 1990); Makarova and Khokhlov (1985); Kataev (1999); 11, Polezhaev and Popov (1990); 12, Fertikov et al. (1983); 13, Kupriyanov and Belikov (1986); Khakhin 1998; 14, Pinegin (1932); Egorov and Krechmar (1967); Kishchinsky (1971); Safronov et al. (1999); 15, Zagorodskikh and Reimers (1984); 16, Smirnov (1994); 17, Pavlov et al. (1993); Kolpashchikov (2000); 18, Yakushkin et al. (2001); 19, Lomanov (pers. comm.); 20, Lavov (1976); 21, Kazmin (1986); Baskin and Skogland (1997); 22, Zyryanov and Zyryanov (1976); Popov et al. (1980); Egorov and Popov (1984); Mordosov et al. (1990); Safronov et al. (1990, 1999); 23, Kupriyanov and Belikov (1986).

[b]The decline of reindeer in the Lapland Nature Reserve was strongly connected with over-exploitation on all the Kola Peninsula. In 1970s, about 10,000 reindeer were harvested with use of helicopters and snowmobiles.

[c]Counted from air on days of high insect harassment, these estimates can be uncertain. These data are under discussion.

reindeer (Druri 1949; Makridin 1962; Egorov 1965; Polezhaev 1976; Bromley and Kucherenko 1983; Sokolsky 1984; Larin and Krashevsky 1989a; Krashevsky 1989; Zyryanov 1990; Zheleznov 1990).

8.7.3
Diseases, Helminths and Ectoparasites (Appendix 2, Tables 4 and 5)

Anthrax (in the past) and necrobiosis (hoof disease) are the most important. Also, foot-and-mouth disease, and inflammation of the lungs (especially harmful for calves after crossing cold rivers) have been reported (Zabrodin 1984).

8.8
Population Structure and Densities (Tables 8.16 and 8.17)

Male to female ratio of embryos in the Kola Peninsula was 1:1 (Makarova and Khokhlov 1985), the male to female ratio of newborns in Kamchatka was 1.09–1.12:1 (Fil 1976). Among adult animals, females were more numerous: in all areas 1 male per 1.35–3.2 females (Pavlov et al. 1985a).

Table 8.19. *Rangifer tarandus.* Official data on population size. (Borisov et al. 1992; Novikov 1996; Paponov 2000)

Locality	Area of suitable habitat (×1,000 km^2)	Population size (×1,000)			
		1981	1990	1995	1999
Amur oblast	50	4.1	5.2	3	3.5
Arkhangelsk oblast	50	24	15.8	17	20
Buryatia	100	3.1	3.5	2.2	2.3
Chita oblast	52	8	10	10	10
Irkutsk oblast	165	20	13.8	12	15
Kamchatka oblast	60	3.5	4.9	5	3
Karelia	6.4	8	4.9	4	3.6
Khabarovsk kray	343	15	17	10	16
Krasnoyarsk kray + Taymyr okrug	2,000	580	595	550	615
Magadan oblast + Chukchi okrug	100	17.0	30	35	163.6
Murmansk oblast	31	2.2	4.0	4	7
Tomsk oblast	152	3.0	4.0	4.5	3
Tuva			4.0	1	0.4
Tyumen oblast	180	16.0	17.2	18	5
Yakutia	2,000	180.0	300.0	240	250
Yamal okrug					35
Russian Federation (total)					1,152.4

8.8.1 Population Density and Size (Tables 8.18 and 8.19).

In many areas (Taymyr, Novosibirskie Islands) enlarged winter grounds and expanded migratory routes followed after the increase in population size (Baskin 1990).

Table 8.20. *Rangifer tarandus.* Wild reindeer harvested (legal harvest with licenses)

Locality[a]	Year	Population size (×1,000)	Numbers harvested (×1,000)	Meat obtained (×1,000 tons)
Kola Peninsula[1,6]	1965	10.5	0.06	
	1970	22.0	0.4	
	1971	7.1	0.3	
	1995	4	0.27	
Lapland[1,b]	1976	2.8	1.8	
Kola Peninsula, Tersky Bereg[1]	1971	7.0	0.5	
	1976	1.5	1.1	
Evenky okrug[c,2,6]	1971–1975		57	0.07
	1995	50	3.4	
Taymyr[3,5,d]	1971–1999		1,700	
	1960	118	15	
	1965	215	15	
	1970	349	15	
	1975	449	57.2	
	1980	477	66	
	1985	575	101	
	1988	570	104	
Yakutia[4,6]	1965	117–122	2.5	
	1969		3.0	
	1986	186–191	24.1	
	1988		27	2.6
	1981–1990		230	114
	1995	240	16.9	
Total Russian Federation[5,6]	1971–1975		140	4.6
	1976–1980		352.2	13.8
	1980–1981		75.6	2.9
	1981–1982		98.6	5.1
	1995	971.7	22.5	

[a]The superscript numbers denote the following references: 1, Semenov-Tien-Shansky (1977); 2, Karelov and Bulavkin (1976); 3, Mikhailov et al. (1990); 4, Novikov (1983b); Egorov and Popov (1984); Safronov et al. (1990, 1999); 5, Kolpashchikov (2000); 6, Novikov (1996).
[b]At the end of hunting only some hundreds of animals had survived. It is a classical case of overexploitation.
[c]Including 860 forest reindeer.
[d]In Taymyr, in 1970–1988, 23–26% of harvested animals were calves.

8.8.2 Impact on Vegetation

During migration, when 100,000 reindeer passed an area of 30 km^2, 10–15% of the food resources were used (Kolpashchikov 1979). In summer, on Taymyr, where 30,000 reindeer were pasturing (3.4 animals/km^2 for 40 days), leaves of willows were reduced by 5–11%, birch by 4%, grass by ca. 2%, total green plants (leaves of shrubs and grass) were reduced by 2%. This removal equals about 203 tons/km^2.

8.9 Relations to Man (Table 8.20)

Tales that wild reindeer herds remove tame reindeer from domestic herds are strongly overstated. In Chukchi Peninsula, only about 5% of tame reindeer were present in wild herds during the autumn migration (Zheleznov 1990). When the gas pipeline "Messoyakha-Norilsk" was constructed, and regular traffic of icebreakers along the Yenisey River started, a shift of migratory routes eastward was observed; reindeer moved around the pipeline and the town of Norilsk. Some well-known crossings of reindeer through the Pyasina, Dudypta and Kheta Rivers have been used since 1969 to arrange slaughtering on an industrial scale, but by late 1980, they had become unprofitable because of the shift of migratory flows to the east. Then, a necessity for arranged hunting on the Kheta and Khatanga Rivers increased (Kolpashchikov and Bolonin 1990; Kolpashchikov 2000). Practically all parts of the reindeer body are used by native people: meat, blood, brains, tongue, liver, kidneys, skins, sinews, even stomach contents as a source of vegetables, and calf excrements for softening skins. Products used for sale are meat, skins, and velvet antlers.

9 Goitred Gazelle – *Gazella subgutturosa*

Gazella subgutturosa Güldenstaedt, 1780.
Artiodactyla, Bovidae, Antilopinae

9.1
Status and Distribution (Fig. 9.1)

9.1.1
Taxonomic Remarks

Only one subspecies is found: *G. s. subgutturosa* Güldenstaedt, 1780 (Bannikov 1989).

9.1.2
Status

Lower risk - near threatened (LR/nt) (Hilton-Taylor 2000). A tendency to decrease in numbers (Flint and Prisyazhnyuk 1986; Bekenov et al. 1992). Protected in nature reserves and by hunting regulations.

9.1.3
Distribution (Fig. 9.2)

Distribution Area. Estimated historical range (Bannikov 1989) ca. 1,880,000 km^2; modern range 300,000 km^2 (Flint and Prisyazhnyuk 1986).

Occurrence in Ecoregions. Tropical/Subtropical Steppe (315), Tropical/ Subtropical Desert (321), Temperate Desert (343).

9.2
Measurements (Table 9.1, Fig. 9.3)

9.3
Habitat

Prefer flat planes, and sand hills but also inhabit low mountains, capable of living in mountain valleys and can even go upwards onto the slopes.

Fig. 9.1. Goitred gazella

Adapted to deserts, presumably with hard soil (Kostin 1955; Heptner 1956; Reimov 1981; Zhevnerov et al. 1983).

Terrain Form. Seashores, shores of fresh water, depressions and flood plains in deserts, flat planes, sand hills, ravines (in Kyzylkumy), plateau (top and foothill), mountain foothill and valleys, even slopes, as in Dzungarian Alatau and Badkhyz (Gorelov 1972; Ishadov 1972; Sludsky 1977; Rustamov et al. 1986; Annenkov 1992; Sokolov and Tembotov 1993).

Altitude. Ustyurt Plateau (0–400 m a.s.l.), Badkhyz (400–1,000 m a.s.l.), Ketmen Range, Dzungarian Alatau Range (1,000–3,000 m a.s.l.), Tajikistan (400–1800 m a.s.l.) (Heptner 1956; Gorelov 1972; Zhevnerov et al. 1983; Annenkov 1992; Sokov 1993).

Soil Texture. Prefer hard ground (gravel, clay), also sandy deserts consolidated by plants.

Snow. A snow depth >10 cm is limiting.

Water. Need water in summer (stay within 10–15 km from water) and need snow when water freezes. Drink mostly in the morning, evening and night. In spring, fresh plants provide most of the required water (Prisyazhnyuk 1986). Needs 3–4 l of water/day (Zhevnerov et al. 1983). Can use salt water up to 20 g/l, a greater salt content in the water leads to mortality among the animals, especially the young (Gorelov 1972; Sludsky 1977).

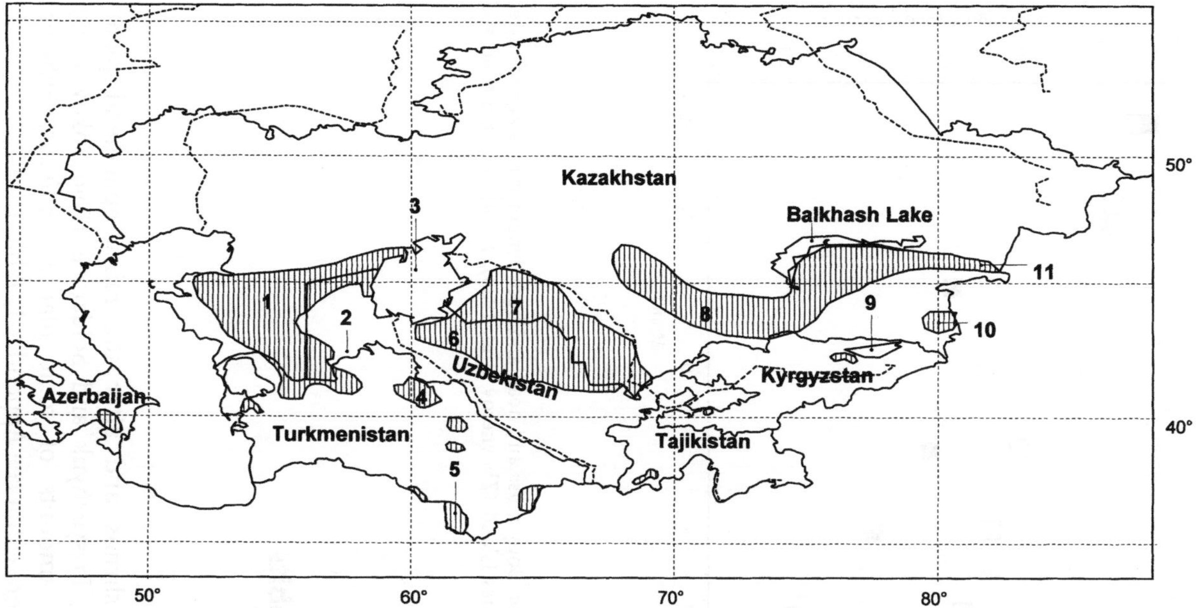

Fig. 9.2. *Gazella subgutturosa*. Distribution map (Zhevnerov et al. 1983; Blank and Kovshar 1986; Sabilaev and Koptsev 1986; Rustamov et al. 1986; Gorbunov and Zarkhidze 1986; Bannikov 1989; Sokolov and Tembotov 1993). *1* Ustyurt; *2* Sarykamysh Lake; *3* Aral Sea, Barsakelmes Island; *4* Karakum Desert; *5* Badkhyz Nature Reserve; *6* Karakalpakstan; *7* Kyzylkum Desert; *8* Betpaqdala Desert; *9* Issyk-Kul Lake; *10* Ketmen Range; *11* Dzungarian Alatau Range

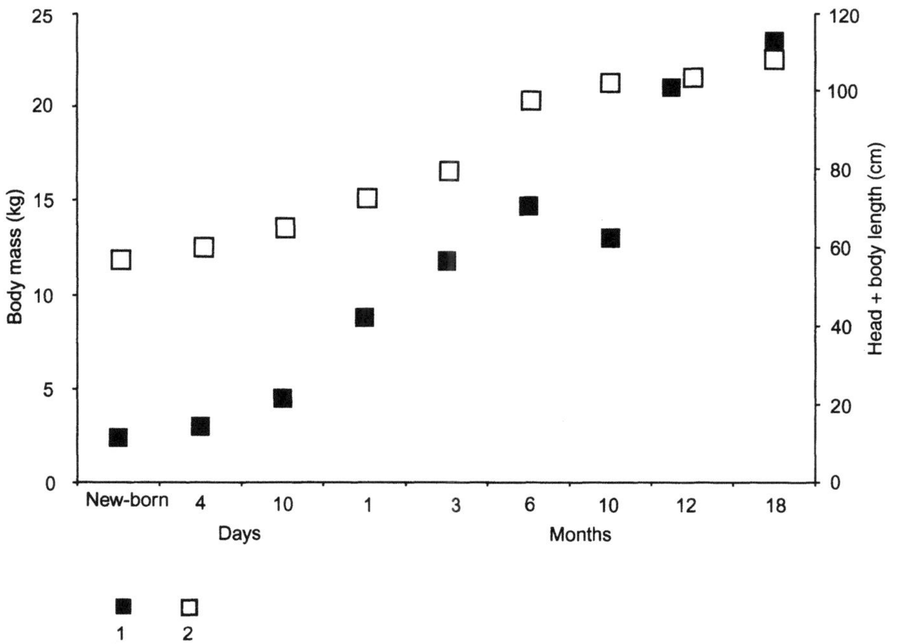

Fig. 9.3. *Gazella subgutturosa.* Body mass and head and body length versus age of males and females from Barsa-kelmes Island (Zhevnerov et al. 1983). *1* Body mass; *2* head and body length

9.3.1 Biotope (Table 9.2)

9.4 Diet and Feeding Habits

9.4.1 Diet

Leaves and twigs of shrubs and after the rains, ephemeral plants (Zhevnerov et al. 1983; Prisyazhyuk and Soldatova 1984; Sokov 1993). Consume considerable amounts of toxic plants: *Anabasis*, *Ephedra*, *Dodartia*. For plant genera eaten see Appendix 1, Table 10).

9.4.2 Life Forms/Plant Parts Eaten (Table 9.3)

Consumption of Food/day. 6 kg of fresh food is consumed (Zhevnerov et al. 1983), according to Prisyazhyuk and Soldatova (1984) the figure is 4.5 kg. The average, available crop in a typical gazelle pasture is 300–400 kg/ha (Kurochkina et al. 1986). At a density of 24 animals/km^2, gazelles consume 30–37% of the vegetation mass (Mardonov 1997).

Table 9.1. *Gazella subgutturosa.* Measurements

Locality[a]	Longitude (°E)	Latitude (°N)	Gender	n	Body mass (kg)	Head and body length (cm)	Tail length (cm)
Ustyurt Plateau[1]	56–58	41–45	Male	7	27.2 (24.2–29.6)	115.0 (105–120.0)	16.6 (12–20.0)
			Female	9	24.2 (20.2–33.2)	112.0 (106–113.0)	16.0 (10–20.0)
Barsakelmes[2]	59	45	Male	32	25 (22–26)	118 (107–126)	16 (15–18)
			Female	10	22 (20–23)	108 (101–119)	16 (14–17)
Betpaqdala[3]	68–74	43–47	Male	18	30.0 (27–33.8)	119 (111–125)	(16–21)
				50	24 (21–33)	109 (102–120)	(10–23)

[a]The superscript numbers denote the following references: 1, Mambetzhumaev (1970); 2, Zhevnerov et al. (1983); 3, Sludsky (1956b).

Table 9.2. *Gazella subgutturosa.* Habitat (x indicates the habitat is used)

Habitat type	Season[a]	Locality[b]							
		Azerbaijan (1)[1]	Ustyurt (2)[2]	Near Aral sea (1)[3], Barsakelmes (2)[4], Betpaqdala (3)[4]			Karakum (1)[5], Badkhyz (2)[6], Tajikistan (3)[7]		
		1	2	1	2	3	1	2	3
Clay-gravel desert			x						
Clay desert with dwarf shrubs of *Artemisia* + *Anabasis* + *Salsola*	S		x		x	x		x	x
Clay desert with shrubs and sedges near rivers and tugai				x			x		x
Sand desert with *Haloxylon* and *Calligonum*	S		x	x	x		x		
Sand desert with *Haloxylon* and *Calligonum*	W		x		x		x		
Sand desert (hilly) with *Salsola* + *Artemisia* vegetation	W	x				x			
Fringes of tugai (*Populus diversifolia* tickets)	W			x					x
Savanna (*Pistacia*) with *Carex* and *Poa*	S							x	x
Grassland								x	

[a]W–Winter, S–Summer.
[b]The superscript numbers denote the following references: 1, Sokolov and Tembotov (1993); 2, Sludsky (1977); Zhevnerov et al. (1983); 3, Ishadov (1972); Rustamov et al. (1986); 4, Sludsky (1977); Zhevnerov et al. (1983); 5, Rustamov et al. (1986); 6, Heptner (1956); Gorelov (1972); Kuznetsov (1986); 7, Sokov (1993).

Table 9.3. *Gazella subgutturosa.* Life forms/plant parts eaten (x indicates plant part used)

Life forms	Plant parts eaten							
	Whole plant	Leaves	Stems/ twigs	Bark	Roots/ bulbs	Flowers	Fruits	Seeds
Shrubs		x	x			x	x	
Dwarf shrubs		x	x			x	x	
Herbs		x	x		x[a]	x	x	x
Grasses/sedges		x	x					
Horsetails						x		
Lichens	x							

[a]Rare.

Table 9.4. *Gazella subgutturosa.* Group size in populations (%)

Locality[a]	Season	Number of animals per group				
Group size		1	2–5	6–10	10–20	>20
Barsakelmes[1]	All year	22	63	13	2	
Badkhyz[2]	Autumn	37		23	23	17

[a]The superscript numbers denote the following references: 1, Zhevnerov et al. (1983); 2, Gorelov (1972).

Feeding Niche. In bottom, field and shrub layers.

9.5 Social Organization and Behavior

Small groups without personal connections, nomadic way of life, significant migrations, harems with territorial males-dominants during rutting season.

9.5.1 Social Life (Table 9.4)

Live in mixed groups in winter; in spring many males and females live as singles and in summer they live in small groups. Adult females are group leaders, in all seasons except during rut. They move 40–50 m ahead of the other animals in the group. All year-round gazelles make areas for defecation, which have an informative function, and during rut are used as a form of territorial marking (Gorelov 1972; Zhevnerov et al. 1983).

9.5.2 Home Range

In winter, gazelles roam in a range of a few tens of km, while in other seasons they are more mobile.

9.5.3 Mode of Movement

Excellent runner. Speed 55–65 km. Dig to find roots and to prepare defection areas; signal by jumping and stamping when disturbed (Mambetzhumaev 1970; Zhevnerov 1984; Sokov 1993).

9.5.4 Dispersal/Migration

During most of the year gazelles are nomadic; only in winter do groups live in a given area. In autumn (rut), males defend territories with markings along borders (Sludsky 1956b; Krivosheev et al. 1960; Mambetzhumaev 1970). In summer, the distribution pattern is determined by availability of water. Earlier migrations were about 450–700 km. Decreased population size has shortened the migrations, which are now usually less than 50–60 km (Zhevnerov et al. 1983; Blank 1990).

9.5.5 Diurnal Activity (Table 9.5)

In winter, polyphasic activity, the animals change between bouts of feeding and resting, during the day. Daily movements between pastures and watering places are about 6–7 km. In winter, gazelles do not visit watering places every day; in summer, they visit them several times a day (Zhevnerov et al. 1983; Blank 1990).

9.5.6 Shelters

Before gazelles lie down they dig out snow and sand. They often use the same places, which become deeper over time. In summer, they use the shade under shrubs. Each 40–50 min they change lying places to avoid the heat. In winter they lie at foothills, in shrubs, and in depressions. Digging hollows in sandy hills has also been observed during very cold

Table 9.5. *Gazella subgutturosa*. Diurnal activity

Locality	3–6	6–9	9–12	12–15	15–18	18–21	21–22.30
Barsakelmes[1]							
Summer							
Tadijkistan[2]							
Summer							
Eastern Kazakhstan[3]							
May							
July							

Shading indicates that animals are active.
[a]The superscript numbers denote the following references: 1, Zhevnerov et al. (1983); 2, Sokov (1993); 3, Blank (1990).

winters. Lying places are situated 5–20 m from each other (Gorelov 1972; Zhevnerov et al. 1983).

9.5.7 Escape Behavior

Very vigilant. Recognize danger at 2 km by sight, and from 300–400 m by sound. Can stay "fixed" for 5–8 min. Approach watering places very slowly, they can sometimes shy without obvious reason. Select open banks to drink water and often look around. Disturbed animals sneeze, strike the ground with a hoof, and make some "signal jumps". Females try to draw attention away from their newborns (Gorelov 1972; Zhevnerov et al. 1983).

9.6 Reproduction (Table 9.6)

There are two rutting seasons: spring and autumn, but only the latter season leads to mating (Tsaplyuk 1972). Before rut, males make "rutting latrines" by digging in the ground and defecating there for 2–3 days. Then harems are established, which include 1 male and 2–5 females. The dominant male keeps the harem together. There are also mixed groups, with one or more males, females and yearlings. There is no sexual activity in the mixed groups. Before calving females separate from the group, they prefer semi-closed habitats, such as shrubs and ravines, where there is protection from wind. Lactation continues for 3 months. A female produces about

Table 9.6. *Gazella subgutturosa.* Breeding parameters

Locality[a]	Age at sexual maturity[a] (months)		Age at participation in breeding (years)	Length of pregnancy (days)	Barrenness[a] (%)	Number of newborns/ female (From all females, %)		
	Male[b]	Female	Male			1	2	3
Ustyurt Plateau[1]	17–18[3]	7–8		165–180[c]		27	73	
Barsakelmes[2]				148–164		30	70	
Tajikistan[3]				148–164[2], 164–174[3]				
Betpaqdala[4]			3–4		9–11	61	39	0.4

[a]The superscript numbers denote the following references: 1, Zhevnerov (1984); 2, Pereladova and Pereladov (1986); 3, Sokov (1993); 4, Sludsky (1956b); Tsaplyuk (1972).
[b]Actually take part in breeding in 3–4 years (Sludsky 1977).
[c]In captivity 149–151 days (Zhevnerov et al. 1983).

400 g of milk, with 4% fat. Lambs can stand up after 17–25 min, suck after 45–60 min and follow the mother after 4–5 days. During the first 2 weeks, lambs hide, twins always lie 40–100 m apart. Mothers feed them three times in a 24-h period. After 5–10 days lambs eat grass and after 3–3.5 months they eat only vegetation and can live alone (Zhevnerov et al. 1983; Pereladova and Pereladov 1986).

9.6.1 Seasonal Breeding Events (Table 9.7)

Maximum Length of Life. Males live for 5–6 years, females live for 8–12 years. (Zhevnerov et al. 1983).

9.7 Mortality

Mass mortality was observed in severe winters with snow depth ≥50 cm and strong, cold winds. Mortality can reach 80% or more. Mortality of males is significantly higher than for females. On Barsakelmes Island, adult male mortality was 37–58%, female mortality was 9–17%, lamb male mortality was 22–33%, lamb female mortality was 11–25% (Zhevnerov et al. 1983). During droughts, mortality of the young can be connected with consuming too much salt water, as they are not experienced enough to recognize salt concentrations by taste (Gorelov 1972). In Kazakhstan, mortality of calves in May was 11%, 17% in June, 33% in July and 39% in September (Sludsky 1956b).

Table 9.7. *Gazella subgutturosa.* Seasonal breeding events

Locality[a]	Rutting season		Calving season		
	Preparation to rut	Basic rut	First calves	Main calving	Last calves
Ustyurt Plateau[1]		1–15 November		April–first part of May	
Barsakelmes[2]	September–October	Early November–mid December	2–5 May		4–5 July
Badkhyz[3]	5 October	November		End of April–early May	
Tajikistan[4]	End of October	End of November–December	End of April	5–15 May	
Betpaqdala[5]		End of October–early December		10–20 May	
Eastern Kazakhstan[6]	Mid October	Second part of November	Mid April	15–25 May	Mid June

[a]The superscript numbers denote the following references: 1, Kostin (1955); 2, Zhevnerov (1984); 3, Gorelov (1972); 4, Sokov (1993); 5, Afanasev and Sludsky (1947); 6, Blank (1990).

Table 9.8. *Gazella subgutturosa.* Gender and age composition of populations (%)

Locality[a]	Males	Females	Lambs
Barsakelmes[1]	13 (5–22)	40 (31–69)	46 (38–58)
Betpaqdala[2]	5–21	79–95	

[a]The superscript numbers denote the following references: 1, Zhevnerov (1984); 2, Afanasev and Sludsky (1947); Zhevnerov et al. (1983).

Table 9.9. *Gazella subgutturosa.* Group composition early and late in the rutting season in the Barsakelmes in 1969–1974, percent of total. (Zhevnerov et al. 1983)

November				December			
Only females	Only males	Mixed groups	Harems	Only females	Only males	Mixed groups	Harems
15	40	20	25	9	31	9	51

Table 9.10. *Gazella subgutturosa.* Population density and size in different localities

Locality[a]	Year	Population density (animals/km$^{2,b)}$)	Population size (×1,000)
Azerbaijan[1]	1940		5.5–6.0
	1960		0.13
	1970		1.5
	1975		2.7
	1983		5
	1986		5
Ustyurt Plateau, the Kazakhstan part[2]	1940		100
Kazakhstan[2]	1940		200
	1960	0.01–0.8	9.3
	1983		12.5
	1986	0.3–2	12.5
	1989	0.1–2.3	30
Ustyurt Plateau, southern	1954		2.7
(Turkmenian) part[3]	1983–1984	0.1	0.25
Sarykamysh basin shores[4]	1985	0.3–0.7	
Ustyurt Plateau, the Karakalpakiya part	1956–1959	(0.6–0.7)	4.5
(24,000 km^2)[5]	1963–1965	0.3–0.4	2.5
	1970	0.2–0.3	1.5
	1981–1984	0.04–0.5	0.4–0.5
Barsakelmes[6] (Data	1929		9[c]
presented in singles)	1936		40
	1940		120
	1947		900
	1949		60[d]
	1970		180
	1976		330
Uzbekistan[7]	1983–1984	0.4–0.5	1.6–2.0
Uzbekistan[7]	1986		5
Kyzylkum Desert	1954–1959	3–4	12–14
Turkmenistan[8]	1910		300[2]
	1960		15[3]
	1981		10[3]
	1983		6
	1986		8
Badkhyz Nature Reserve[9]	1970		1.5
	1978	50–60	6
	1985		2.1–2.3
	1998		1.7
Tajikistan[10]			0.2
Kyrgyzstan[10]			0.1
Former USSR	1986		ca. 30
territory, total[10]	1991		ca. 60

[a]The superscript numbers denote the following references: 1, Vereshchagin (1947); Shchadilov (1986); 2, Sludsky (1977); Zhevnerov et al. (1983); Blank (1990); 3,8, Rustamov et al. (1986); 4, Gorbunov and Zarkhidze (1986); Chernov (1986); 5,7, Sabilaev and Koptsev (1986); 6, Vasenko (1950); Zhevnerov (1984); 9, Kuznetsov (1986); Gorelov and Ishanov (1999); 10, Flint and Prisyazhnyuk (1986); Flint et al. (1991).
[b]Authors estimated the density by dividing the numbers of observed gazelles by the length of car or plane routes in km.
[c]Seven females and 2 males were introduced on the island.
[d]In the severe winter of 1948/1949, most of the animals died.

9.7.1
Predators

Wolf, fox, caracal, dog, jackal, sand cat, golden eagle, raven, crow and gulls. Ca 39% of hyena (*Hyaena hyaena*) excrements contain gazelle remnants, 10% of wolf excrements and 3–8% of fox excrements also contain gazelle remnants (Zhevnerov et al. 1983; Lukarevsky 1995).

9.7.2
Diseases, Helminths and Ectoparasites (Appendix 2, Table 6)

9.8
Population Structure and Densities (Tables 9.8 and 9.9)

9.8.1
Population Density and Size (Table 9.10)

In the Badkhyz area in 1887, hunters counted about 1,000 gazelles in 1 h; in 1924 5,000 gazelles visited the watering place on the Egrichek river daily; in 1958, herds of 100–200 gazelles were observed (Gorelov 1978).

9.9
Relations to Man

A protected species. As a result of heavy poaching, protection in the nature reserves is very important (Bannikov 1979). There are attempts to breed gazelles in a large enclosure in Uzbekistan (Flint and Prisyazhnyuk 1986).

10 Saiga Antelope - *Saiga tatarica*

Saiga tatarica Linnaeus, 1766.
Artiodactyla, Bovidae, Antilopinae

10.1
Status and Distribution (Fig. 10.1)

10.1.1
Taxonomic Remarks

One subspecies occurs: *S. t. tatarica* Linnaeus, 1766 (Bannikov 1989).

10.1.2
Status

Lower risk - conservation-dependent (LR/cd) (Hilton-Taylor 2000). CITES II. The Kalmykia population has strongly declined (Lushchekina and Struchkov 2001; Milner-Gulland et al. 2001), other populations are numerous; hunted. Protected by hunting regulations (Zhirnov et al. 1998b).

10.1.3
Distribution (Fig. 10.2)

In Kalmykia, irrigation forced the steppe saiga population to semi-desert and desert zones. When sheep husbandry collapsed there, due to overgrazing, new pastures for saiga appeared. These areas, "Chernye Zemli", are now the main habitat for the Kalmykia saiga population, and cover 18,000–28,000 km^2, compared with 60,000–70,000 km^2 in the previous saiga range in Kalmykia (Bliznyuk 1977, 1980; Leshchenko 1997; Zhirnov et al. 1998a).

Distribution Area. Ca. 2,560,000 km^2.

Occurrence in Ecoregions. Temperate Desert (342, 343).

Fig. 10.1. Saiga

10.2 Measurements (Table 10.1 and Figs. 10.3, 10.4)

10.3 Habitat

Plains, sometimes undulated, with or without snow. Grass-wormwood and grass semi shrubs (*Kochia - Salsola*), steppes and semi-deserts.

Terrain Form. Plains, in extremely dry years saiga visit mountains (Bannikov et al. 1961; Baskin 1976; Fadeev and Sludsky 1983).

Altitude. 0–200 m a.s.l. (Kalmykia, Volga - Ural); 0–600 m a.s.l. (Ustyurt Plateau, Betpaqdala Desert - Turgay Steppe) (Bannikov et al. 1961; Baskin 1976; Fadeev and Sludsky 1983).

Snow. Snow >20 cm limiting (Rashek 1965). Weight pressure of hoof is 480 g/cm^2 (Sludsky 1963).

Temperature. Winter, with temperatures below −30 °C, causes high mortality.

Water. In Kazakhstan, saiga have a great need for water when the ground is not covered by snow, or when they feed on dry plants; under such conditions saiga visit fresh water daily; crossing long distances to reach areas with water (Fadeev and Sludsky 1983). In Kalmykia, watering is not necessary in favorable years with frequent rains and moist vegetation. Drinking is needed when the water concentration in their plant food becomes <60% (Zhirnov 1982).

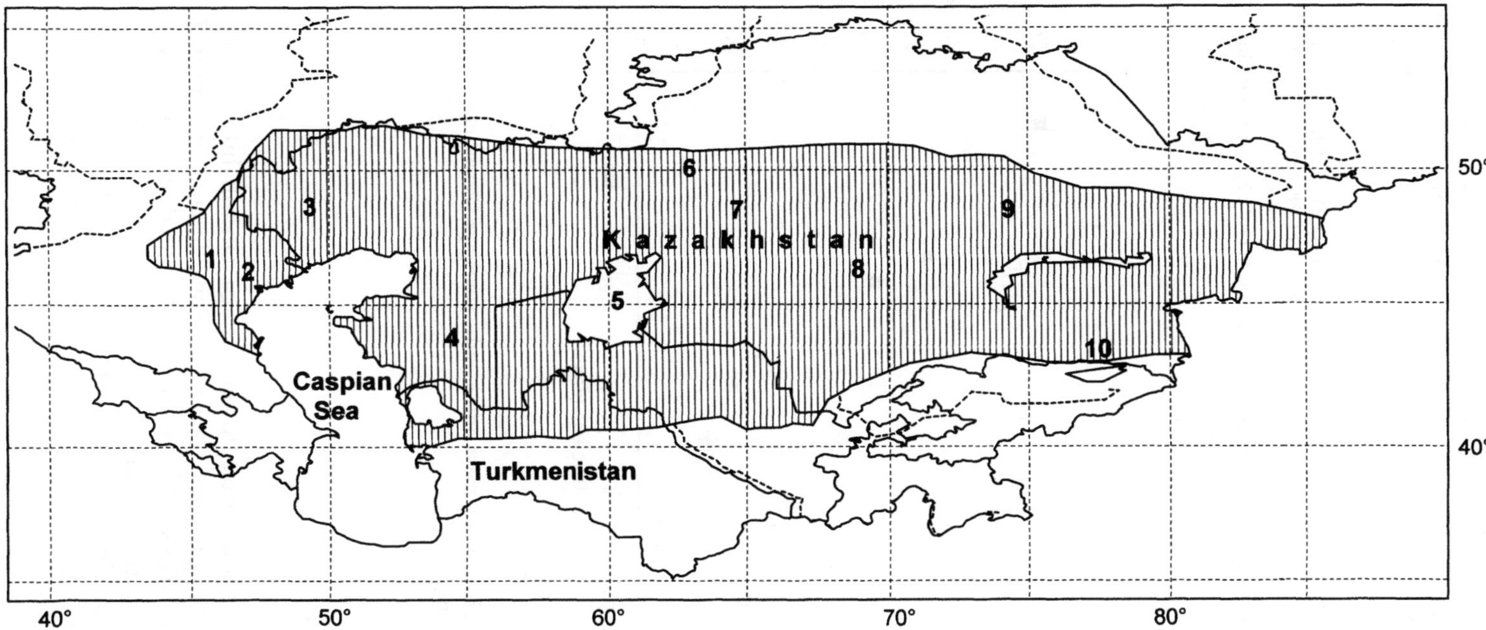

Fig. 10.2. *Saiga tatarica.* Distribution map (Zhirnov et al. 1998a). *1* Kalmykia; *2* Chernye Zemli; *3* Volga – Ural area; *4* Ustyurt; *5* Aral Sea, Barsakelmes Island; *6* Turgay Lowland; *7* Ulutau Mountains; *8* Betpaqdala semi-desert; *9* Kazakh Upland; *10* Zailiyskoe Alatau Range

Table 10.1. *Saiga tatarica.* Measurements

Locality[a]	Longitude (°E)	Latitude (°N)	Gender	n	Body mass (kg)	Head and body length (cm)	Tail length (cm)
Kalmykia[1]	45–47	45–47	Male	96	25.8–43.5		
			Female	75	15.2–28.2		
Betpaqdala[2]	67–73	45–47	Male	50	44.0 (37–49)	174.0 (152–190)	11.7 (9–14)
			Female	50	31.0 (22–37)	149 (138–160)	12.0 (9–15)

[a]The superscript numbers denote the following references: 1, Leshchenko (1997); 2, Fadeev and Sludsky (1982).

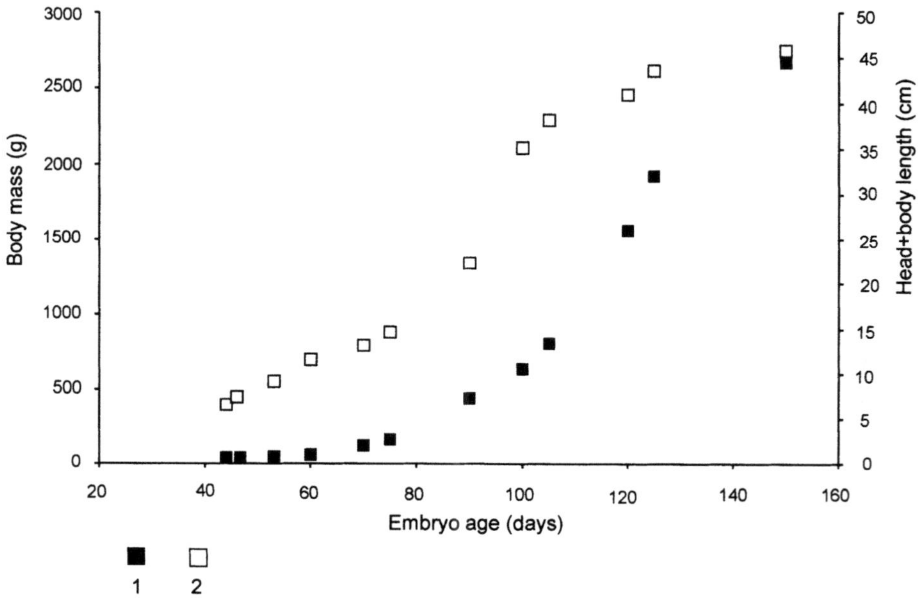

Fig. 10.3. *Saiga tatarica.* Embryo body mass and length versus age (Petrishchev 1998). *1* Body mass; *2* head + body length

10.3.1 Biotope (Table 10.2)

Plain dry steppes and semi-deserts. However, saiga can inhabit undulated plains, sandy hills; lambing occurs even in the low Ulutau Mountains; in extreme situations saiga also search for food in the subalpine meadows of Trans-Ili Alatau Range and Kazakh Upland. During 1975–1995, saiga in Kalmykia were forced out into semi-deserts, with sandy hills (Kapitonov 1980; Fadeev and Sludsky 1983; Bekenov 1988; Zhirnov et al. 1998a).

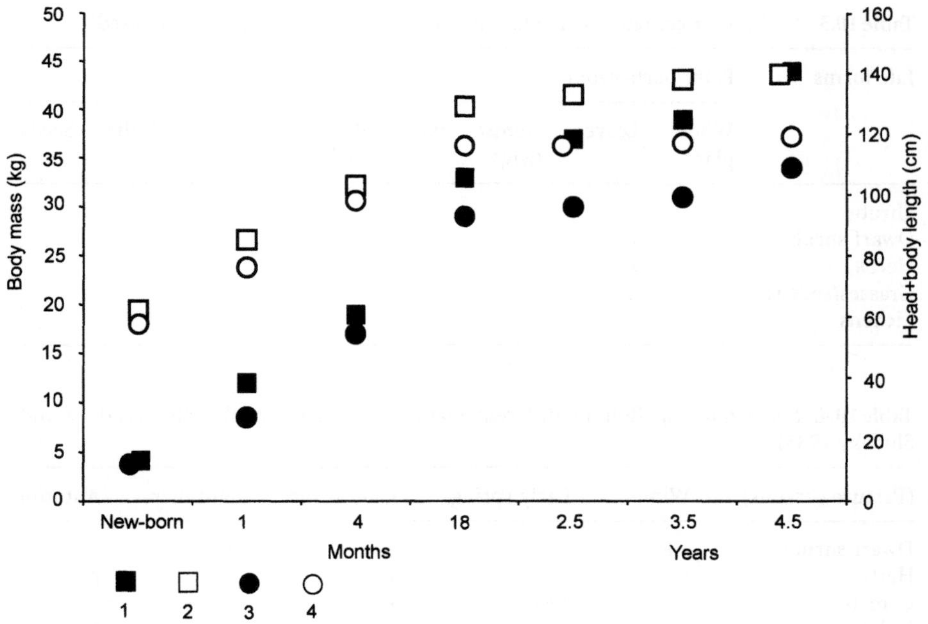

Fig. 10.4. *Saiga tatarica.* Body mass versus age of the Kazakhstan saiga (Fadeev and Sludsky 1983). *1* Male mass, *2* male head + body length; *3* female mass; *4* female head + body length

Table 10.2. *Saiga tatarica* Habitat[a]

(Ecoregion)[b] (Locality)	342[1] Kalmykia	343[2] Volga-Ural	343[2] Ustyurt Plateau	342, 343[2] Betpaqdala – Turgaj
Steppe	+++	+++		+++
Semi-desert	++	+	+++	+++
Desert				
Clay			+	+
Sand	+			r

[a]Symbols indicate percent of habitat use was >30% (+++); 20–30% (++): 10–20% (+); <10% (r); x indicates habitat use noted, no numerical data reported.
[b]The superscript numbers denote the following references: 1, Zhirnov et al. (1998a); 2, Fadeev and Sludsky (1983).

Table 10.3. *Saiga tatarica.* Life forms/plant parts eaten (x indicates plant part used)

Life forms	Plant parts eaten							
	Whole plant	Leaves	Stems/ twigs	Bark	Roots/ bulbs	Flowers	Fruits	Seeds
Shrubs		x						
Dwarf shrubs		x	x					
Herbs		x	x					
Grasses/sedges		x	x					
Lichens	x							

Table 10.4. *Saiga tatarica.* Role of different plant forms in seasonal diets.[a] (Fadeev and Sludsky 1983)

(Pasture seasons)	Winter	Early spring	Late spring	Summer	Autumn
Dwarf shrubs	+++	+	++	+++	+++
Herbs	r		+++	+	r
Cereals	+	+++	++	++	+
Lichens	+				r

[a]Symbols used indicate food type consumed: heavily and is one of main foods (+++); often (++); occasionally (+); rarely (r).

10.4 Diet and Feeding Habits

10.4.1 Diet

Main foods are shrubs, lichens, herbs and grasses (Lebedeva 1959, 1960; Bliznyuk 1980). For a list of plant genera eaten see Appendix 1, Table 11. Mineral licks are not used.

10.4.2 Life Forms/Plant Parts Eaten (Table 10.3)

10.4.3 Seasonal Aspects of Diet (Table 10.4)

Feeding Niche. In field and shrub layers.

Consumption of Food/Day (Table 10.5). In captivity, consume 1.2–1.5 kg of dry matter in the summer, 0.7 kg in the winter (Abaturov et al. 1982; Kholodova 1989).

Table 10.5. *Saiga tatarica.* Consumption of food by adult animals/day

Locality[a]	Season	Adult food consumption/day				Food consumption – all ages	
		Male (kg fresh weight)	Female	Both sexes		As proportion of total biomass[3] (%)	As proportion of food plant[4] (%)
				(kg fresh weight)	(kg dry weight)		
Kalmykia[1]	Summer			1.5–4.0	0.5–2.0	5 (3–10)	10 (5–20)
Kazakhstan[2]	Summer	1.5–6.9	3.1–5.8			1–3	
	Winter	2.0–5.7	1.5–4.5				

[a]The superscript numbers denote the following references: 1, Kholodova (1989); 2, Fadeev and Sludsky (1982); 3, Abaturov (1984); 4, Abaturov et al. (1982).

Table 10.6. *Saiga tatarica.* Herd sizes in different seasons (percent of observed groups)

Locality[a]	No. of animals per group														
	1–50					51–500				>501					
	December		May	July	October	December		May	July	October	December		May	July	October
	Early	Late				Early	Late				Early	Late			
Kalmykia[1]	61	19	22	49	12	23	56	68	42	53	16	25	10	9	35
Kazakhstan[2]	81	28	14	41	15	15	60	27	53	57	4	12	59	6	28

[a]The superscript numbers denote the following references: 1, Zhirnov et al. (1998c); 2, Fadeev and Sludsky (1983).

10.5 Social Organization and Behavior

10.5.1 Social Life (Table 10.6)

Live in herds. The herd leader is usually an adult female or an adult male (in the rutting period and during the spring migration, when males and females move separately). There is free exchange of individuals between herds (Baskin 1974). During rut, males establish harems (1 male and about 30 females in Kazakhstan (Fadeev and Sludsky 1983), 10–12 females on Barsakelmes Island (Vasenko 1950), 10–15 females in the Volga - Ural between-river-area (Rakov 1956), and 2–6 females in Kalmykia (Bannikov et al. 1961). Males force the females of the harems to stay within their territories.

10.5.2 Home Range

Nomadic way of life. Males defend "a herd distance" around their harems; in late summer and autumn during rut they establish plots of territories (Baskin 1976).

10.5.3 Mode of Movement

Ambling is the normal pace. Adults run 60–65 km/h, lambs (8–15 days old) run at 40–50 km/h. Dig snow. Good swimmers (Fadeev and Sludsky 1983).

10.5.4 Dispersal/Migration

There are permanent summer, lambing and winter pasture grounds (in the scale of hundreds of kilometers), as well as migratory routes for each geographical population. Saiga move between pasture grounds depending on snow depth (20 cm limiting), drought, and rains, which stimulate growth of fresh vegetation. Lambing grounds are usually situated near water and in places with sufficient food. Plain valleys with sparse wormwood-feather grass vegetation, slightly undulating sands, *Salsola* shrubs, clay fields with small ponds are usual biotopes where lambing takes place. In the 1960s–1970s, the lambing grounds in Kazakhstan consisted of 1,544 (110–9,000) km^2. 15,000–200,000 females gathered there. Animal density during lambing reaches 600/km^2, 160/km^2 on average. Density on summer pastures is lower, 15–250 animals/km^2 in Kazakhstan, 50–130 animals/km^2 in Kalmykia, 11–20 animals/km^2 in Ustyurt (Fadeev and Sludsky 1982;

Table 10.7. *Saiga tatarica.* Diurnal activity[a]

Locality[b]	Time of day (h)							
	0–3	3–6	6–9	9–12	12–15	15–18	18–21	21–24
Kalmykia[1]								
Spring	+	+++	+++	++	+	++	+++	+++
Kazakhstan[2]								
Spring			x	x	x	x		
Summer		x	x			x	x	
Autumn			x	x	x	x	x	

[a]Symbols indicate % of animals active was >30% (+++); 20–30% (++): 10–20% (+); x indicates active animals observed, no numerical data reported.
[b]The superscript numbers denote the following references: 1, Zhirnov (1982); 2, Fadeev and Sludsky (1983).

Fadeev and Ivanov 1988; Bekenov and Grachev 1998). The main direction of spring and autumn migrations is north to south, and are driven by climate and vegetation cycles on different latitudes. Extent of migrations in Kalmykia is 300–800 km, Volga - Ural 250–300 km, Ustyurt 300–400 km, Betpaqdala - Turgay 500–600 (maximum 1,200 km). Speed of migration is mostly 10–15 km/day, maximum 40–70 km/day. Increased migration speeds can occur due to harsh cooling, snowfall and the drying of watering places (Fadeev and Sludsky 1983). In spring, males and females migrate in different herds and take different routes; males migrate about one month earlier than females. During migrations there is an exchange of individuals between herds from different parts of the range, but natural connections between populations (Kalmykia, Volga–Ural, Ustyurt and Betpaqdala–Turgay) is not interrupted (Fadeev 1980b; Bekenov and Grachev 1998; Zhirnov et al. 1998a).

10.5.5 Diurnal Activity (Table 10.7)

10.5.6 Shelters

Saiga rake the snow before the lying down. When hot, animals paw the ground before lying down; animals change lying places from time to time.

10.5.7 Escape Behavior

In Kazakhstan, saiga detected men at a distance of 390 ± 58 m, the first flight occurred at 276 ± 26 (SE)m, the first stop at 325 ± 28 (SE)m (Baskin 1976).

Table 10.8. *Saiga tatarica.* Breeding parameters

Locality[a]	Age at sexual maturity (months)		Participation in breeding (years)		Pregnancy (days)
	Male	Female	Male	Female	
Kalmykia[1]	18–30	6–7	1.5	0.5	140–145
Betpaqdala[2]	18–19	6–7	1.5	0.5	138 (131–141)[b]

[a]The superscript numbers denote the following references: 1, Bannikov et al. (1961); Zhirnov (1982); 2, Tsaplyuk (1962, 1966); Rashek (1963, 1974).
[b]In captivity.

10.6 Reproduction (Tables 10.8 and 10.9)

Males at rut develop several special skin and hair features, and scent glands produce abundant secretions, with a very strong smell. At the beginning of rut males combat fiercely while still in large herds. The stronger males force away weaker males and establish harems with 2–15 (maximum 30) females. Males control territories 15–40 m around the females, and these territories move with the roaming herds (Dal et al. 1958; Bannikov et al. 1961; Rashek 1963; Baskin 1976). Fecundity is low during years with dry summers, and winters with deep snow.

Females give milk (at 05:00–06:00 h in the morning, 12:00–14:00 h in the afternoon, and 19:00–21:00 h in the evening) only to their own calves, even at high densities of calves at calving grounds, searching them out by their footprints. Feeding takes only 20–30 s (sucking is very intensive), then the female leaves the calves and they spend their time hidden. Without the mother, the calves can lie in the same place for 2 days. A calf sucks 40–50 ml, while 2–3 days old; 350–400 ml, when 7 days old, and 0.9–1.0 ml, when 30 days-old (Adolf 1957; Bannikov et al. 1961). Milk contains about 6% fat, 6% protein and 3% lactose (Demeuov 1971). Calves suck for 2–3 months. Grass is eaten for the first time at an age of 3–4 days, and after 15 days, grass makes up 95% of their diet (Zhirnov 1982).

10.6.1 Seasonal Breeding Events (Table 10.10)

The time of migration changes from year to year, depending on the time of the snow melting, and development of vegetation. The very short periods of rutting and lambing are probably adaptations to avoid predation. There are two periods of sexual activity for males (in June and December), but only the second rut leads to breeding (Tsaplyuk 1962).

Table 10.9. *Saiga tatarica.* Fecundity of females of different age (%)

Locality, years[a]	Barren females (%)		Pregnant females with 1–3 embryos (%)						Embryos/female	
(Age, years)	<1	>1	<1			>1			<1	>1
(No. of embryos)	0	0	1	2	3	1	2	3		
Kalmykia, 1958–1959[1]	?	?	95	5	0	26	74	0		
Volga-Ural, 1986–1994[2]	29 (0–36)	7 (3–11)	61 (36–75)	9 (4–25)	1 (0–4)	15 (4–19)	76 (70–89)	2 (0–4)	0.8 (0.7–1.2)	1.7 (1.6–1.9)
Ustyurt Plateau, 1986–1994[2]	31 (10–47)	2 (0–4)	51 (29–84)	18 (0–62)	0	26 (19–38)	72 (58–81)	0	0.9 (0.5–1.5)	1.7 (1.5–1.8)
Betpaqdala, 1964–1978[2]	35 (1–40)	12 (2–32)	88 (64–100)	7 (0–12)	0	19 (6–36)	71 (57–92)	2 (0–3)	1.0 (0.7–1.2)	1.8 (1.6–1.9)
Betpaqdala, 1986–1994[2]	20 (0–35)	4 (0–10)	63 (19–100)	17 (0–59)	0	21 (14–39)	75 (54–85)	0	1.0 (0.6–1.2)	1.7 (1.5–1.8)

[a]The superscript numbers denote the following references: 1, Bliznyuk (1980); Zhirnov et al. (1998c). 2, Fadeev and Sludsky (1982).

Table 10.10. *Saiga tatarica.* Seasonal breeding events

Locality[a]	January	February		March		April		May		June		July	August		September		October		November		December	
Kalmykia[1]		M	M	M	M	M	N	N	N				M	M	M	M	M	M			R	R
Volga-Ural[2]								N	N												R	
Ustyurt Plateau[2]					M	M	M	MN	M													
Betpaqdala[2]			M	M	M	M	M	M	MN	MR[b]	MR[b]		M	M	M	M	M	M	M	M		R

M, Migration. R, rut; N, newborns.
[a]The superscript numbers denote the following references: 1, Bannikov et al. (1961); 2, Tsaplyuk (1966); Fadeev and Sludsky (1983); Rashek (1974).
[b]Sexual activity of males without actual mating.

Maximum Length of Life. 7.5 years. However, about 94% of the males die before 2.5 years of age, and most males participate in breeding only once.

10.7 Mortality

Deep snow, and drought cause high mortality (Dal et al. 1956). In Kazakhstan, the bad winters for saigas were in 1826/1827, 1848/1849, 1851/1852, 1856/1857, 1879/1880, 1888/1889, 1891/1892, 1917/1918, 1927/1928, 1944/1945, 1945/1946, 1949/1950, 1950/1951, 1971/1972, 1975/1976, 1976/1977, 1983/1984, 1984/1985, 1987/1988, and 1993/1994 (Sludsky 1963; Zhirnov et al. 1998a). In 1971/1972, in Kazakhstan, deep snow (20–35, up to 50 cm) resulted in the death of 400,000 saiga (Fadeev 1975b). Most males are exhausted after the rut and die during winter, only 3–5% survive (Bannikov et al. 1961; Sludsky 1963; Rashek 1978). Other causes of mortality are embryo abortions after winters with deep snow (up to 20–30% of females abort), overcooling of newborns after rains and frosts, drowning when crossing water, and being killed by cars. Rate of calf mortality in Kazakhstan and Kalmykia, during their first year, is 50–60% (Bannikov et al. 1961; Fadeev and Sludsky 1983).

10.7.1 Predators

Wolf, dog, and raptors (Sludsky 1962). A wolf generally kills 2–4 saiga per month. Calving of saiga takes place where wolves have their cubs and saiga lambs are frequently killed. The wolf families protect the territory of the lambing grounds against other wolves (Filimonov 1979). Dogs kill up to 50% of the lambs in Kalmykia (Bibikov and Zhirnov 1975). On lambing grounds, eagles (*Aquila nipalensis* and *A. heliaca*) gather to eat carrion. Golden eagles can attack lambs.

10.7.2 Diseases, Helminths and Ectoparasites (Appendix 2, Table 6)

Pallasiomyia antilopum (skin gadfly) was important during the 19th century, but after strong decline of the saiga population it became extinct (Grunin 1962).

Foot-and-mouth disease is severe (up to 76% of calves and many adults die). Brucellosis was found in 1.2% of the studied animals. Pastrilles killed 70,000 animals in 1981 within 4 days. Necrobiosis and toxoplasmosis have been recorded (Rementsova 1962; Galuso et al. 1963; Petrov et al. 1979; Berkinbaev 1992). Because of the migratory way of life, saiga transmit diseases and helminths were obtained from domestic animals over large areas (Fadeev and Sludsky 1983).

Table 10.11. *Saiga tatarica.* Gender and age composition of populations (%)

Locality[a]	Year	Male : female ratio		Population structure (%)[b]			
		Embryos	Newborns	Adults (≥1 year)		Calves	
				Males	Females	Males	Females
Kalmykia[1]	1978–1990	(0.73–1.07) : 1	(0.77–1.04) : 1	20 (18–21)	42 (33–55)	15 (7–28)	23 (17–28)
Kazakhstan[2]	1965–1974	0.9 : 1	(1.0–1.12) : 1	20	39	41	
Kazakhstan[2]	1988–1994						
Volga-Ural		(0.75–1.18) : 1	(0.81–0.97) : 1	13 (8–17)	56 (51–62)	31 (21–37)	
Ustyurt Plateau		(0.89–1.22) : 1	(1.09–1.15) : 1	11 (2–18)	59 (51–78)	30 (18–40)	
Betpaqdala[2]		(0.67–1.2) : 1	(1.05–1.14) : 1	7 (5–12)	58 (48–66)	35 (23–47)	

[a]The superscript numbers denote the following references: 1, Bannikov et al. (1961); Prisyazhnyuk and Pronyaev (1981); Bukreeva (2002); 2, Fadeev and Sludsky (1982, 1983).
[b]Results of catching by net corrals.

Table 10.12. *Saiga tatarica.* Population size in different localities (Bekenov et al. 1990; Grachev and Bekenov 1993; Bekenov and Grachev 1998; Zhirnov et al. 1998b; Grachev et al. 1999; Sidorov 2000; Bukreeva 2002)

Locality	Years	Population size (×1,000)
Kalmykia, sections to NW Caspian Sea	1950	100
	1955	270
	1960	290
	1965	120
	1970	190
	1972	213
	1974	425
	1975	500
	1980	380
	1981	430
	1985	222
	1990	160
	1994	142
	1997	270
	1998	180
	1999	55
	2000	25
	2001	18
Volga-Ural	1970	50
	1974	70
	1980	120
	1981	160
	1985	50
	1990	138
	1993	250
	1998	104
Ustyurt Plateau	1965	64
	1970	56
	1974	70
	1980	170
	1981	190
	1985	190
	1990	202
	1993	216
	1998	246
Betpaqdala	1955	663
	1993	Strong decline
	1994	282
	1998	120
Turgay	1960	690
	1965	412
	1970	995
	1972	850
	1974	956
	1980	400
	1981	470
	1985	400
	1990	361
	1993	510

Table 10.13. *Saiga tatarica.* Harvest. (Fadeev and Sludsky 1982; Bekenov et al. 1992; Zhirnov et al. 1998b; Grachev et al. 1999; A. Lushekina, pers. comm.)

Locality	Years	No. of animals (×1,000)		Harvest during periods (tons)	
		Total annually	Hunted during periods	Meat (×1,000)	Horns
Kalmykia	1951–1955	180–270	103	1.4	
	1956–1960	290–400	612	8.5	
	1961–1965	120–510	347	4.8	
	1966–1970	190–215	52	0.7	
	1971–1975	150–500	195.7	2.7	
	1976–1980	380–650	353.1	9.1	
	1981–1985	222–430	253	3.5	
	1986–1990	160–200	26.3	0.36	
	1991–1995	142–168			
Kazakhstan	1960	600	138	2.5	
	1965	480	72	1.3	5.4
	1971	1100	194	3.7	9.1
	1976	750	321	5.5	15.0
	1981	820	193.1	3.0	35.0
	1986		87		
	1991	825	112		
	1993–1997	470–810	13–62[a]		
	2000	25			

[a]Annually.

10.8
Population Structure and Densities (Table 10.11)

A decline in male numbers has been observed since 1989, because of selective hunting for horns. In the Kalmykia population, 7% of the males were ≥2.5 years, while 23% of the females were ≥2.5 years old (Zhirnov et al. 1998c).

10.8.1
Population Density and Size (Table 10.12)

Since the 1960s, two full cycles of fluctuations in saiga numbers, in Kalmykia, have been observed (Sidorov and Bukreeva 1999).

10.8.2
Impact on Vegetation

By the 1970s, there were many complaints that the saiga damage crops. However, this damage is very minor and occurs only in the dry years, when saiga migrate northwards and come into agricultural areas.

10.9 Relations to Man (Table 10.13)

In Kalmykia, irrigation canals have destroyed the migration routes. Since 1975 saiga have been forced to use the same area "Chernye Zemli" all year-round and now have a semi-settled way of life (Zhirnov et al. 1998b). Meat and horns (used in Chinese medicine) are important products (Table 10.13). During 1840–1861, annually 7,000–345,000 horns were exported from Kazakhstan (Nebolsin 1855). A couple of horns weigh 240 g (190–345 g) (Fadeev and Sludsky 1983). Saiga meat contains proteins (20–22%), and energy (246 kcal/100 g) (Zhitenko 1974).

11 European Bison Wisent – *Bison bonasus*

Bison bonasus Linnaeus, 1758.
Artiodactyla, Bovidae, Bovinae

11.1 Status and Distribution (Fig. 11.1)

11.1.1 Taxonomic Remarks

Two subspecies occur: *B. b. bonasus* Linnaeus, 1758 (European part of the former USSR); *B. b. caucasicus* Turkin and Satunin, 1904 (NW Caucasus) are found (Heptner 1989).

11.1.2 Status

Endangered (EN – A2ce, C2a) (Hilton-Taylor 2000). CITES I. In Russian Federation endangered (Rozhnov 2001). Pure *B. b. bonasus* only exists in the Belovezhskaya Forest Nature Reserve, the Tsumanskoe hunting husbandry (Volyn, Ukraine), the Prioksko-Terrasny Nature Reserve and the Cherga Farm in Altai. Hybrids of the cross between *B. b. bonasus* and *B. b. caucasicus* exist in most areas. Hybrids of European and American bison exist in the Kavkazsky Nature Reserve and in its neighboring area (with the share of pure European bison blood being about 60/64) (Kalugin 1968). Among free-living European bison there are now 760 *B. b. bonasus* and 1,100 hybrids of *B. b. bonasus* – *B. b. caucasicus* (Kudryavtsev and Belousova 1999).

11.1.3 Distribution (Fig. 11.2)

Distribution Area. Ca. 948,000 km^2 (estimated historically) (Heptner 1989).

Occurrence in Ecoregions. Warm Continental (212), Prairie (252), Prairie Regime Mountains (M252), Temperate Steppe (332).

Fig. 11.1. European bison

11.2
Measurements

Body mass of males is about 500 kg, maximum 1,000 kg; body length, maximum 300 cm; tail length 80 cm (Heptner 1989). Newborns: male 28 kg (16–35 kg), female 24 kg (15–33 kg) (Krasiñski et al. 1999).

11.3
Habitat

In winter bison are found in forests with clearings, forest edges, flood plains; in summer they are found on fringes of meadows and sub-alpine grassland (Baskin 1979; Krasiñska et al. 1987; Nemtsev and Kiseleva 1987).

Terrain Form. Plains and flood plains in Belovezhskaya Forest, and foothills, valleys, slopes, and high plains in mountains of Caucasus.

Altitude. In Caucasus, 400–600 m a.s.l. (winter), 600–1,300 m a.s.l. (summer).

Snow. Normally found in areas with snow depth <50 cm. When feeding, bison can remove snow with their head when the snow is 30–40 cm deep.

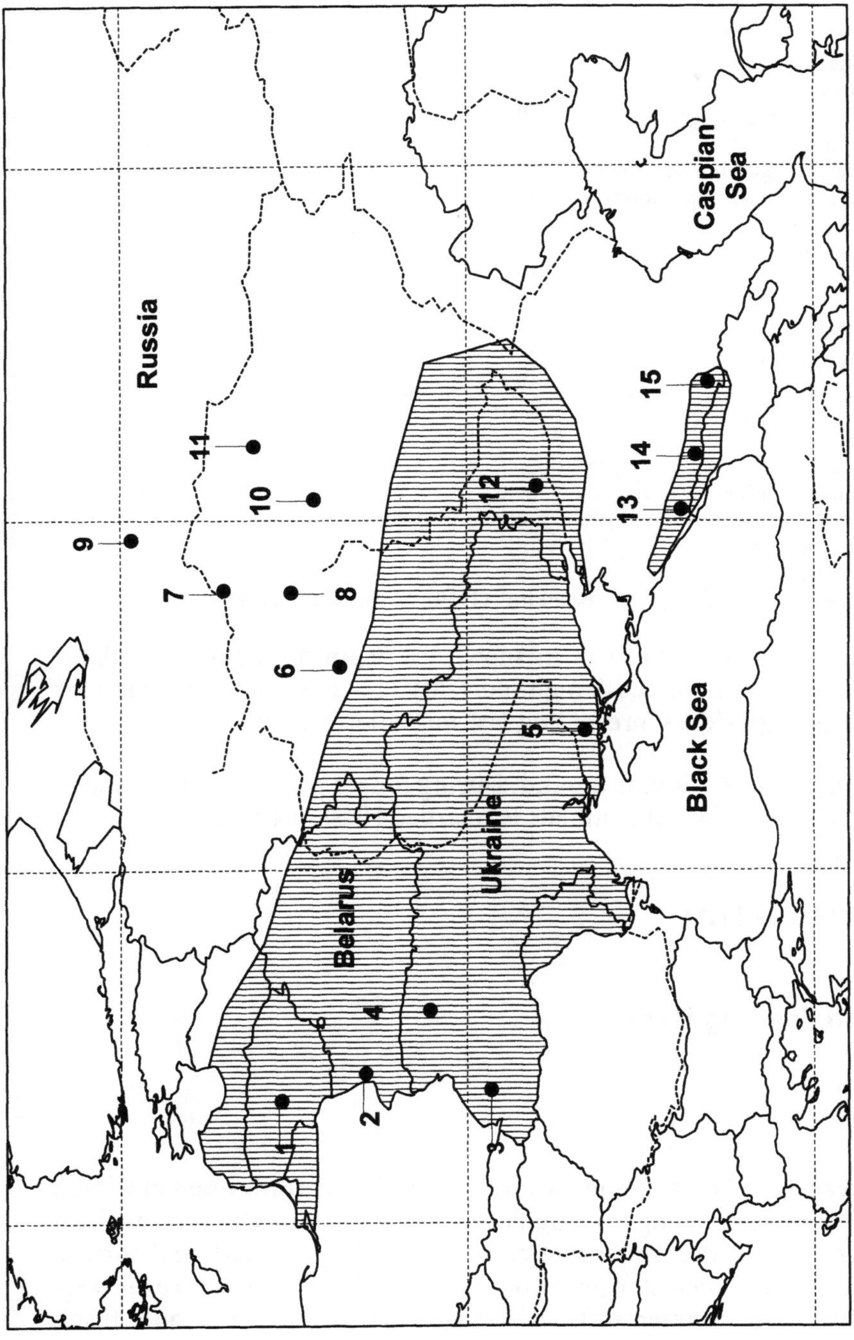

Fig. 11.2. *Bison bonasus.* Distribution map (historical range) (Heptner 1989). *1* Lithuania; *2* Belovezhskaya Forest Nature Reserve; *3* Carpathian Mountains; *4* Tsumanskoe Game Husbandry; *5* Askania-Nova Nature Reserve; *6* Oryol oblast; *7* Sknjatino Game Husbandry; *8* Prioksko-Terrasny Nature Reserve; *9* Vologda oblast; *10* Oksky Nature Reserve; *11* Vladimir oblast; *12* Rostov oblast'; *13* Kavkazsky Nature Reserve; *14* Teberdinsky Nature Reserve; *15* N Ossetia

Table 11.1. *Bison bonasus.* Habitat (x indicates use of habitat observed)

	Season[a]	Belovezhskaya Forest[b]	Caucasus[c]
(Ecoregions)		212	M252
Mature forest			
Coniferous with good grass cover	W		x
Coniferous - broadleaf	W		x
Broadleaf with good undergrowth	W		x
Deciduous forest near upper forest line	S		x
Fresh water shore			
Meadow	S	x	
Forest	S	x	
Flood plain	S	x	
Bog, boggy forest	S	x	
Clearings, edges	W	x	
	S	x	
Grassland			
Meadows	S	x	
Grassland subalpine	S		x

[a]W, winter; S, summer.
[b]Korochkina (1973).
[c]Vinogradov (1870); Filatov (1912); Bashkirov (1939); Nasimovich (1939); Naniev (1956).

When disturbed, they can run in snow, at a depth of about 1 m (Filatov 1910). Strong snowfalls cause high mortality (12–23%) in the populations enclosed by high ridges, preventing fast migration (Nemtsev 1985).

Water. Drink water all year around. During longer periods without water, the development of young animals is retarded (Baskin 1979).

11.3.1 Biotope (Table 11.1)

11.4 Diet and Feeding Habits

11.4.1 Diet

In Belovezhskaya Forest, bison have been forced to use only grass in winter, because of overbrowsing by ungulates (Wroblewski 1927). Normally, a significant part of the winter diet consists of twigs, tree bark and shrubs, as well as plants which stay green over winter (*Rubus casesius, Festuca* sp. in Caucasus). In summer, soft herbs and grass are the main food. Bison use salt licks. For list of plant genera eaten see Appendix 1, Table 12 (Vinogradov 1870; Nasimovich 1939; Gębczynska et al. 1991).

Table 11.2. *Bison bonasus*. Plant types/plant parts eaten.[a] (Kalugin 1968; Gębczynska et al. 1991)

Plant types	Plant parts eaten							
	Whole plant	Leaves	Stems/ twigs	Bark	Roots/ bulbs	Flowers	Fruits	Seeds
Trees		++	+++	+++			++	++
Shrubs		+++	+++				++	x
Dwarf shrubs		+++	+++					
Herbs		+++	+++			x		
Grasses/sedges		+++	+++					
Lichens	+							
Fungi	r							

[a]Symbols indicate percent of plant use was >30% (+++); 20–30% (++): 10–20% (+); <10% (r); x indicates plant use noted, no numerical data reported.

11.4.2
Life Forms/Plant Parts Eaten (Table 11.2)

Consumption of Food/Day. In captivity, consumption of food/day (fresh weight): herbs 36 kg (30–45 kg), aspen twigs 0.9–2.4 kg, willow twigs 1.1–2.2 kg, elm 1–3.2 kg, *Euonymus* 1.4–2.7 kg (Kalugin 1968). Under normal conditions in C Caucasus, in early winter, bison eat ca. 22 kg as fresh weight (ca. 12 kg as dry weight) per day; in late winter, they eat ca. 13 kg as fresh weight (7.4 kg in dry weight) per day (Kazmin 1992). In captivity, in winter, young, 1.5–3 year old animals consume 5–8 kg each (dry weight), 4–8 year old animals consume 10–15 kg each (dry weight). In summer, the young animals consume 20–35 kg (fresh weight), adults consume 40–50 kg (Kholodova 1989; Kholodova and Belousova 1989). In C Caucasus, the bison population consumes ca 4% of meadow biomass. However, when the animal density is 20/km^2 the consumption amounts to 13–30% (Kazmin 1992; Kazmin et al. 1992).

Feeding niche. In field and shrub layers.

11.5
Social Organization and Behavior

11.5.1
Social Life (Table 11.3)

Mixed groups (females, young animals and males), male groups and solitary bulls (in Belovezhskaya Forest solitary bulls account for 60% of males) are the basic units of the population. From July, mixed herds are formed,

Table 11.3. *Bison bonasus.* Seasonal group size and composition changes in Belovezhskaya Forest. (Bunevich and Kochko 1988; Krasińska and Krasiński 1995; Krasiński et al. 1999)

Months	No. of animals per group			
		Mixed groups	Male groups	Female groups
April–July		18–20	3–4	3–4
August–September	Open area	23 (2–140)	2 (1–3), max. 9	3–4
	Forest	11 (1–74)		

when males join female groups. The size and structure of mixed groups are changeable (Table 11.3; Krasiński et al. 1994, 1999). Group size: 15–18 animals (Belovezhskaya Forest, in the past), 4–8 animals, maximum 60 (aboriginal Caucasian population) (Usov 1865; Kartsov 1903; Filatov 1912). Animals with highest status are those with the greatest weight (Deryagina 1972). Females (usually with a calf) become group leaders (Filatov 1912; Krasiński et al. 1999).

11.5.2 Home Range

Settled way of life. In the Kavkazsky Nature Reserve four local subpopulations are known, with little exchange of individuals (Nemtsev 1998). Bison heavily concentrate their activities to a minor part of their home range. In the Belovezhskaya Forest, mixed groups demonstrate fidelity to some areas (Krasińska et al. 1987). During winter, the size of home ranges of radio-collared bulls was 10.7 (0.8–33.2) km^2, cows 7.9 (0.7–34.3) km^2. In the snow-free season, bulls had home ranges of 69.5 (29–151.6) km^2, cows 68.8 (45.4–97.5) km^2. Younger bulls had significantly smaller home ranges than older bulls (Krasiński et al. 1999, 2000). There is overlapping of the home ranges, both sexes do not defend their home ranges. In the home range there are clear, visible trails (especially to salt licks and watering places), stalls (shelters from blood-sucking insects), scratching places (trees, stamps, hammocks), and places for rolling. Fidelity of animals to the home range impedes settling in new places during reintroduction (Baskin 1979).

11.5.3 Mode of Movement

Step and diagonal gallop are usual (Gambaryan 1974). Incapable of digging in snow, although scratching movements of legs were observed in excited bulls.

11.5.4
Dispersal/Migration

Seasonal migrations of aboriginal Caucasian bison were mostly correlated to snow conditions, and reached 10–30 km (Filatov 1912). Modern Caucasian bison make vertical migrations in autumn and spring within 400–3,000 m a.s.l. (Kalugin 1968). Sometimes single animals (more often young males) leave the population range and move 70–180 km (Korochkina 1973; Baskin 1979).

11.5.5
Shelters

Stalls in thickets of undergrowth of spruce and fir to escape blood-sucking insects.

11.5.6
Escape Behavior

Bison have excellent smell, good hearing and poor vision. It is impossible to approach a bison group down wind (flight distance is about 65–100 m), but up wind hunters can approach to within 10–15 m in open places. Disturbed animals flee down the slope. Frequently disturbed groups abandon their home range and move to an area situated some tens of km away (Dinnik 1910; Filatov 1910; Baskin 1979). Defend their calves from predators (brown bear), and from humans. Against wolves bison protect calves by arranging themselves in a circle with their heads pointing outwards (Usov 1865).

11.6
Reproduction (Table 11.4)

Males become mature sexually at 3 years of age. However, they do not participate in reproduction until the age of 6–7 for behavioral reasons. Bulls mostly participate in breeding between the ages of 6–12 years. Females usually become mature in the third year of life, giving birth to the first calf in the fourth year, but frequently in the fifth or sixth year. The gestation period of a cow lasts for 264 (254–277) days. After calving, estrus occurs after 40–80 days (minimum 22), and again after 20–30 days. Generally, female bison breed every second year, up to 20 years, sometimes to 25 years of age. Age of maturity and intervals between births are affected by habitat conditions and being kept in captivity (Severtsov 1940; Pererva and Kiseleva 1991; Krasiñska and Krasiñski 1995; Krasiñski et al. 1999). After the release of bison in the Kavkazsky Nature Reserve, females gave birth at 5–6 years of age and thereafter every 2–3 years (Nemtsev 1985). Fertil-

Table 11.4. *Bison bonasus*. Breeding parameters

	Male participation in breeding (years)		Age of calving (years)		Pregnancy (days)	Intervals between birth (years)
	Maturing	Mating	First	Last		
Kavkazsky, aboriginal population[1]		6–?		20		2–3
Belovezhskaya Forest, population in early of 20 century[2]	3	6–?	5			2
Kavkazsky, modern population[3]	1.3–3	5–6	4–6		261–283	2–3
Belovezhskaya Forest, with supplemental food[4]	2–3	5–10	3–4	13–14	257–272	1.3
Prioksko-Terrasny, in captivity[5]	2–3	5–11	3–4	19	262–267	1

[a]The superscript numbers denote the following references: 1, Filatov (1912); 2, Severtsov (1940); 3, Kalugin (1968); 4, Korochkina (1971); 5 Zablotsky (1957).

ity of females in the Belovezhskaya Forest during 1971–1996 was 39% (21–55), barrenness 59% (45–65); calves/adult female 0.41% (0.35–0.55); gender ratio of newborn, (males : females) 1.01 : 1(0.6 : 1–1.7 : 1); newborns in the population 15% (6–21) (Bunevich and Kochko 1988; Bunevich 1999).

Bison usually have one calf, two calves are rare. During rut, males reunite with female groups. Hierarchy is strong. Dominant bulls continuously keep other bulls away from females in estrus, and try to control their movements. Two weeks before calving, females pasture separately. For the first few days after calving the mother stays with the newborn and joins the herd when the calf can follow. The calf can stand 1 h after birth, and after 1.5 h it can follow the mother. Milk is high in fat content (8–9%), but low in quantity. Lactation is for 8–10 months, sometimes up to 2 years. Usually, the female chase away the older calf at calving. Calves start to feed on grass when they are 19–20 days old, and tree leaves at 2 months (Zablotsky 1957).

Table 11.5. *Bison bonasus.* Seasonal breeding events

Locality[a]	Rut Basic rut	Calving		
		First calves	Basic calving	Last calves
Belovezhskaya Forest[b,1]	August–September		April–May	
Belovezhskaya Forest[c,2]	All year excluding May	1 April	April–August	December
Caucasus[3]	Last August–Early September		5 May–15 June	

[a]The superscript numbers denote the following references: 1 Kartsov (1903); Wroblewski (1927); 2 Korochkina (1971); Bunevich and Kochko (1988); 3 Vinogradov (1870); Filatov (1910); Dinnik (1910).
[b]In the last of 19th century.
[c]Present.

Table 11.6. *Bison bonasus.* Gender and age composition of populations (%)

Locality[a]	Males		Females	Yearlings	Calves
	Adult	2–4 years			
Belovezhskaya Forest[1]	26	12	34	28	
Kavkazsky[2]		31		48	12 9

[a]The superscript numbers denote the following references: 1, Wroblewski (1927); 2, Nemtsev (1985).

11.6.1 Seasonal Breeding Events (Table 11.5)

Maximum Length of Life. 25 years.

11.7 Mortality

Causes of mortality in the Belovezhskaya Forest Nature Reserve: old age 49% (in the 19th century), 6–13% (modern period); diseases (modern period), including diseases of the genital organs of male bison (43%) and blindness (9%); predators 11% (in the 19th century); poachers 20% (in the 19th century); killed by other bison 9–16% (modern period); drowning 3% (modern period) (Kartsov 1903; Wroblewski 1927; Krasiñski et al. 1994, 1999). Annual mortality of the wild Belovezhskaya Forest population in

Table 11.7. *Bison bonasus.* Wild animals in the Russian Federation in 1996 (Pererva 1997)

Locality	Genetic origin	Number
NW Caucasus	Lowland-Caucasian	570[a]
N Ossetia	Lowland-Caucasian	135
Teberdinsky Nature Reserve	Lowland-Caucasian	43
Tver oblast	Lowland-Caucasian	31
Vladimir oblast	Lowland-Caucasian	13
Vologda oblast	Lowland-Caucasian	8
Rostov oblast	Lowland-Caucasian	9
Oryol oblast	Lowland-Caucasian	9
Russian Federation (totally, including farms)[b]	Lowland-Caucasian and pure Lowland	343
Kavkazsky Nature Reserve	Hybrids: European × American bison	280
Hunting Reserves around the Kavkazsky Nature Reserve	Hybrids: European × American bison	180
Nalchik game farm	Hybrids: European × American bison	70

[a]In 1998.
[b]In 1996.

Table 11.8. *Bison bonasus.* Population density and size in Lithuania, Ukraine and some nature reserves of Belarus and Russian Federation

Locality[a]	Years	Population density (animals/10 km^2)	Population size
Lithuania[1]	2001		42
Ukraine (total)[2]	1965		23
	1970		75
	1975		102
	1980		202
	1985		388
	1990		582
	1995		656
	1999		489
Carpathian[2]	1995		260
	1998		225
Tsumanskoe Game husbandry[2]	1995		170
Belovezhskaya Forest[3]			
Polish and Belarus parts together	1857		1,898
	1862		500
	1880		550
	1890		410
	1902	5.4	665
	1914	5.3	727
	1918		152
	1919		4
	1941		21
	1945		5

Table 11.8. *Continued*

Locality[a]	Years	Population density (animals/10 km^2)	Population size
Belarus part[4]	1960		39
	1970	4.4	63
	1975	6.4	87
	1980	10.4	143
	1985	13.5	206
	1993		308
	1998		238
Kavkazsky Nature Reserve[b,5]	1870		2,000
	1880s		1,000
	1882		500
	1890s		440
	1914		500–600
	1921		10–15
	1941		5
	1960		136
	1965		449
	1970		450
	1975		430
	1980		760
	1992		400
	1996		280
	1999		250
Tseysky Nature Reserve (N Ossetia)[6]	1990	14	200–220
	1998	3	50–60

[a]The superscript numbers denote the following references: 1, Baleišis et al. (2002); 2, Demchuk et al. (1997); Perzanowski and Kozak (1999); I. Kozak (pers. comm.); 3, Kartsov (1903); Wroblewski (1927); Pucek 1991; 4, Bunevich and Kochko (1988); Krasińska et al. (2000); 5, Dinnik (1910); Pfizenmayer (1929) cit. Nasimovich (1989); Bashkirov (1939); Kalugin (1968); Nasimovich (1989); Pucek 1991; A. Kudaktin (pers. comm.); 6, Veinberg and Komarov (1999).
[b]Since 1941 – hybrids.

19th century was 6–8% (Wroblewski 1927), today it is 3.5% on average (Krasiński et al. 1999) Foot-and-mouth disease in 1890, 1907 and 1919 killed the Caucasian bison population (Bashkirov 1939). In the Kavkazsky Nature Reserve, 25% of the animals died in the 1967/1968 winter due to the very deep snow. In cold springs, with a heavy snowfall, up to 31% of newborns can die (Nemtsev 1985, 1998).

11.7.1 Diseases Helminths and Ectoparasites (Appendix 2, Table 7)

11.8 Population Structure and Densities (Table 11.6)

Male: female ratio in the population of free-ranging bison in the Kavkazsky Nature Reserve in 1978 was found to be 0.96:1.

11.8.1 Population Density and Size (Tables 11.7 and 11.8)

11.9 Relations to Man

Protected species, restored from a few animals (Kulagin 1919; Zablotsky 1949, 1957; Nasimovich 1989). There are breeding centers in the Prioksko-Terrasny Nature Reserve, the Oksky Nature Reserve, the Askaniya-Nova Nature Reserve, and the Belovezhskaya Forest Nature Reserve. Efforts to enlarge its present range by introducing animals into new areas, such as Lithuania, Vologda, oblast, Vladimir oblast, Tver oblast, Bryansk oblast, Oryol oblast and Tajikistan are made (Pererva 1997; Gusarov 1999; Sipko et al. 1999).

12 West Caucasian Tur – *Capra caucasica*

Capra caucasica Güldenstaedt and Pallas, 1783. Artiodactyla, Bovidae, Caprinae

12.1 Status and Distribution (Fig. 12.1)

12.1.1 Taxonomic Remarks

No subspecies are recognized (Heptner 1989).

12.1.2 Status

Endangered (EN – A1d+2cde) (Hilton-Taylor 2000). Hunting by licenses (Sokolov and Tembotov 1993).

12.1.3 Distribution (Fig. 12.2)

Distribution Area. Ca. 17,000 km^2.

Occurrence in Ecoregions. Prairie Regime Mountains (M252).

12.2 Measurements (Table 12.1)

12.3 Habitat

Stenotopic. Use cliff-like areas with grasses in coniferous, subalpine and alpine mountain zones.

Terrain Form (Table 12.2). High mountain plains, slopes and tops of mountains (Kotov 1968; Tarasov 1975).

Fig. 12.1. Caucasian goat

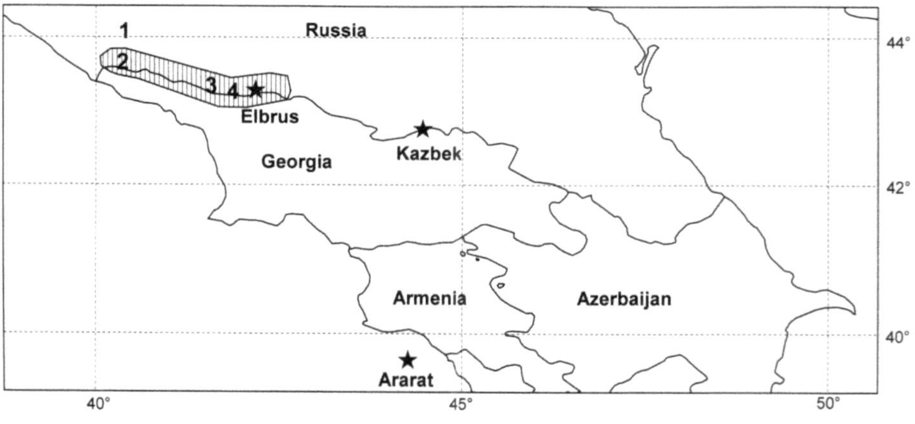

Fig. 12.2. *Capra caucasica.* Distribution map (Sokolov and Tembotov 1993). *1* Krasnodar kray, Adygey; *2* Kavkazsky Nature Reserve; *3* Teberdinsky Nature Reserve; *4* Karachay-Cherkessia

Altitude. (Table 12.3).

Temperature. Inhabit areas with a temperature range from −19 to +26 °C (Kotov 1968).

Snow. Prefer southern slopes with snow <30 cm (Kotov 1968).

Table 12.1. *Capra caucasica.* Measurements (Kotov 1968)

Locality	Longitude (°E)	Latitude (°N)	Gender	*n*	Body mass (kg)	*n*	Head and body length (cm)	Tail length (cm)
Kavkazsky Nature Reserve	40	44	Male	15	142.8 (123.5–155)	23	178.7 (159.0–196.0)	14.3 (12.0–17.0)
			Female	9	65.5 (58.0–70.7)	47	150.0 (136.0–164.0)	12.1 (8.0–14.0)

Table 12.2. *Capra caucasica.* Use of slopes facing in different directions in the Kavkazsky Nature Reserve (%). (Kotov 1968)

Direction of slope	Winter	Spring	Summer	Autumn
N	1	9	15	16
NE	4	0	4	4
NW	1	0	4	5
S	22	25	26	36
SE	58	38	11	1
SW	0	5	10	3
W	4	1	13	7
E	10	22	17	28

Table 12.3. *Capra caucasica.* Distribution of goats (%) in different altitudes of the Kavkazsky Nature Reserve. (Nasimovich 1949a; Kotov 1968)

Altitude (m a.s.l.)	Vegetation zone	Winter	Spring	Summer	Autumn
0–1,000	Forest	0	0	0	0
1,000–1,700	Forest	Singles[a]	Singles	Singles	Singles
1,700–2,000	Forest	Singles	Singles	0	Singles
2,001–2,300	Subalpine	Singles	4	1–2	1–8
2,301–3,000	Alpine	0	96	96–98	92–99
3,000–3,300	Glaciers	0	Singles	0.4–3	Singles

[a]Singles – very small number of animals.

12.3.1 Biotope

In the Kavkazsky Nature Reserve, during winter, some of the goats use alpine and subalpine grassland, others live in coniferous and deciduous forests near the forest line. In spring, both subpopulations concentrate in subalpine grassland near the forest line. In summer, goats use subalpine

Table 12.4. *Capra caucasica.* Life forms/plant parts eaten (x indicates plant part used. Nasimovich 1949a; Kotov 1968)

Life forms	Plant parts eaten							
	Whole plant	Leaves	Stems/ twigs	Bark	Roots/ bulbs	Flowers	Fruits	Seeds
Trees			x	x				
Shrubs		x	x					
Herbs		x	x			x		
Grasses/sedges		x	x			x		
Mosses	x							
Lichens	x							
Fungi	x							

and alpine grassland and alpine tundra. In early autumn, once again, they prefer grassland near the forest line, where there is fresh vegetation (Nasimovich 1938; Kotov 1968).

12.4 Diet and Feeding Habits

12.4.1 Diet

Herbs, twigs of shrubs and trees, and grass. Require salt and very actively use salt licks, salt water, and find artificial salt grounds very quickly (Nasimovich 1938). For a list of plant genera eaten see Appendix 1, Table 13.

12.4.2 Life Forms/Plant Parts Eaten (Table 12.4)

Feeding niche. In field and shrub layers.

12.5 Social Organization and Behavior

12.5.1 Social Life (Table 12.5)

Individuals aggregate in feeding places and follow the leader when disturbed. Composition of groups is often variable, i.e. the same animals live together only for a few days (Baskin 1976). Smaller groups are often more constant (Nasimovich 1949a; Kotov 1968). The size of groups increases with

Table 12.5. *Capra caucasica.* Seasonal changes in group size. (Kotov 1968)

Seasons	Average number of animals per group		
(Years)	1947	1957	1963
Population size	3,637	5,603	11,864
Winter	12	11	12
Spring	14	15	17
Summer	18	17	20
Autumn	16	15	17
Average	16	17	19

the population size; the maximum group size found was close to 300 animals (Table 12.5).

12.5.2 Home Range

Distinct ranges have not been demonstrated. Goats frequently use trails through cliffs, which are 4–5 km in length, the same passes through ridges; cavities and roof-like rocks are used as shelter, and patches for wintering. Patches where goats spend the winter are 0.08–0.55 km^2, in summer they are 0.02–1.9 km^2 (Kotov 1968).

12.5.3 Mode of Movement

Step and run in an undisturbed state, and heavy gallop when disturbed. Speed of running along flat surfaces is low; a dog or wolf can easily catch up with goats. Excellent jumpers on cliffs and through firm snow. A thick skin pad on hooves and the sharp edges of hooves give a better grip. Can slide along snow covered slopes. In narrow gorges, they leap from one wall to another (Kotov 1968).

12.5.4 Dispersal/Migration

There are no migrations, but there are displacements up and down mountains, depending on snow cover. Some groups can move from one mountain range to another situated 10–12 km away (Nasimovich 1949a).

12.5.5 Diurnal Activity (Tables 12.6 and 12.7)

Table 12.6. *Capra caucasica.* Feeding activity in the Kavkazsky Nature Reserve. Hours 21:00–03:00 not covered by observations (percent of all animals observed at a given time). (Kotov 1968)

Time (h)	3–5	5–7	7–9	9–11	11–13	13–15	15–17	17–19	19–21
Winter	0	93	91	97	92	82	77	100	0
Summer	92	88	79	52	47	39	70	83	84

Table 12.7. *Capra caucasica.* Summer daily activity in the Kavkazsky Nature Reserve (percent of all observed animals)

Activity	Time of day (h)[a]						
	5–11[2]	8–10[1]	10–14[1]	14–18[1]	11–17[2]	18–20[1]	17–20[2]
Feeding	41	65	7	40	40	72	78
In salt licks	28	4	2	19	9	25	1
Walking	3	16	18	16	8	0	18
Resting	10	15	73	25	17	3	0
Rest or walk on snow	18	0	0	0	26	0	3

[a]The superscript numbers denote the following references: 1, Nasimovich (1949a); 2, Kuznetsov (1969).

12.5.6 Shelters

Use cavities in rocks and roof-like cliffs to avoid bad weather. Use the shade of mountains and trees, they also use snowfields, and move to glacier lakes to avoid blood-sucking insects (Nasimovich 1949a; Kotov 1968).

12.5.7 Escape Behavior

Frequency of looking around for danger is low, but in herds some animals keep a look out. They have excellent vision and detect danger from a great distance. Very little attention is paid to bears (minimum distance 25 m), since they indicate a small predation risk. Goats are aggressive to chamois and deer. They are also unconcerned by avalanches and falling stones. In the reserves, hikers can approach goats to within some tens of meters. Aircraft evokes panic, and the goats run away (Kotov 1968; Baskin 1976).

12.6 Reproduction

Breeding parameters: age of maturity of males 2.5 years, females 1.5–2.5 years; males participate in breeding at 6 years of age, females at 3 years of age; length of pregnancy is 150–160 days (Inyakova 1957b; Kotov 1968).

12.6.1 Seasonal Breeding Events

Rutting season is in the second half of November–December; lambing is from May to the first half of June.

Table 12.8. *Capra caucasica.* Gender and age composition of populations in 1941–1966 (percent of all observed animals). (Ekvtimishvili 1952; Kotov 1968; Tarasov 1973)

Locality	Gender ratio Males : females	Males	Females	Males + females		Unknown gender and age
		Older than 2 years of age		Yearlings	Kids	
Kavkazsky	1.2 : 1	35	29	8	15	13
Teberdinsky	1.3 : 1				15	

Table 12.9. *Capra caucasica.* Fluctuations of population density (in typical habitats) and size. (Kotov 1968; Tarasov 1973; Bobyr 1999; Trepet 1997)

Locality	Year	Population density (animals/10 km^2)	Population size (×1,000)
Kavkazsky	1947		3.6
	1952		5.5
	1957		5.6
	1963		11.9
	1967	51–414	
	1972		15
	1982		10
	1992		5.2
	1999		3.2
Teberdinsky	1952		2.3
	1957		3.9
	1963		2.7
	1967		3.6
	1972	50–60	2.7
	1981	40–58	1.8–2.5
	1999	30	1.3

12.7 Mortality

In the Kavkazsky Nature Reserve, annual mortality is about 10% (Kotov 1968). Mortality of kids during their first month is ca. 50% (Kotov 1968). Snow avalanches result in the highest mortality and kill about 300 animals each winter in the Kavkazsky Nature Reserve. Many animals are injured when they fall from rocks (Nasimovich 1938; Naniev 1956; Kotov 1968; Tarasov 1975).

12.7.1 Predators

In the Kavkazsky Nature Reserve goat remnants were found in 14–43% of wolf scats in winter, 12–45% in spring, 27–34% in summer, 18–30% in autumn (Kudaktin 1986).

Wolf is the main predator; 12–36% of scats (depending on altitude) contain goat hairs (Teplov 1938; Kudaktin 1998), but wolves are rare in high mountains. In the past, the leopard was an important predator (Dinnik 1914). 22–60% of brown bear scats and 2–17% of lynx scats contain goat remnants (Kudaktin 1998).

12.7.2 Diseases, Helminths and Ectoparasites (Appendix 2, Table 8)

12.8 Population Structure and Densities (Table 12.8)

Ca. 14–15% of the population in August were kids, 1937–1977 (Kotov 1968).

12.8.1 Population Density and Size (Table 12.9)

12.9 Relations to Man

Almost all individuals live within nature reserves, outside of which, they are killed for horns and meat.

13 East Caucasian Tur – *Capra cylindricornis*

Capra cylindricornis Blyth, 1841.
Artiodactyla, Bovidae, Caprinae

13.1 Status and Distribution (Fig. 13.1)

13.1.1 Taxonomic Remarks

No subspecies are recognized (Heptner 1989).

13.1.2 Status

Vulnerable (VU – A1d+2de, C1) (Hilton-Taylor 2000). Hunting with licenses (Sokolov and Tembotov 1993).

13.1.3 Distribution (Fig. 13.2)

Distribution Area. Ca 27,000 km^2.

Occurrence in Ecoregions. Prairie Regime Mountains (M252).

13.2 Measurements

Male body mass increases with age: 4–6 years ca. 71 kg, 6–11 years ca. 100 kg, 12–16 years ca. 140 kg. Female body mass is ca. 56 kg (Vereshchagin 1938). In Dagestan, the head and body length of males is 190 cm, in females it is 138 cm; in N Ossetia, males are 178 cm, females are 146 cm (Abdurakhmanov 1973; Kuliev 1981).

Fig. 13.1. Dagestan goat

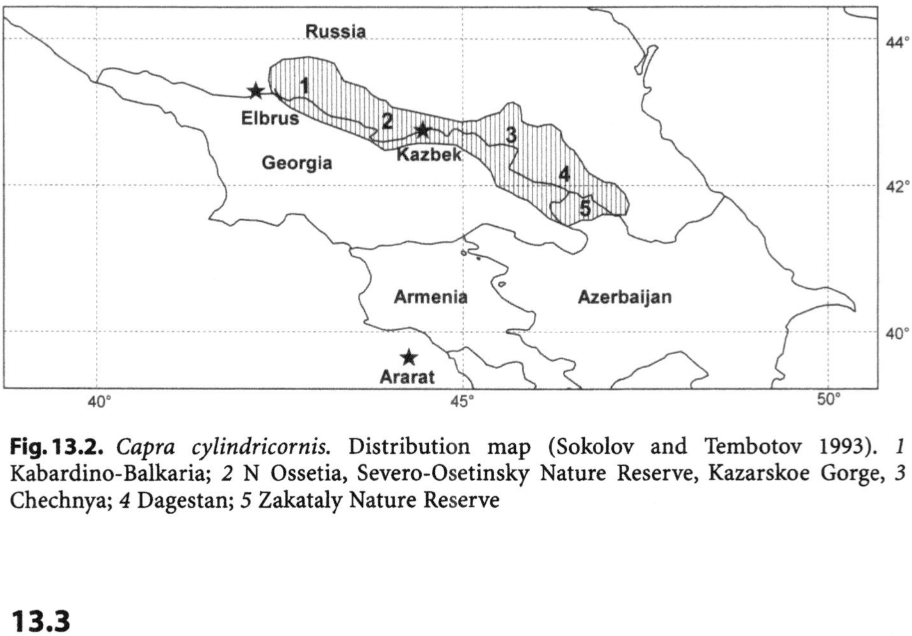

Fig. 13.2. *Capra cylindricornis.* Distribution map (Sokolov and Tembotov 1993). *1* Kabardino-Balkaria; *2* N Ossetia, Severo-Osetinsky Nature Reserve, Kazarskoe Gorge, *3* Chechnya; *4* Dagestan; *5* Zakataly Nature Reserve

13.3 Habitat

Rocky slopes of mountains and cliffs along river valleys. In winter, patches with a high plant standing crop (more than 2,000 kg/ha) are important because the deep snow covers the slopes. The optimal slope is about 36° for goats of any age and sex. However, in summer, females with kids suffer badly from predation and use habitats with more cliffs, but with less food. Females predominate at medium altitudes, where the composition of cliffs and pasture plots are more favorable. During summer, males suffer less from predation and need more food. They use larger food patches, situated

Table 13.1. *Capra cylindricornis.* Use of slopes (%) facing in different directions. (Magomedov and Akhmedov 1994)

Slope direction	Summer		Winter		
	Females	Males	Females	Males	Mixed groups
SE	8	6	26	58	55
S	9	28	36	21	19
SW	8	6	33	17	16
W	13	3	2	0	10
NW	29	21	3	4	0
N	21	19	0	0	0
NE	7	10	0	0	0
E	5	7	0	0	0

Table 13.2. *Capra cylindricornis.* Distribution of animals at different altitudes (%). (Magomedov and Akhmedov 1999)

Altitude (m a.s.l.)	Summer (May–September)		Winter (October–April)	
	Females	Males	Females	Males
<2,500	0.9	0.1	10.7	3.3
2,500–2,750	15.7	3.2	48.0	49.4
2,750–3,000	52.2	19.9	38.5	33.8
3,000–3,250	24.2	23.9	2.8	13.5
3,250–3,500	6.5	46.7	0	0
<3,500	0.5	6.2	0	0

on more gentle slopes in the subalpine zone near glaciers (Akhmedov and Magomedov 1995; Magomedov and Akhmedov 1999). Veinberg (1984) stresses that this species is more closely associated with forested cliffs, in the upper part of mountains, than other goat species.

Terrain Form (Table 13.1). Valleys, slopes and tops (in Dagestan) of mountains (Vereshchagin 1938; Veinberg 1984; Magomedov and Akhmedov 1994; Magomedov et al. 2001).

Altitude. (Table 13.2) 1,300–4,000 m a.s.l. (N Ossetia); 1,500–5,000 m a.s.l. (Dagestan) (2,600–3,200 m a.s.l. are preferred); 1,300–1,500 m a.s.l. (Azerbaijan) (Veinberg 1984; Magomedov and Akhmedov 1994; Akhmedov 1995).

Table 13.3. *Capra cylindricornis.* Habitat[a]

Habitat type	Location[b]	
	N Ossetia[1]	Dagestan[2]
Mature forest[c]		
Coniferous (pine) with grass cover	x	
Broadleaf (curved forest near forest line, with grass and shrub cover)	x	
Broadleaf		+
Shrubland	x	
Grassland		
Valley side		++
Mountain slope	x	+++
Alpine	x	+
Tundra (alpine)	x	r

[a]Symbols indicate percent of habitat use was >30% (+++); 20–30% (++): 10–20% (+); <10% (r); x indicates habitat use noted, no numerical data reported.
[b]The superscript numbers denote the following references: 1, Veinberg (1984); 2, Magomedov and Akhmedov (1994).
[c]Use preferably during wintertime.

Snow. Snow depth of >50 cm is limiting. However, goats are capable of crossing snowfields about 1 m deep (Veinberg 1984). Pressure of hooves, on average, is 600 g/cm^2 (Nasimovich 1955).

Water. Regularly go down to water courses to drink.

13.3.1 Biotope (Table 13.3)

In three main areas, inhabited by goats of the Severo-Ossetinsky Nature Reserve (Ossetia), the animals prefer to use unforested zones, but in the Kasarskoe gorge (N Ossetia) 70–100% of goats live in the forest zone during spring. Snow conditions heavily affect the distribution of goats. During summer, the majority of females with newborns, and animals <3 years of age continue to live in the forest zone (Veinberg 1984).

13.4 Diet and Feeding Habits

13.4.1 Diet

Mostly cereals and sedges, also herbs in summer, tree and shrub twigs constitute a minor part of diet (Ekvtimishvili 1952; Abdurakhmanov 1977;

Table 13.4. *Capra cylindricornis.* Life forms/plant parts eaten (x indicates plant part used)

Life forms	Plant parts eaten							
	Whole plant	Leaves	Stems/ twigs	Bark	Roots/ bulbs	Flowers	Fruits	Seeds
Trees		x	x	x				
Shrubs		x	x					
Dwarf shrubs		x	x					
Herbs		x	x					
Grasses/sedges		x	x					
Fungi	x							

Kuliev 1981). In early summer, goats use salt licks near mineralized springs, and travel long distances to reach them (Veinberg 1984). For a list of plant genera eaten see Appendix 1, Table 14.

Life Forms/Plant Parts Eaten (Table 13.4).

Feeding Niche. In field and shrub layers.

Consumption of Food/Day. Males, under normal conditions (with live weight ca. 80 kg), eat 2.5–3 kg (dry weight) in summer. Per kilogram body weight, young goats consume 1.5–2 times more food than an adult. In captivity, an animal consumes 816–1,297 g of food in summer, and 375–410 g (dry weight) in winter. Depending on the availability of fodder – measured at 3,650, 2,600, 640 kg/ha – goats grazed for 310, 367, 511 min, and consumed 599, 657, 612 g/animal (dry weight), respectively (Yarovenko 1997; Magomedov et al. 2001).

13.5 Social Organization and Behavior

13.5.1 Social Life (Table 13.5)

Female groups with kids and yearlings, groups of old males and groups of young males, and mixed groups. There is personal recognition between individuals and an established hierarchy between animals inhabiting the patches. There is no territoriality, not even males protect their patches during rut, but marking trees by horn scratching is usual (Veinberg 1984).

Table 13.5. *Capra cylindricornis.* Seasonal changes in group size and composition. (Veinberg 1984)

Months	Average number of animals per group			
	Adult males	Young males	Females	Mixed herds
November–January	3	2	5	10
February–April	9	2	6	3
May–July	16	3	8	10
August–October	10	4	9	21
Average	10	3	7	10

13.5.2 Home Range

Fidelity to a given area is present in animals (Veinberg 1984). Males are more connected with some patches of the habitats, while females are probably more connected to particular habitat types (Akhmedov and Magomedov 1995).

13.5.3 Mode of Movement

The hard front of the hooves protrudes down and forms a crest that keeps animals secure on hard snow and ice. A skin pad on the back of hooves helps animals to walk and jump on stony surfaces (Vereshchagin 1938; Sokolov 1959). Careless movement through stone fields results in stones falling; trails in snowfields can release avalanches. Dig snow, rise on back legs to reach twigs of trees.

13.5.4 Dispersal/Migration

Male groups occupy more elevated parts of the mountains than female groups. Regular, short vertical migrations depend on snow cover and rain (N Ossetia and Dagestan). In Ossetia, there are also longer migrations from the Great Caucasus Range to the gorges of the northern slopes. Significant migrations were observed in winter, from Azerbaijan to Dagestan, to avoid deep snow, and back again in summer, to avoid blood-sucking insects and poachers. Males are more migratory than females (Vereshchagin 1938; Abdurakhmanov 1973; Kuliev 1981; Veinberg 1984; Magomedov and Akhmedov 1999).

Table 13.6. *Capra cylindricornis.* Diurnal activity (percent of active animals among observed animals during the time period). (Veinberg 1984)

	Time of day (h)								
	3–5	5–7	7–9	9–11	11–13	13–15	15–17	17–19	19–21
Winter		87	98	96	81	80	89	88	
Summer	100	85	62	27	15	33	55	87	100

13.5.5 Diurnal Activity (Table 13.6)

Prefer to be active during the evening, night and morning (Markov 1938; Ekvtimishvili 1952; Zalikhanov 1967; Veinberg 1984).

13.5.6 Shelters

Always rest on rocks or on stony cover near rocks. They clean and flatten a surface by scraping with their legs before lying down (Veinberg 1984).

13.5.7 Escape Behavior

Goats run straight to rock cover to escape predators, but this behavior is not successful against hunters.

13.6 Reproduction

Breeding parameters: age at maturity of males is 3.5 (sometimes 2.5) years, females 2.5 years; males participate in breeding at 6 years of age; the length of pregnancy is 160–165 days; there is usually 1 kid/female, 2 kids/female are rare (Vereshchagin 1938; Sokolov and Tembotov 1993).

Rut is preceded by a period when males search for females. A remarkable feature of rut is the raising of tails by males ≥3 years. Typical male sexual behaviors are guarding females, and characteristic demonstrations, such as a low stretch of the neck, twisting of the head, tongue flicking, sniffing of the female vulva, lip curls, and urinating on the muzzle. Battles between mature males are fierce but rare, however, they attack young males very often (Veinberg 1984).

13.6.1 Seasonal Breeding Events (Table 13.7)

Table 13.7. *Capra cylindricornis*. Seasonal breeding events

Locality[a]	Jan		Febr	Mar	Apr		May		June		July		Aug		Sept	Oct	Nov	Dec	
N Ossetia[1]	R[b]							N	N	N	N	N	N					R	R
Kabardino-Balkaria[2]	R						N	N										R	R
Azerbaijan[3]	R					N	N											R	R

[a]The superscript numbers denote the following references: 1, Veinberg (1984); 2, Zalikhanov (1967); 3, Kuliev (1981).
[b]R, Rutting season; N, newborn.

Table 13.8. *Capra cylindricornis.* Mortality in Dagestan. (Akhmedov and Magomedov 1995)

Season	Proportion of animals of different gender and age that died (%)			
	Adult males	Adult females	Yearlings	Kids
Summer	14	25	0	25
Autumn	0	0	29	0
Winter	57	25	42	50
Spring	29	50	29	25

13.7 Mortality (Table 13.8)

Mortality of adults in the Severo-Ossetinsky Nature Reserve is about 5–6% (Veinberg 1984). Avalanches are the predominant cause of mortality (Zalikhanov 1967). Goats suffer from avalanches in autumn, when they migrate to winter habitats (Naniev 1956), and in spring, on southern slopes (Veinberg 1984).

13.7.1 Predators

Predators are lynx (28.6% of scats with goat hairs), brown bears (8.3%), wolf (7.1%), golden eagle, *Gypaetus barbatus* (Zalikhanov 1967; Abdurakhmanov 1973, 1977; Veinberg 1979).

13.7.2 Diseases, Helminths and Ectoparasites (Appendix 2, Table 8)

13.8 Population Structure and Densities (Table 13.9)

13.8.1 Population Density and Size (Table 13.10)

In Dagestan, appropriate goat habitats cover 4,665 km^2 and there were 20,420 goats counted (Magomedov and Akhmedov 1994).

The annual population increase in the Severo-Ossetinsky Nature Reserve was 7–8.5% (Veinberg 1984).

Table 13.9. *Capra cylindricornis.* Gender and age composition of populations (percent of all observed animals)

Locality[a]	Male:female	Males, adult	Females, adult	Yearling	Kids
Severo-Osetinsky[1,b]	1.3–1.4:1		55	13	32
Dagestan[2]		24	37	21	18

[a]The superscript numbers denote the following references: 1, Veinberg (1984); 2, Abdurakhmanov (1973).
[b]Percent of all observed females, yearlings and kids.

Table 13.10. *Capra cylindricornis.* Population density and size in different localities

Locality[a]	Year	Population density (animals/10 km²)		Population size (×1,000)
		Summer[b]	Winter	
Nature Reserves of N Caucasus[1]		100–120	240, max. 660	
N Caucasus[2]	1992			17–22
Chechenia	1990			5.5
Dagestan (hunting grounds)	1994		51 (18–73)	20.4
	1999			20
N Ossetia	1999			2.6
Azerbaijan + Georgia[3]				
Zakataly Nature Reserve + Lagodekhi Nature Reserve	1990			3.2
Ismailinsky Nature Reserve	1988		24	

[a]The superscript numbers denote the following references: 1, Veinberg (1984), Kudaktin (pers. comm); 2, Zalikhanov 1967; Tochiev 1968; Magomedov and Akhmedov 1994; Akhmedov 1995; Sidorov (official data) (2000); 3, Kuliev 1981; Kudaktin (pers. comm.).
[b]Alpine zone.

13.8.2 Impact on Vegetation

In winter habitats, grass cover is heavily grazed and there is a high frequency of browsing on willow twigs. In Dagestan alpine meadows, goats utilize 600–1,400 kg/ha (dry weight) of plant biomass, which amounts to 29–54% of the total annual crop (Yarovenko 1997).

13.9 Relations to Man

Goats survive mostly in the nature reserves. On hunting grounds only licensed hunting is allowed, but poaching for horns, meat, and skin is common.

14 Markhor – *Capra falconeri*

Capra falconeri Wagner, 1839.
Artiodactyla, Bovidae, Caprinae

14.1 Status and Distribution (Fig. 14.1)

14.1.1 Taxonomic Remarks

One subspecies *C. f. heptneri* Zalkin, 1945 (Uzbekistan, Turkmenistan) is found (Heptner 1989).

14.1.2 Status

Endangered (EN–A2cde) (Hilton-Taylor 2000). CITES I.

14.1.3 Distribution (Fig. 14.2)

Distribution Area. Ca. 25,000 km^2.

Occurrence in Ecoregions. Tropical/Subtropical Desert Regime Mountains (M321).

14.2 Measurements

In Uzbekistan, body mass of males was 86–109 kg, females 32–41 kg; head and body length of males 155–170 cm, females 145 cm (Ishunin 1961).

14.3 Habitat

Mountain cliffs and the slopes of gorges. They avoid glaciers at high altitudes , and have become adapted to the warmer parts of the mountains.

Fig. 14.1. Screw-horned goat

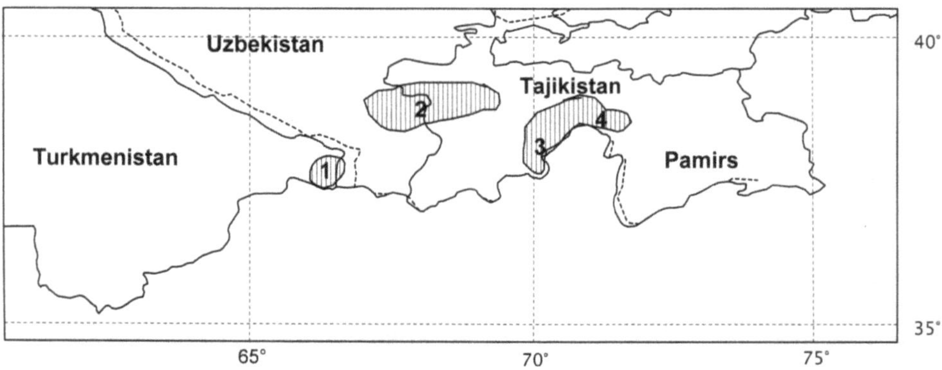

Fig. 14.2. *Capra falconeri.* Distribution map (Bannikov 1978). *1* Kugitangtau Range, Surkhan Nature Reserve; *2* Gissar Range; *3* Khazratishokh Range; *4* Darvaz Range

C. sibirica inhabits the same mountains, but occupies more elevated and colder areas (Ishunin 1987). Avoid areas with deep snow; leave areas that have had snowfall.

Terrain Form. Foothills (on fallen stones), cornices on slopes of gorges, slopes (cliffs, which are ten to several hundred of meters in height) and the tops of mountains (cliffs near the top of ridges) (Sultanov 1953; Ishunin 1961).

Table 14.1. *Capra falconeri.* Habitat[a]

Habitat type	Locality[b]		
	Tajikistan[1]	Uzbekistan[2]	
		Gissarsky	Kugitangtau
Forest			
Coniferous (*Juniperus*) in gorges and slopes	++	+	+
Mixed (islets along gorges)	+++		
Broadleaf sparse forest on slopes (*Pistacia, Acer*)	++		
Savanna (in and on edge of gorges)		+++	+++
Shrub-land (islets along gorges)	+++		
Grassland			
Cornices of slopes	r	++	++
Alpine (clearings in upper part of *Juniperus* savanna)			+++

[a]Symbols indicate % of habitat use was >30% (+++); 20–30% (++): 10–20% (+); <10% (r); x indicates habitat use noted, no numerical data reported.
[b]The superscript numbers denote the following references: 1, Sapozhnikov (1975); Sokov (1993); 2, Ishunin (1961, 1987).

Altitude. 1,500–2,800 m a.s.l., in summer males visit the alpine zone at an altitude of 3,000 m a.s.l. (Sultanov 1953; Ishunin 1987; Sokov 1993).

Water. Need water in summer, and visit watering places twice a day (Sokov 1993).

14.3.1
Biotope (Table 14.1)

14.4
Diet and Feeding Habits

14.4.1
Diet

Twigs of trees and shrubs, herbs, and grass. For a list of plant genera eaten see Appendix 1, Table 15.

14.4.2
Life Forms/Plant Parts Eaten (Table 14.2)

Table 14.2. *Capra falconeri.* Life forms/plant parts eaten (x indicates the type eaten). (Sultanov 1953; Sapozhnikov 1984)

Life forms	Plant parts eaten							
	Whole plant	Leaves	Stems/ twigs	Bark	Roots/ bulbs	Flowers	Fruits	Seeds
Trees		x	x				x	x
Shrubs		x	x					
Herbs		x	x					
Grasses/sedges		x	x					

Consumption of Food/Day. An adult female consumes 7–11 kg (fresh weight), a kid consumes 2.4–4 kg (Sapozhnikov 1984).

Feeding Niche. In field and shrub layers.

14.5 Social Organization and Behavior

14.5.1 Social life

In summer, mostly groups of 3–5 males (maximum 10) or females with kids were observed, but mixed groups or single animals were also observed. There are mixed groups from the time of the rut and throughout the winter.

14.5.2 Home Range

Settled way of life. Use the same feeding places, as well as the same resting places under rocks, during winter. In the Khazratishokh Range, the home range of goats occupies 7,000 ha; in the western part of Darvaz Range 3,800 ha; in Kushvoriston 2,000 ha (Ishunin 1961; Sapozhnikov 1984; Sokov 1993).

14.5.3 Mode of Movement

By using small ledges, goats can climb very steep walls of rock. In an extreme case, a male was observed to climb 100 m by jumping off and rebounding from near-vertical walls of a gorge; the distance between the walls was 4–5 m (Ishunin 1961).

14.5.4 Dispersal/Migration

Ishunin (1987) suggested that goats migrate seasonally between mountains, but Sapozhnikov (1984) did not make these observations. Sokov (1993) found seasonal and daily movements of 2–5 km. To escape deep snow they move to southern slopes. In summer they are forced, by domestic sheep and dogs, to hide in gorges and on cliffs; in winter, they occupy pastures rich in vegetation. No seasonal or daily migrations were observed (Sapozhnikov 1984).

14.5.5 Diurnal Activity

Active feeding in the morning, from about noon, they rest during hot days. In winter, they feed with breaks during the day (Sokov 1993).

14.5.6 Shelters

In summer, goats will lie in open places, in the shadow of mountains or under small shrubs, they never lie on rocks or flat places. In winter, they lie under roof-like cliffs, often near the foot of cliffs and above stony scree. Shelters always have a good view.

14.6 Reproduction

Breeding parameters: age at maturity of males is 2.5 years, for females it is 1.5 years; males participate in breeding at ≥3.5 years of age, females at >1.5 years of age; length of pregnancy is 153–170 days; there are 1–2 kids/female.

In the period before rut, males roam over large areas to search for females. Later, males that are ≥3.5 years old, form harems. Fierce battles between males occur. During rut males become thin. Kids are able to follow the mother from the second day of life. They suck until December, but stay with the mother for 2–3 years (Sokov 1993).

14.6.1 Seasonal Breeding Events

Rutting season from November–December; lambing from late May–early June (Zhirnov 1977; Sokov 1993).

Table 14.3. *Capra falconeri.* Population size in different localities

Locality[a]	Years	Population size
Uzbekistan		
Kugitangtau[1]	1951	135
	1976	180
Surkhansky Nature Reserve[2]	1992	37
	1993	86
Tajikistan[3]	1967	1,000
Tajikistan[4]	1967	500
	1983	400
Former USSR territory[5]	1977	1,000–1,200

[a]The superscript numbers denote the following references: 1, Zheleznyakov (1952); Sokov (1993); 2, Chernogaev et al. (1996); 3, Sultanov (1953); Sapozhnikov (1976); 4, Sokov (1993); 5, Zhirnov (1977).

14.7 Mortality

31% of snow leopard scats contain goat remnants (Sultanov 1953).

14.8 Population Structure and Densities

In Tajikistan, among 50 goats observed in autumn, 10% were adult males, 26% were adult females, 24% were yearlings, and 40% were kids (Sokov 1993).

14.8.1 Population Density and Size (Table 14.3)

14.9 Relations to Man

Protected species. The main reason for the diminishing number of goats is the loss of habitat to domestic sheep.

15 Wild Goat - *Capra hircus*

Capra hircus Linnaeus, 1758.
Artiodactyla, Bovidae, Caprinae

15.1 Status and Distribution (Fig. 15.1)

15.1.1 Taxonomic Remarks

In the Russian literature and Red List 2000 the synonym *C. aegagrus* is used. Two subspecies: *C. a. aegagrus* Erxleben, 1777 (Caucasus), *C. a. turcmenica* Zalkin, 1950 (syn. *C. a. blythi* Hume, 1875) (Turkmenistan) are recognized (Heptner 1989).

15.1.2 Status

Vulnerable (VU–A2cde) (Hilton-Taylor 2000). In the Russian Federation they are decreasing in number (Rozhnov 2001).

15.1.3 Distribution (Fig. 15.2)

Distribution Area. Ca. 75,000 km^2.

Occurrence in Ecoregions. Prairie Regime Mountains (M252), Mediterranean Regime Mountains (M262), Tropical/Subtropical Steppe Regime Mountains (M313).

15.2 Measurements

Body mass of males is 35–38 kg, body mass of females is 26–30 kg; head and body length of one male was 150 cm (Sokolov 1959; Gasparyan 1975).

Fig. 15.1. Bearded goat

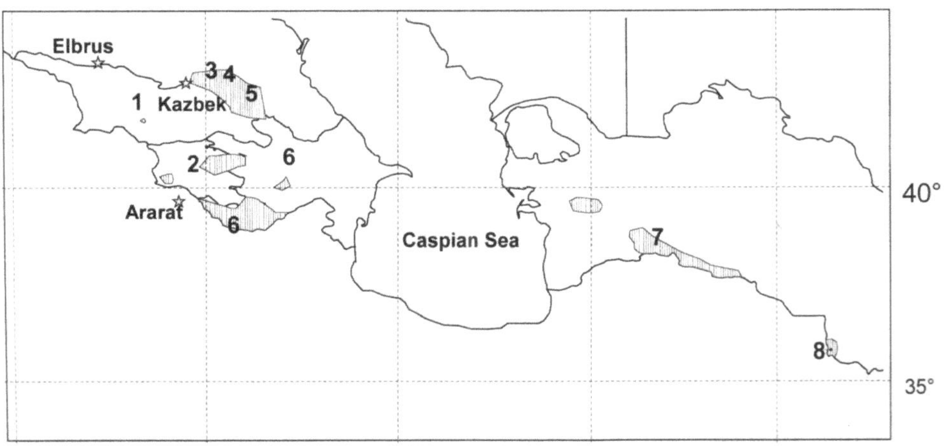

Fig. 15.2. *Capra hircus.* Distribution map (Kuliev 1981; Heptner 1989; Sokolov and Tembotov 1993). *1* Georgia; *2* Armenia; *3* Ingush Republic; *4* Chechen Republic; *5* Dagestan; *6* Azerbaijan; *7* Kopet-Dag Range; 8 Badkhyz Nature Reserve

15.3 Habitat

Inhabits lower parts of mountains than *Capra cylindricornis*; often lives on dry slopes and ridges, close to sheltering rocks (Dinnik 1910; Heptner and

Table 15.1. *Capra hircus.* Habitat[a]

Habitat type	Locality[b]		
	Chechen, Ingush[1]	Armenia[2]	Kopet-Dag[3]
Forest			
Coniferous	+++		
Shrubland (on slopes)	++	+++	
Grassland (desert type)			
Slopes	r	++	+++
Plateau		++	r

[a]Symbols indicate percent of habitat use was >30% (+++); 20–30% (++): 10–20% (+); <10% (r); x indicates habitat use noted, no numerical data reported.
[b]The superscript numbers denote the following references: 1, Gasparyan and Mezhlumyan (1969); 2, Dal (1949); 3, Korshunov (1986).

Formozov 1941; Vereshchagin 1959; Shamardanov and Ramazanov 1980; Zakariev 1982; Batkhiev 1989; Prilutskaya and Pishvanov 1988).

Terrain Form. Hills, plateau tops, mountain slopes.

Altitude. 500–3,900 m a.s.l. (Armenia) (Gasparyan 1975).

Water. Need water, except during spring, when they eat green plants.

15.3.1 Biotope (Table 15.1)

15.4 Diet and Feeding Habits

15.4.1 Diet

During winter and the dry seasons, twigs from trees and shrubs are the predominant food. In spring, green herbs and grass play the same role. For a list of plant genera eaten see Appendix 1, Table 16. Use salt licks, especially in spring (Gasparyan 1964).

15.4.2 Life Forms/Plant Parts Eaten (Table 15.2)

Feeding Niche. In field and shrub layers, sometimes below ground.

Table 15.2. *Capra hircus.* Life forms/plant parts eaten (x indicates plant part used)

Life forms	Plant parts eaten							
	Whole plant	Leaves	Stems/ twigs	Bark	Roots/ bulbs	Flowers	Fruits	Seeds
Trees		x	x	x				
Shrubs		x	x					
Herbs		x	x		x			
Grasses/sedges		x	x					

Table 15.3. *Capra hircus.* Seasonal changes of group size in Kopet-Dag expressed as percent of observed groups. (Korshunov 1988b)

Season	No. of animals per group							
	1	2	3–5	6–10	11–20	21–50	51–100	>100
Winter–spring	4	2	21	23	20	22	8	0
Lambing	0	9	28	18	17	25	3	0
Watering period	1	6	22	27	22	17	4	1
Rutting period	6	6	25	17	14	32	0	0

Table 15.4. *Capra hircus.* Seasonal changes in group composition in Kopet-Dag (% of observed groups). (Korshunov 1988b)

Season	Males	Females	Male + females	Males + females + juveniles	Females + kids
Winter–spring	3	16	0	65	16
Lambing	8	12	14	42	24
Watering period	7	0	6	47	40
Rut	3	13	6	62	16

15.5 Social Organization and Behavior

15.5.1 Social Life (Tables 15.3 and 15.4)

In groups, there is a hierarchy based on the size of horns and the body mass of animals. Dominant males are usually also the leaders of groups during traveling and approaching watering places, older females are leaders of mixed groups (Korshunov 1988a).

15.5.2
Dispersal/Migration

Migrate down mountains and move to other slopes in order to avoid deep snow. Due to the low population density, migrations are not significant anywhere.

15.5.3
Shelters

To escape danger goats hide among rocks, in cavities, and under roof-like cliffs (Yavruyan et al. 1971).

15.6
Reproduction

Breeding parameters: age at maturity of males and females is 1.5 years, males participate in breeding from ≥3–4 years of age, females from ≥2–3 years of age; barrenness can reach 20%; usually 1 kid/female, 2 kids/female are rare.

15.6.1
Seasonal Breeding Events (Table 15.5)

Maximum Length of Life. Males live for about 6 years on average, maximum 11 years (Korshunov 1988a).

15.7
Mortality

15.7.1
Predators

Irbis (*Uncia uncia*) is an important predator in Kopet-Dag (Korshunov 1985). In Caucasus, wolf and lynx are probable predators (Batkhiev 1989).

15.8
Population Structure and Densities (Table 15.6)

In 1984, in the Kopet-Dag Range (Turkmenian part of the species range), 6,600 goats were estimated to inhabit 107,000 ha (Korshunov 1986). In 1990, 250–300 goats were found in Chechnya (Sokolov and Tembotov 1993). In 1987, there were 700 goats in Armenia (Movsesyan and Ayrumyan 1987).

Table 15.5. *Capra hircus*. Seasonal events

Locality[a]	January		February	March	April		May		June		July		August		September		October	November		December	
Kopet-Dag[1]						N[b]	N	N	N	N	N	N	W	W	W	W		R	R	R	
Chechen, Ingush[2]								N	N											R	R
Georgia[3]	R					N	N	N	N	N									R	R	R
Armenia[4]						N	N	N													R

[a]The superscript numbers denote the following references: 1, Korshunov (1988b); 2, Dinnik (1910); Tochiev (1973); Batkhiev (1989); 3, Ekvtimishvili (1954); 4, Sarkisov (1944); Gasparyan (1975).
[b]W, Watering; R, rut; N, newborns.

Table 15.6. *Capra hircus.* Population structure in different populations

Locality[a]	Years	State of population	Adult male/ female ratio	Total population (%)		
				Males	Females	Kids
Kopet-Dag[1]		Growth	1:2.27	22	50	28
		Decline	1:1.69	29	48	23
	1986	In nature reserve	1:2.09	24	51	25
	1986	In hunting grounds	1:1.78	28	51	21
Chechen, Ingush[2]	1988			25	65	10
Dagestan[3]	1995	In hunting grounds		22	32	30[b]
Armenia[4]	1939			60	40	
	1950			50	50	

[a]The superscript numbers denote the following references: 1, Korshunov (1988b); 2, Batkhiev (1980, 1989); 3, Akhmedov (1999); 4, Dal (1944, 1951).
[b]Plus 16% of yearlings.

15.9 Relations to Man

An important species for hunting in some areas (Armenia, Turkmenistan), goats have been almost exterminated. Nowadays, hunting is permitted only with special licenses.

16 Asiatic Ibex – *Capra sibirica*

Capra sibirica Pallas, 1776.
Artiodactyla, Bovidae, Caprinae

16.1 Status and Distribution (Fig. 16.1)

16.1.1 Taxonomic Remarks

Two subspecies occur: *C. s. sibirica* Pallas, 1776 (Altai); *C. s. alaianna* Noak, 1902 (Tien Shan Mountains, Pamirs - Alay) (Heptner 1989).

16.1.2 Status

Protected in the nature reserves and by hunting regulations (Fedosenko and Savinov 1983; Sokov 1993; Smirnov 1994).

16.1.3 Distribution (Fig. 16.2)

Distribution Area. Ca. 704,000 km^2.

Occurrence in Ecoregions. Subarctic Regime Mountains (M133), Prairie Regime Mountains (M251).

16.2 Measurements (Table 16.1 and Fig. 16.3)

16.3 Habitat

Cliffs and rocks, often on steep slopes, but at very different altitudes (600–5,000 m a.s.l.) and within patches of pastures with high available plant biomass (>2,000 kg/ha) where goats aggregate during winter. Males prefer the upper, gentler slopes; females with young live in more rocky areas

Fig. 16.1. Siberian goat

(Dmitriev 1938; Sobansky 1992). Grassland (dry or meadow-like), in Middle Asia they often live on grassland with sparse shrubs (*Juniperus*), subalpine grassland, and tundra (Sokov 1993). Avoid tall grass, closed shrub areas and forest.

Terrain form. Foothills (Altai), slopes of gorges (Tien Shan, Altai), mountain slopes and tops (everywhere) (Shnitnikov 1936; Skalon 1936; Meklenburtsev 1949; Zhiryakov 1975; Sobansky 1992).

Altitude. 400–5,000 m a.s.l.; prefer heights of 1,700–3,000 m a.s.l. (Dmitriev 1938; Antipin 1941; Meklenburtsev 1949; Fedosenko and Savinov 1983).

Snow. Avoid snow >20–30 cm. Because of high hoof pressure (in Altai 500 g/cm^2 Dulkeit, cited in Nasimovich 1955), in Tien Shan 848 (551–996) g/cm^2 (Zimina 1953) and short legs, goats are careless in deep snow. Snow-rich winters are disastrous.

Water. Usually do not suffer from water shortage. In a few localities they constantly visit watering places (Fedosenko and Savinov 1983).

16.3.1 Biotope (Tables 16.2 and 16.3)

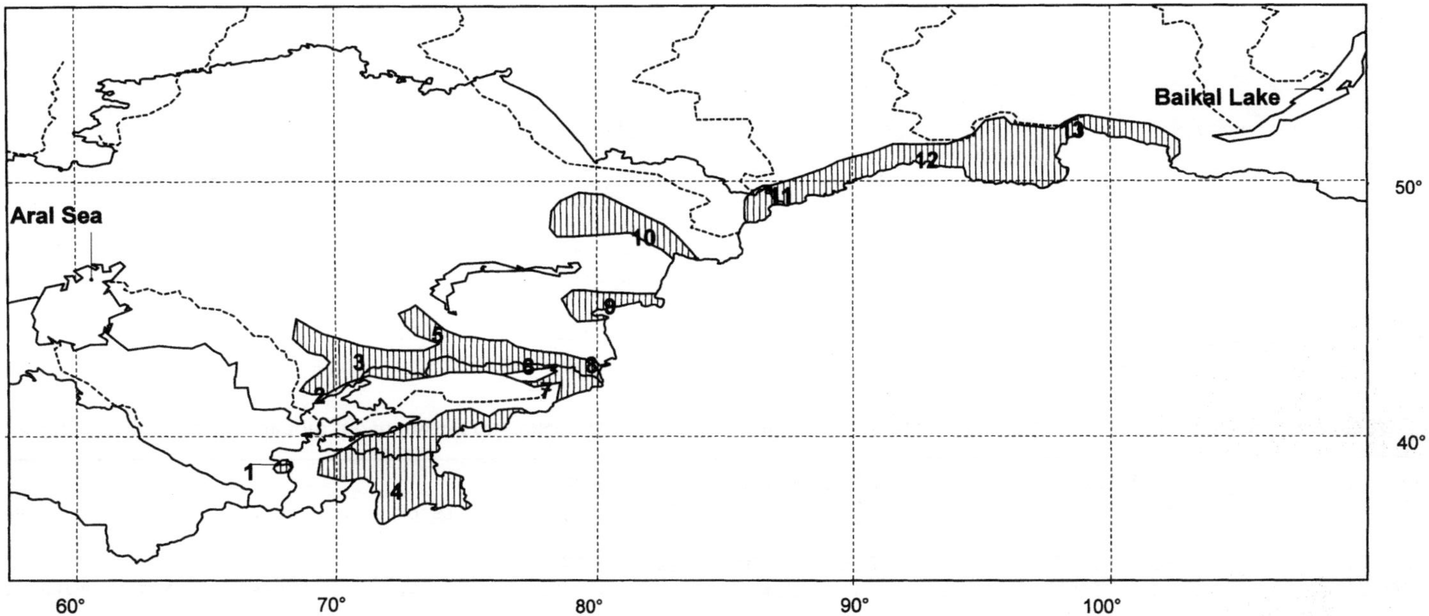

Fig. 16.2. *Capra sibirica*. Distribution map (Yanushevich et al. 1972; Fedosenko and Savinov 1983; Sokolov 1988). *1* Gissar Nature Reserve; *2* Ugam Range; *3* W Tien Shan Mountains (Talas Alatau Range); *4* Pamirs, Alay Range; *5* Kyrgyz Range; *6* C Tien Shan (Zailiyskoe Alatau Range, Kungey Alatau Range); *7* C Tien-Shan (Terskey Alatau Range); *8* Ketmen Range; *9* Dzungarian Alatau Range; *10* Tarbagatay Range; *11* Altai Mountains; *12* W Sayan Mountains, Tuva; *13* E Sayan Mountains, Buryatia

Table 16.1. *Capra sibirica.* Measurements

Locality[a]	Longitude (°E)	Latitude (°N)	Gender	*n*	Body mass (kg)	*n*	Head and body length (cm)	Tail length (cm)
Kyrgyzstan[1]	70–80	39–43	Male	6	67.7 (50–80)	6	149.6 (131–175)	15 (14–16)
			Female	6	45.3 (41–49)	6	132.3 (122–140)	14 (13–15)
Kyrgyzstan[1]	70–80	39–43	Male	15	77.1 (60–95)	15	153.8 (133–172)	16.2 (15–20)
			Female	6	42.0 (40–50)	6	130 (109–150)	14.7 (13–17)
Dzungarian Alatau[2]	79–82	45	Male	28	70.4 (39.6–114)	28	151.1 (124–171)	18.2 (17–18.5)
			Female	51	41.2 (30.0–55.2)	51	125.3 (108–140)	15.8 (12.0–19)
Altai[3]	84–90	49–52	Male	8	85.7 (70–103)	8	151.6 (138–160)	14.7 (13–17)
			Female	5	56.0 (50–64)	5	141.0 (132–145)	15.1 (14.5–16)

[a]The superscript numbers denote the following references: 1, Yanushevich et al. (1972); 2, Fedosenko and Savinov (1983); 3, Sobansky (1992).

Table 16.2. *Capra sibirica.* Distribution of animals in the Talas Alatau Range (percent of observed animals/month. (Grachev and Smirnova 1980)

	Altitude (m)	January	March	May	July	September	November
Grassland							
Savanna	1,300–1,800	6	14	3	0	0	0
With sparse shrubs, *Juniperus sabina*	1,800–2,300	93	24	36	5	3	51
Subalpine	2,300–2,900	1	53	54	70	61	44
Alpine	2,900–3,400	0	9	7	25	36	5
Glaciers	3,400–4,200	0	0	r	0	0	0

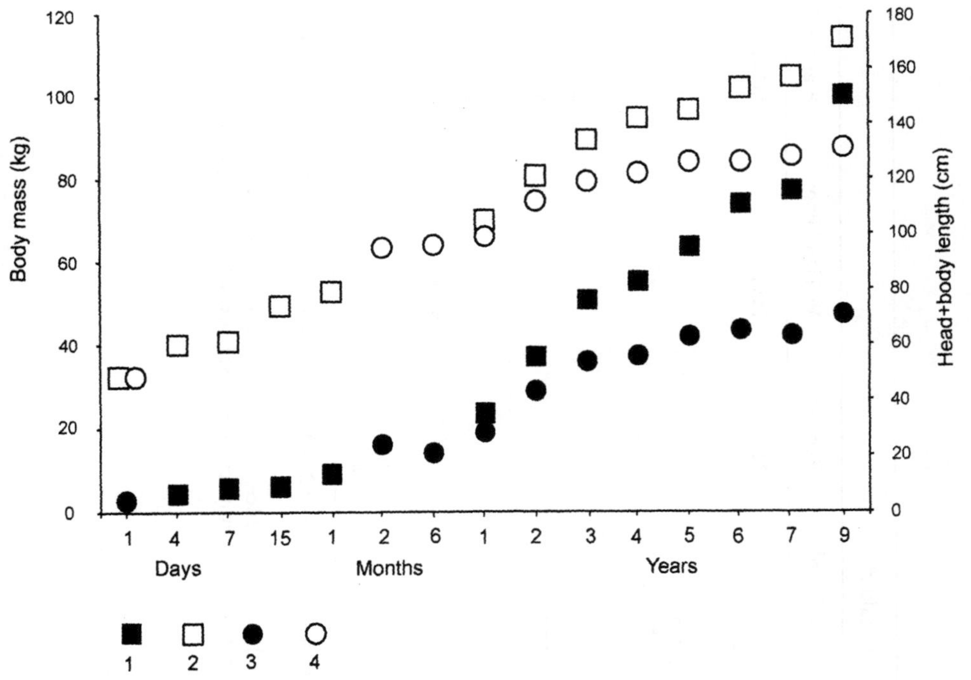

Fig. 16.3. *Capra sibirica.* Body mass and head and body length versus age of goats from the Dzhungarian Alatau Range. *1* Male mass; *2* male head and body length; *3* female mass; *4* female head and body length. (Fedosenko and Savinov 1983)

16.4 Diet and Feeding Habits (Table 16.4)

16.4.1 Diet

Herbs, grass and twigs of shrubs. Use salt licks, and sometimes drink water from salt streams (Fedosenko and Savinov 1983). For a list of plant genera eaten see Appendix 1, Table 17.

Table 16.3. *Capra sibirica.* Habitat[a]

Habitat	Locality[b]					
	Tajikistan[1]	Pamirs[2]	W Tien Shan[3]	E and C Tien Shan[4]	Altai[5]	Sayan[6]
(Ecoregion)		M321	M341	M341	M251	M133
Mature coniferous forest (sparse)	+					
Valley slope					+	
Shrubland (sparse *Juniperus*)	+++					
Slope of gorge	+++				+++	x
Shore of fresh water						r
Grassland						
Semi-desert		x		x		
Steppe	+++	x		x		
Shore of fresh water					r	r
Valleys				x		
Slope of gorge				x	++	x
Slopes			+++	x	+++	x
Sub-alpine	+++	x				
Alpine	x	x		x	+++	x
Tundra (alpine)		x			+	x

[a]Symbols indicate percent of habitat use was >30% (+++); 20–30% (++): 10–20% (+); <10% (r); x indicates habitat use noted, no numerical data reported.

[b]The superscript numbers denote the following references: 1, Sokov (1993); 2, Shnitnikov (1936); Meklenburtsev (1949); 3, Zhiryakov (1975); 4, Antipin (1941); Fedosenko and Savinov (1983); 5, Dmitriev (1938); Sobansky (1992); 6, Skalon (1936); Fedosenko and Savinov (1983).

Table 16.4. *Capra sibirica*. Feeding and resting. (Fedosenko and Savinov 1983)

Month	Proportion of day used for feeding (%)	Length of feeding periods (min)	Length of resting periods (min)
March	68	88 (50–120)	42 (7–120)
May	66	56 (40–106)	55 (25–81)
June	47	45 (22–91)	52 (16–84)
November	54	102 (70–155)	28 (52–145)

Table 16.5. *Capra sibirica*. Life forms/plant parts eaten (x indicates the type used). (Fedosenko and Savinov 1983)

Life forms	Plant parts eaten							
	Whole plant	Leaves	Stems/ twigs	Bark	Roots/ bulbs	Flowers	Fruits	Seeds
Trees		x[a]						
Shrubs		x	x			x	x	
Herbs		x	x			x		
Grasses/sedges		x	x					
Horsetails			x					

[a]x indicates plant part used.

16.4.2 Life Forms/Plant Parts Eaten (Table 16.5)

Consumption of Food/Day. Males eat up to 16 kg (fresh weight); females eat 8–10 kg, but can eat up to 14 kg (Egorov 1955; Sapozhnikov 1984).

Feeding Niche. In field and shrub layers.

16.5 Social Organization and Behavior

16.5.1 Social Life (Tables 16.6, 16.7 and 16.8)

The composition of groups is not fixed, and there is with an exchange of animals between groups. Leaders (older females in the female groups, and the oldest or largest males in the male groups) play an important role during escape and movement of the group.

Table 16.6. *Capra sibirica.* Group composition. (Kovshar and Yanushko 1965; Fil 1970; Fedosenko and Savinov 1983; Sobansky 1992)

Group types	Winter	Spring	Summer	Late autumn-early winter
(Seasonal events)	Wintering	Lambing	Summering	Rut
Singles (old males)	x			
Females with kids			x	
Mixed groups (all sexes and ages)	x	x		x
Adult males			x	

Table 16.7. *Capra sibirica.* Seasonal changes in group size (percent of observed groups)

Locality[a]	Season	No. of animals per group					
		1	2–5	6–10	11–20	20–100	200
Zailiyskoe Alatau Range[1]	Summer	3	38	31	24	4	0
Altai[2]	May–September	8	38	26	23	5	0
Altai[2]	Snow period	12	43	23	11	9	2

[a]The superscript numbers denote the following references: 1, Zhiryakov (1975); 2, Sobansky (1992).

Table 16.8. *Capra sibirica.* Seasonal group size and composition changes in different localities. (Fedosenko and Savinov 1983)

Locality	Seasons	Number of animals per group		
(Group composition)		Females	Males	Mixed groups
Talas Alatau	Spring	5–6	3–11	18–20
	Summer	22	10	38
Zailiyskoe Alatau Range	Winter			9–13
	Spring	5–7	4–6	8–15
	Summer	7–8	2–3	7–18
	Autumn		3	
Terskey Alatau Range	Spring			11
	Summer	5	4	3
Dzungarian Alatau Range	Spring	6	6–9	24
	Summer	26	3–13	32
	Autumn	2–8	3	10–20
W Sayan	Spring	12–46	8–20	
	Summer		2–25	
	Autumn		3–20	

16.5.2 Home Range

Trails, lying places and droppings are found within the home range of goats. Mostly settled, they slowly re-occupy territories where they have been exterminated. Move up and down slopes at an altitude of 700–800 m a.s.l. (Sobansky 1992). Male groups are more agile, and have home ranges up to 2 km in diameter (Fedosenko and Savinov 1983). In Tajikistan, summer groups move within areas of 5,000–15,000 ha ; home ranges of some groups can overlap (Sokov 1993). However, severe snow conditions can force goats to migrate up to 30 km in Tien Shan, and Pamirs (Savinov 1964).

16.5.3 Mode of Movement

Slow movements are characteristic. Usually, move by stepping or trotting. Goats have an impressive capacity to run on steep cliffs, and jump from near-vertical rock walls. They have highly developed coordination. Have soft skin pads under the hooves. However, they rarely fall on their horns when they jump (Shnitnikov 1936; Tsalkin 1950). Females can climb 300 m in 4–5 min, stopping twice, and adult males climb the same height in 11–15 min with 14 stops (Fil 1970). They dig in the snow and the ground with their front legs, but not as frequently as other ungulates (Nasimovich 1955; Sobansky 1992).

16.5.4 Dispersal/Migration

Everywhere there are seasonal migrations to avoid deep snow. When the snow melts, or at the first signs of new growth, the goats return to summer pastures. In the Dzungarian Alatau Range, seasonal migrations extend over 60–100 km (Fedosenko and Savinov 1983) They also make short-term moves downwards before intense snowfalls; they move into the center of mountains to avoid strong wind, and human disturbance. On 14 December 1998, in the Sayano-Shushensky Nature Reserve, 1,800–2,000 goats (almost all the goats of the reserve) formed a column 1.5 km long and moved along a slope of the Yenisey basin (Sashko and Smaznov 1999).

16.5.5 Diurnal Activity (Table 16.9)

Polyphase diurnal feeding activity.

Table 16.9. *Capra sibirica.* Diurnal activity

Locality	6–8	8–12	12–15	15–17	17–19	19–21	21–22.30	Late night
Tadijkistan[1]								
Summer	■	■			■	■	■	■
Kazakhstan[2]								
Middle, high mountains (2200–2400 m)								
Spring	■	■	■	■	■	■		
Summer	■				■	■		
Late autumn-winter	■	■	■	■	■	■		
Low mountains								
Summer	■					■	■	■
Altai[3]	■		■		■	■		

Shading indicates that animals are active. The superscript numbers denote the following references: 1, Sokov (1993); 2, Fedosenko and Savinov (1983); 3, Sobansky (1992).

16.5.6 Shelters

Use cavities, roof-like cliffs, big stones, canyons, slopes protected from the wind, and even old houses, as shelters from bad weather. They use open, windy places or the shadow of rocks and trees to escape from blood-sucking insects (Fedosenko and Savinov 1983). To rest they need flat areas >0.5 m^2, which are difficult to find on steep slopes and are subject to competition. Scratch the snow or ground before they lie down.

16.5.7 Escape Behavior

Before escaping, goats stand still and try to recognize the danger. In contrast to sheep, frightened goats scatter in all directions, then gather again and move in a line up a slope, hide among cliffs or the steep walls of canyons (Sokov 1993). To avoid avalanches they run away and jump to the nearest stone; they pass dangerous places with great speed. Predators (leopard,

Table 16.10. *Capra sibirica.* Breeding parameters. (Tsalkin 1950; Kovshar and Yanushko 1965; Savinov 1964; Fedosenko and Savinov 1983; Sobansky 1992)

Locality	Age at sexual maturity (years)		Pregnancy (days)	Barrenness (%)		No. of kids/ 100 females	Females with 2 lambs (%)
	Males	Females		2 years old	≥3		
Talas Alatau							11
Dzungarian Alatau	2.5[a]	1.5	180	75	0		13
Altai						60.1 (39–83)	30–35

[a]1.5-year-old males have a low intensity of spermatogenesis (Tsaplyuk and Savinov 1973).

bear) evoke an acute escape reaction. In nature reserves goats let approaching humans come within 25 m; out of curiosity they approach man to a distance of 30–40 m and follow behind (Fedosenko and Savinov 1983). Capable of recognizing man by sight at 2.5–3 km; but are indifferent to human smell and the sound of falling stones.

16.6 Reproduction (Table 16.10)

Two peaks of spermatogenesis are found (March–May and June–November) and the sexual behavior of males can then be observed. Actual mating takes place in late summer–November. Females can have repeated heats if not fertilized (Egorov 1953; Tsaplyuk and Savinov 1973). Many males (old and some young) surround a female during rut and try to copulate, but mating is successful when the pair has separated from the group. A female gives birth to 1–2 kids, very rarely does a female have 3 kids (1 out of 25 cases). Kids suck for 11 months. In May, kids suck for about 35 s at each occasion, in August around 20 s, in November only about 8 s (Fedosenko and Savinov 1983).

16.6.1 Seasonal Breeding Events (Table 16.11)

Maximum Length of Life. 14 years in the wild (Yanushevich et al. 1972); 22 years in captivity (Sokolov 1959).

16.7 Mortality

Significant mortality during snow-rich winters, which occur at intervals of 5–15 years. Goats die under avalanches and when crossing large rivers.

Table 16.11. *Capra sibirica*. Seasonal breeding events

Locality[a]	January	February	March	April		May		June		July	August	September	October		November		December	
Talas Alatau[1]																		
High						N	N								R	R	R	
Moderate														R	R	R	R	
Zailiyskoe Alatau[2]							N	N										
Dzungarian Alatau[3]					N	N							R	R	R	R		
Altai[4]					N	N	N									R	R	
Sayan[5]					N									R	R	R	R	R

R, Rutting season; N, newborns.
[a]The superscript numbers denote the following references: 1, Savinov (1962); 2, Fedosenko and Savinov (1983); Zhiryakov (1975); 3, Fedosenko and Savinov (1983); 4, Dmitriev (1938); Sobansky (1992); 5, Skalon (1936).

Table 16.12. *Capra sibirica.* Gender and age composition of populations (%)

Locality[a]	Adult males		Females		Kids
	Total	≥5 years of age (of all males)	Adult	Yearlings	
Pamirs[1]	41	28	46	13	13–16
Talas Alatau[2]	28	69	40	15	17
Zailiyskoe Alatau[2]	22–42	44	40–44	12	14–26
Dzungarian Alatau[2]	15	57	33	25	27

[a]The superscript numbers denote the following references: 1, Egorov (1955); 2, Fedosenko and Savinov (1983).

16.7.1 Predators

Wolf (5–13% of scats contained goat remnants; 13–40% of the goat carcasses found were killed by wolf), lynx (25% of scats with goat remnants), snow leopards mostly hunt older, lone males), brown bear (<1% scats with goat remnants), and golden eagle (Shnarevich 1948; Shulpin 1948; Meklenburtsev 1949; Heptner and Sludsky 1972; Fedosenko and Sludsky 1981; Fedosenko 1982, 1986; Fedosenko and Savinov 1983; Zavatsky 1986). In W Sayan, of the males killed by wolves, 26% were 7–10 years and 51% were ≥10 years, of the females killed by wolves, 60% were 2–4 years, and 40% were 7–12 years (Zavatsky 1986).

16.7.2 Diseases, Helminths and Ectoparasites (Appendix 2, Tables 8 and 9)

Between 1917–1920, there was high mortality due to mange. This disease still causes high mortality today (Chernogaev et al. 1996). Necrobiosis, pasteurellosis and helminths also occurred (Appendix 2, Tables 8 and 9). Larva of *Oestrus* spp. often fill the nose and other head cavities (Grunin 1957). Fleas are also very numerous (Shnarevich 1948; Doszhanov 1980).

16.8 Population Structure and Densities (Table 16.12)

16.8.1 Population Density and Size (Table 16.13)

Table 16.13. *Capra sibirica*. Population density and size in typical habitats

Locality[a]	Year	Population density (animals/10 km²)	Population size (×1000)
Gissarsky Nature Reserve[1]	1984	1.7	0.5
	1987	1.9	0.54
	1990	2.2	0.65
	1993	1.6	0.48
Tajikistan[2]	1983	5–10, max. 110	41.0
Kazakhstan[3]	1992		15–20
	1999		20–25
Ugam Range	1980		0.5
Talas Alatau	1980	10.0	1.0
Kyrgyz Alatau	1980	5–7, max. 19	0.5
Zailiyskoe Alatau Range	1980	10–13	1.4
Kungey Alatau Range	1980	4.0	0.4
Terskey Alatau Range	1980		1.3
Ketmen Range	1980	4.0	0.3
Dzungarian Alatau Range	1980		10.0
S Altai	1980		1.0
Kyrgyzstan[4]	1966	0.4–16	13.9
Altai[5]		7.7 (5–16)	
	1975		2.5
	1980		2.0–2.5
	1990		4.5
W Sayan[6]		4.0, max. 300	
	1988		1.5–2.0
	1994		3.0
E Sayan[6]		1.0	
	1975		1.0
	1994		1.8
Russian Federation[c]	1999		5.9

[a]The superscript numbers denote the following references: 1, Chernogaev et al. (1996); 2, Sokov (1993); 3, Ayzin (1969); Fil (1970); Savinov (1975); Fedosenko and Savinov (1983); Zhiryakov (1975); Bekenov et al. (1992); Bekenov (1999); 4, Yanushevich et al. (1972); 5, Sobansky (1992); 6, Moskvitin and Smirnov (1975); Sokolov (1988); Smirnov (1980, 1994); 7, Sidorov (2000).

16.9 Relations to Man

Hunting species. Used for meat, horns and skin. In Kazakhstan, the annual harvest is about 60 animals (Bekenov et al. 1992), in Tajikistan, about 1,500–2,000 (Sokov 1993).

17 Chinese Goral – *Naemorhedus caudatus*

Naemorhedus caudatus Milne-Edwards, 1867.
Artiodactyla, Bovidae, Caprinae

17.1 Status and Distribution (Fig. 17.1)

17.1.1 Taxonomic Remarks

In Russia, a synonym *Naemorhedus goral caudatus* (syn. *raddeanus*) Milne-Edwards, 1867 is used (Heptner 1989). One subspecies *N. c. raddeanus* is found.

17.1.2 Status

Vulnerable (VU–A2 cd) (Hilton-Taylor 2000). CITES I In Russian Federation endangered (Rozhnov 2001). Hunting is banned.

17.1.3 Distribution (Fig. 17.2)

Distribution Area. Ca. 382,000 km^2.

Occurrence in Ecoregions. Warm Continental Regime Mountains (M212).

17.2 Measurements (Table 17.1 and Fig. 17.3)

17.3 Habitat

Habitats are connected with granite intrusions. There are two types of habitats: cliffs at the seashore (usually 100–300 m high) and cliffs in mountains, at some distance from the sea. The last type of habitat is usually situated in small, forested areas (Bromley and Kucherenko 1983; Nesterov 1992;

Fig. 17.1. Goral

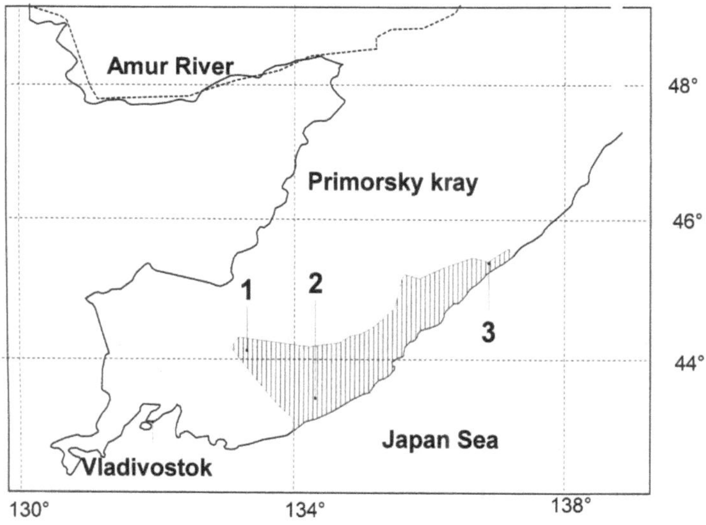

Fig. 17.2. *Naemorhedus caudatus.* Goral distribution (Bromley and Kucherenko 1983; Voloshina and Nesterov 1992). *1* Tumannaya Mountain; *2* Lazovsky Nature Reserve; *3* Sikhote-Alinsky Nature Reserve

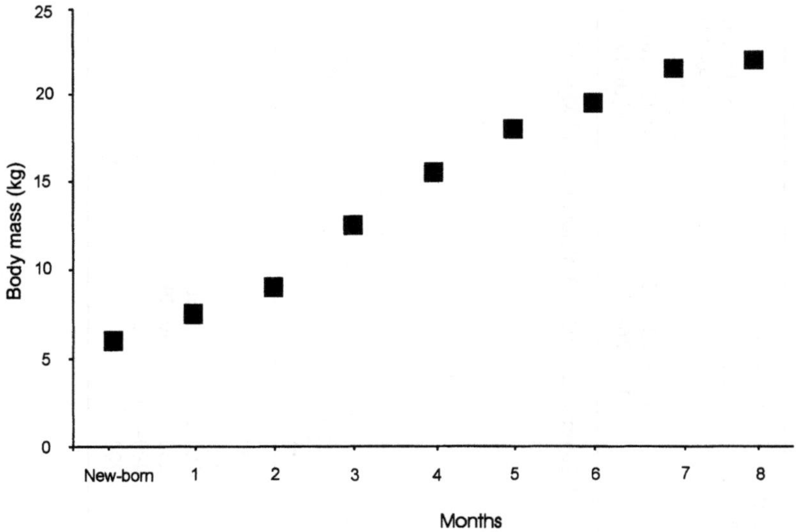

Fig. 17.3. *Naemorhedus coudatus.* Body mass versus age (in enclosures within the natural habitat). (Solomkina 1992)

Voloshina and Nesterov 1992). Cliff areas >5–10 km long are the preferred habitats (Voloshina and Nesterov 1992).

Terrain Form (Table 17.2). Seashores (cliffs, rare terraces), cliffs along streams in moderately high mountains (Bromley and Kucherenko 1983; Nesterov 1992)

Temperature. Inhabits areas with a temperature range from −10 to 13 °C (minimum −34 to 38 °C) to +14–16 °C (coastal); from −21 °C (minimum −42 °C) to +21 °C (inland) (Nesterov 1992).

Snow. Snow depth >35–40 cm is limiting. Eat snow to obtain water.

Water. Regularly drink water; in winter they eat snow several times per day (1–8 min each bout). They obtain salt by eating plants covered by salt spray and seaweed found on the shore. Whether they drink seawater is under discussion (Bromley and Kucherenko 1983; Myslenkov and Voloshina 1989).

17.3.1 Biotope (Table 17.3)

Table 17.1. *Naemorhedus caudatus.* Measurements

Locality[a]	Longitude (°E)	Latitude (°N)	Gender	*n*	Body mass (kg)	Head and body length (cm)	Tail length (cm)
Tumannaya Mountain[1]	134	44	Male	5	34 (30–39)	113.5 (109–118)	15.9 (15–18)
			Female	3	30 (24.2–32)	112 (106–117)	15.5 (15–16)
Lazovsky Nature Reserve[2,b]	134	43	Male	8	40.3 (36.8–47)		43.7 (40–48)[c]
			Female	7	40.3 (38–42)		40.5 (36–45)[c]
Sikhote-Alinsky Nature Reserve[3,b]	137	45	Male	11	33 (26–41)	128 (123–132)	15 (14–18)
			Female	5	32 (27–45)	130 (126–135)	16 (16–17)

[a]The superscript numbers denote the following references: 1, Bromley and Kucherenko (1983); 2, Solomkina (1992); 3, Voloshina and Myslenkov (1992).
[b]In enclosures.
[c]With hairs.

Table 17.2. *Naemorhedus caudatus.* Characteristics of cliff habitats (Voloshina and Nesterov 1992)

Locality	Longitude (°E)	Latitude (°N)	Altitude (m)	Extent from foothill to top (m)	Steepness (°)	Square of cliff area (ha)	Cliff area of the total area (%)	Density of goral/km² of cliffs
Chernie Skali	135°47′	44°16′	20–250	400	45–80	200	70	5–6
Abrek	135°47′	44°16′	50–500	600	80–90	400	40	12–50
Kamennye Vorota	135°30′	44°30′	400–500	500–600	40–80	200–220	45–50	7–8
Klyuch Skalistyi	134°20	43°38′	300–800	500–600	60–70	150	30–40	8–10
Uppers of Arsenevka River	133°07′	43°30′	400–800	600	80–90	400–500	5	4–6
Izvilinka River	134°0′	43°50′	350–500	250–400	45–60	200		8–10

Table 17.3. *Naemorhedus caudatus.* Habitat.[a] (Myslenkov and Voloshina 1989; Voloshina and Nesterov 1992)

Habitat type	Seashore	Inland
Mature forest		
Pine		+
Larch	r	
Mixed pine-oak	+	+++
Mixed pine-birch		+
Oak	+++	+++
Grassland		
Slope of valley		r
Slope of sea shore	+++	

[a]Symbols indicate percent of habitat use was >30% (+++); 20–30% (++): 10–20% (+); <10% (r); x indicates habitat use noted, no numerical data reported.

17.4 Diet and Feeding Habits

17.4.1 Diet

In the Sikhote-Alinsky Nature Reserve they consume 269 species (58%) of the 446 plant species occurring in the habitats (Shaulskaya 1992). In winter, trees and shrubs constitute up to 40% of their diet. For a list of plant genera eaten see Appendix 1, Table 18.

17.4.2 Life Forms/Plant Parts Eaten (Tables 17.4 and 17.5)

Feeding Niche. In field and shrub layers.

17.5 Social Organization and Behavior

17.5.1 Social Life

Parcel groups consist of 2–3 males, 5–7 adult females, 4–6 kids, 2–4 male yearlings, and 2–5 female yearlings. The subgroups can be 2 females with kids, or 2–5 animals <3 years of age, or bachelor groups (1 adult male and

Table 17.4. *Naemorhedus caudatus.* Life forms/plant parts eaten (x indicates plant part used)

Life forms	Plant parts eaten							
	Whole plant	Leaves	Stems/ twigs	Bark	Roots/ bulbs	Flowers	Fruits	Seeds
Trees		x	x	x				
Shrubs	x	x	x	x		x		
Herbs	x	x	x			x		
Grasses/sedges	x							
Mosses	x							
Lichens	x							
Fungi	x							

Table 17.5. *Naemorhedus caudatus.* Seasonal changes in food composition. (Shaulskaya 1980)

Food	Winter	Spring	Summer	Autumn
Trees	6	4	82	4
Shrubs	16	7	9	12
Grasses/sedges	77	89	6	81
Mosses, lichens, mushrooms	1	0	3	3

1–3 young males), or mixed groups (of up to 16 animals). These subgroups can join in different patterns and for different lengths of time. The distance between individuals is 2–15 m, and the intergroup distance is 100–200 m. Hierarchy is of a linear type and the territorial male is the dominant one. Severe conflicts mostly occur between the dominant male and kids approaching maturity. Combat rarely occurs and usually consists of no more than threatening poses. Leaders are important during escape from danger, they also determine the place of pasturing (Myslenkov and Voloshina 1989).

17.5.2 Home Range

Goral live in groups in which the members recognize each other; they inhabit areas with distinct borders. The home ranges of parcel groups occupy 55–150 ha (1.5–4 km along seashore cliffs). Within these groups there are subgroups of animals, which are personally connected. Home ranges of subgroups are 5.6 ha (4.4–6.8) with a core area of 2 ha (1.2–3).

Table 17.6. *Naemorhedus caudatus.* Feeding activity (percent of the time periods). (Voloshina et al. 1992)

Periods (h)	5–7	7–9	9–11	11–13	13–15	15–17	17–19	19–21
December		58	70	72	54	70	68	
June	55	40	250	37	45	40	20	22

Males occupy territories of 22–55 ha. The home range includes trails, lying places, areas marked by animals and areas of defecation (33–66 areas of defecation/ha) (Myslenkov and Voloshina 1989). Each home range is used by 2–3 adult males and is defended against neighboring males. Marking is done by fraying trees with horns, and by depositing a small volume of fluid from a gland behind the horn (Sokolov et al. 1992a). Adult females have their own home ranges. They do not defend their borders and, therefore, female home ranges overlap (Myslenkov and Voloshina 1989).

17.5.3 Mode of Movement

Adapted to moving in rocky landscapes. They can jump 2.5–3 m in distance and 2 m in height. In 35 cm of snow they can only jump 2 m in distance. They dig through soil and snow when feeding (Bromley and Kucherenko 1983). They move at speeds of 60 m/h (40–120), during winter grazing, and 90 m/h (33–200) in summer (Voloshina and Myslenkov 1992).

17.5.4 Dispersal/Migration

Very settled. Only young males disperse.

17.5.5 Diurnal Activity (Table 17.6)

From 05:00 to 21:00 hours of the day, gorals have 4–5 periods of feeding, and resting. Feeding also occurs during the night. They rest among the rocks and then move to the pastures to feed. In winter, feeding takes 72–84% of the time, chewing takes up 4–13%.

17.5.6 Shelters

Dig in the ground before lying down, as a result the resting places develop into hollows. These are situated near trees and shrubs, on flat rocks, and in cavities (Voloshina and Myslenkov 1992).

Table 17.7. *Naemorhedus caudatus.* Seasonal breeding events. (Myslenkov and Voloshina 1978)

Locality	Rutting period	Lambing	
		First kids	Most of kids
Sikhote-Alinsky	Second part of October–end of December	17–27 May	Mid June

17.5.7 Escape Behavior

The flight distance, when humans approach, is about 90 m (25–200), for single animals it is slightly less, for groups of 2–5 animals slightly more; they do not stop until they reach a distance of 100–150 m. Kids allow humans to approach to 6–10 m (Myslenkov and Voloshina 1978; Bromley and Kucherenko 1983). Looking around takes 5–10% of time.

17.6 Reproduction

Breeding parameters: age at maturity of males and females is 1.5 years; the first participation in breeding for males is at 2.5 years of age, females at 1.5 years of age; the length of pregnancy is 210 days (177–231); there is usually 1 kid/female, 2 kids/female are rare (Solomkina 1983; Myslenkov and Voloshina 1989).

Estrus lasts for 15–30 h and is occurs again after 20 days. During rut, a female can have up to 4 estruses (Myslenkov and Voloshina 1989). Kids hide during the first 20–35 days, they run well at 3–4 days, eat green grass at 5–7 days. Suckling continues for >100 days.

17.6.1 Seasonal Breeding Events (Table 17.7)

The rutting period is during the second part of October–end of December; lambing starts 17–27 May, most kids are born in mid-June (Myslenkov and Voloshina 1989).

Maximum Length of Life. 15 years (maximum 16–18).

17.7 Mortality

The expansion of area is very slow because of high mortality at the margins. Poachers and severe winters are the main causes of mortality.

Table 17.8. *Naemorhedus caudatus.* Gender and age composition (percent of all observed animals)

Locality[a]	Males	Females	Males + females		
	Adult	≥3 years of age	2 years old	Yearling	Kids
Lazovsky[1]	27	33	23		17
Sikhote-Alinsky[2]	13	31	30		26

[a]The superscript numbers denote the following references: 1, Khokhryakov (1997); 2, Myslenkov and Voloshina (1989).

Table 17.9. *Naemorhedus caudatus.* Population size. (Bromley and Kucherenko 1983; Khokhryakov 1997)

Locality	Years	Population size
Chernie Gori	1980	20
Lazovsky	1980	120–150
	1993	210–220
Sikhote-Alinsky		110
Inland sections		250–300
Total in Far East	End of 19th century	2000
	1980	500–550

17.7.1 Predators

Wolves (kill 3% of the population) and lynx are the main predators; bears (two species) are scavengers. Leopard and golden eagle rarely kill gorals.

17.7.2 Diseases, Helminths and Ectoparasites (Appendix 2, Tables 8 and 9)

Few helminths are reported (see Appendix 2, Tables 8 and 9) (Voloshina and Solomkina 1992).

17.8 Population Structure and Densities (Table 17.8)

17.8.1 Population Density and Size (Table 17.9)

Population size is more or less stable, except for 1920–1950 when there was a strong decline due to poaching (Bromley and Kucherenko 1983).

17.9 Relations to Man

Protected species. The size of the existing populations has not changed for many years, despite protection. The very settled way of life may prevent rapid increase in numbers. Although a few individuals have been observed on new sites, they have failed to establish new populations (Voloshina et al. 1986).

18 Musk Ox – *Ovibos moschatus*

Ovibos moschatus Zimmermann, 1780.
Artiodactyla, Bovidae, Caprinae

18.1
Status and Distribution (Fig. 18.1)

18.1.1
Taxonomic Remarks

Evidences by Vereshchagin (1959b) suggest that musk ox was exterminated from Taymyr by the first explorers during the 17th century. In 1947, the species was re-introduced into the Taymyr Peninsula (lower part of Bikada River) from Banks Island (Canada), and in 1975 into Taymyr and Wrangel Island from Nunivak Island (USA, Alaska) (Uspensky 1976, 1984). In 1995–2000, ca. 70 musk ox from Taymyr were introduced into Yakutia (Anabar River, Ust-Lensky Nature Reserve, Indigirka River).

18.1.2
Status

Hunting is banned.

18.1.3
Distribution (Fig. 18.2)

Animals now occupy all suitable habitats on Wrangel Island. In Taymyr, in 1986, descendants of reintroduced animals (175 animals) occupied a range of 4 million ha; by 1996 1,600 animals occupied a large range in E Taymyr (Fig. 18.2; Yakushkin 1998).

Occurrence in Ecoregions. Tundra (121), Tundra Regime Mountains (M122).

18.2
Measurements (Table 18.1)

18.3
Habitat

Fig. 18.1. Musk ox

Terrain Form. Seashores; lowland; upland hills; foothills; valley sides; slopes and plateaus of mountains.

Altitude. 10–650 m a.s.l. (Wrangel Island).

Temperature. Inhabits areas with a temperature range from –52 to +25 °C (Taymyr), from –48 to +18 °C (Wrangel Island) (Yakushkin 1979).

Snow. Prefer areas with snow depth <50 cm in Taymyr, and <40 cm on Wrangel Island.

18.3.1 Biotope (Table 18.2)

18.4 Diet and Feeding Habits

18.4.1 Diet

Shrubs (*Salix* spp.), dwarf shrubs, grass, lichens, and dry grass. For a list of plant genera eaten see Appendix 1, Table 19.

18.4.2 Life Forms/Plant Parts Eaten (Table 18.3)

In Taymyr, an analysis of rumens showed that their contents were 47–86% grass, 10–52% shrubs and twigs, 2–7% lichens and 0.2–12% moss (Yakushkin 1998).

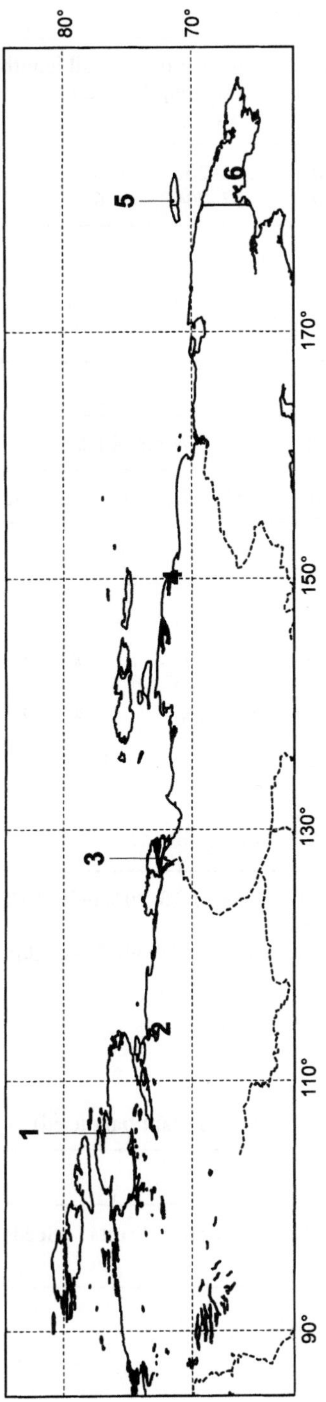

Fig. 18.2. *Ovibos moschatus.* Distribution map (Zheleznov 1990; Yakushkin 1998; N. Solomonov (pers. comm.). *1* Bikada River (Taymyr); *2* Anabar River (Yakutiya); *3* Ust-Lensky Nature Reserve (Yakutia); *4* Indigirka (Yakutiya); *5* Wrangel Island; *6* Chukotka

Table 18.1. *Ovibos moschatus*. Measurements. (Yakushkin 1998)

Locality	Longitude (°E)	Latitude (°N)	Gender	*n*	Body mass (kg)	Head and body length (cm)	Tail length (cm)
Taymyr	106	75	Male	3	280–330	234–262	6
			Female	3	115–160	177–204	6

Table 18.2. *Ovibos moschatus*. Habitat[a]

Habitat	Locality[b]							
	Taymyr Peninsula[1]				Wrangel Island[2]			
(Season)[c]	Wi	Sp	Su	Au	Wi	Sp	Su	Au
Arctic desert								
Stony-grass						x		
Shrubby						x		
Tundra								
Shrubs (*Dryas, Salix*) – *Carex-Eriophorum*, moss (on elevations)		+	r		x		x	x
Shrubs (*Salix, Dryas*), *Carex-Eriophorum*, moss	+++		+++	+++	x		x	x
Grass-shrubs-moss in shores of fresh water		+++	+++					
Shrubs and meadows in flood plain			+++	+++				
Bogs	+		++	++				

[a]Symbols indicate percent of habitat use was >30% (+++); 20–30% (++): 10–20% (+); <10% (r); x indicates habitat use noted, no numerical data reported.
[b]The superscript numbers denote the following references: 1, Pospelova (1989); Yakushkin (1998); 2, Zheleznov (1990).
[c]Wi, Winter; Sp, spring; Su, summer; Au, autumn.

Table 18.3. *Ovibos moschatus*. Life forms/plant parts eaten (x indicates plant part used)

Life forms	Plant parts eaten							
	Whole plant	Leaves	Stems/ twigs	Bark	Roots/ bulbs	Flowers	Fruits	Seeds
Shrubs		x	x					
Dwarf shrubs		x	x					
Herbs		x	x					
Grasses/sedges		x	x					
Lichens	x							

Table 18.4. *Ovibos moschatus.* Size and composition of groups

Locality[a]	Year	Population size	Herd size	Size of male groups
Taymyr[1]	1986	175	23 (7–32)	
	1993	830	17–23	2–7
Wrangel[2]	1985	35	29	2–3
	1990	99	18–48	4–10

[a]The superscript numbers denote the following references: 1, Yakushkin (1998); 2, Kovalev (1991).

Feeding Niche. In bottom, field and shrub layers.

18.5 Social Organization and Behavior

18.5.1 Social life (Table 18.4)

Young females stay in their maternal herd for 2–4 years, then together with their newborns, they abandon the herd and join another one. Males stay with the maternal herd for 2 years. Later, half of them abandon the herd and join a group of 3–4 year old males. The rest of the males spend their third year in the maternal herd; single animals stay for 4 years or more as subdominants, depending on the aggressiveness of dominant males (Kovalev 1991).

18.5.2 Mode of Movement

Good diggers in the snow.

18.5.3 Dispersal/Migration (Table 18.5)

Settled way of life in winter; there are spring and early winter migrations; they roam in summer.

18.5.4 Diurnal Activity (Table 18.6)

Table 18.5. *Ovibos moschatus.* Dispersal and migrations in Wrangel Island. (Kovalev 1991)

Period	Observed behavior	Mean daily range (km^2)	Mean daily route (km)	Maximum range during the period (km^2)	Maximum route length during the period (km)
1 December–15 April	Settled	0.03	0.5	50	20
15 April–20 May	Migration	0.01	7	50	80
20 May–20 June	Settled	0.03	1	20	10
20 June–10 September	Roaming	0.1	2	200	80
10 September–20 November	Settled	0.03	0.5	20	10
20 November–1 December	Migration	0.05	5	50	100

Table 18.6. *Ovibos moschatus.* Diurnal activity

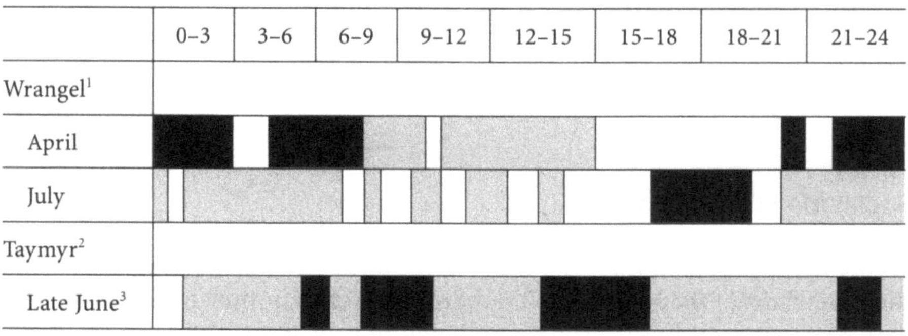

Gray shading indicates feeding; black, rest; white, moving. The superscript numbers denote the following references: 1, Chernyavsky et al. (1980); 2, Yakushkin (1998).
[a]Observations made in captivity in a large enclosure.

18.5.5 Escape Behavior

Recognize man from 1–3 km; the flight distance in mountain areas is 200 m; on the plain it is 500 m (Kovalev 1991).

18.6 Reproduction (Table 18.7)

Newborns can stand after about 30 min and suck within their first hour. Lactation continues for 3–4 months (Yakushkin 1998).

18.6.1 Seasonal Breeding Events (Table 18.8)

Table 18.7. *Ovibos moschatus.* Breeding parameters

Locality[a]	Age at sexual maturity (years)		Participation in breeding (years)		Pregnancy (months)	Litter size
	Male	Female[b]	Male	Female		
Taymyr[1]	2–3	2	4–5	2–3	8–8.5	1 (rare 2)
Wrangel[2]	4	2–3	5–8	2–3	9	1

[a]The superscript numbers denote the following references: 1, Yakushkin (1998); 2, Zheleznov (1990).
[b]A few females of 15–16 months participate in rut.

Table 18.8. *Ovibos moschatus.* Seasonal breeding events

Locality[a]	Rutting period		Lambing		
	Start	Finish	First kids	Most kids	Last kid
Taymyr[1]	10–31 August		17 April	25 April–10 May	15 June
Wrangel[2]	Early July	August–mid-September	Late April		Mid-July

[a]The superscript numbers denote the following references: 1, Yakushkin (1998), 2, Zheleznov (1990); Kovalev (1991).

Maximum Length of Life. 17 years. Before 1990, all 20 animals, which arrived on Wrangel Island in 1975, died.

18.7 Mortality

During 1982–1987, calf mortality was 4.5% in Taymyr (Katsarsky 1989).

18.7.1 Predators

Form a defensive circle with their heads outwards, when man approaches.

18.8 Population Structure and Densities (Table 18.9)

18.8.1 Population Density and Size (Table 18.10)

Table 18.9. *Ovibos moschatus.* Gender and age composition of the populations (%)

Locality[a]	Years	Adult		3 years		2 years	1 year	Calves
		Male	Female	Male	Female			
Taymyr[1]	1985	16	21	5	6	13	16	23
	1993	2	38	6[b]	6[b]		20	28
Wrangel[2]	1980	15	23			15	23	24
	1985	15	18	15	15	15		22
	1991	33	19	6	6	17	19	

[a]The superscript numbers denote the following references: 1, Katsarsky (1989); Yakushkin (1998); 2, Kovalev (1991).
[b]2 and 3 years old.

Table 18.10. *Ovibos moschatus.* Population size

Year	Locality[a]	
	Taymyr Peninsula	Wrangel Island
1975	35	20
1981	51	
1984		27
1986	175	
1990	420	116
1996	1100	325
1998	1560	

[a]The superscript numbers denote the following references: 1, Uspensky (1976); Yakushkin (1998); Tsarev (1999); 2, Chernyavsky et al. (1981); Kovalev (1991); Tsarev (1999).

18.9 Relations to Man

Introduced species are protected. There are plans to introduce musk ox in a number of areas of Yakutia, Chukchi Peninsula and N Ural (Tsarev 1999).

19 Argali, Asian Wild Sheep, Marco Polo Sheep – *Ovis ammon*

Ovis ammon Linnaeus, 1758.
Artiodactyla, Bovidae, Caprinae

19.1 Status and Distribution (Fig. 19.1)

19.1.1 Taxonomic Remarks

This description includes wild sheep populations which were previously described in the Russian and USSR literature as *O. ammon*, *O. argali*, *O. collium*, *O. argali dauricus*, *O. heinsii*, *O.ammon humei*, *O. karelini*, *O. argali mongolica*, *O. nigrimontana*, *O. polii*, *O. ammon przevalski*, *O. sairensis*, and *O. severtzovi* (Nadler et al. 1973; Heptner 1989). Six subspecies: *O. a. nigrimontana* Severtzov, 1873 (Karatau Range), *O. a. severtzovi* Nasonov, 1914 (Nuratau Range, Aktau Range), *O. a. karelini* Severtzov 1873 (Tien-Shan Mountains), *O. a. collium* Severtzov, 1873 (Kazakh Upland), *O. a. polii* Blyth, 1840 (E Pamirs Mountains, Dzungarian Alatau Range), *O. a. ammon* Linnaeus, 1758 (Altai Mountains, Tuva, Transbaikal) are found (Heptner 1989; Lyapunova et al. 1997).

19.1.2 Status

O. a. ammon (Altai, Tuva) is vulnerable (VU–A2cde, C1) CITES II; *O. a. collium* (Kazakhstan Upland) is vulnerable (VU–A2cde, C1) , *O.a. karelini* (Tien Shan) is vulnerable (VU–A2cde, C1 + 2a), CITES II; *O. a. nigrimontana* (Karatau Range) is critically endangered (CR–C2b) CITES I; *O. a. polii* (E Pamirs Mountains, Dzungarian Alatau Range) is vulnerable (VU–A2cde, C1) CITES II; *O. a. severtzovi* (Nuratau Range, Aktau Range) is endangered (EN–A2cde, C2b) (Hilton-Taylor 2000). In Russian Federation *O. a. ammon* is endangered (Rozhnov 2001).

19.1.3 Distribution (Fig. 19.2)

Distribution Area. Ca. 851.000 km^2.

Fig. 19.1. Altai sheep

Occurrence in Ecoregions. Subarctic Regime Mountains (M133, M134), Prairie Regime Mountains (M251), Tropical/Subtropical Desert Regime Mountains (M321), Temperate Desert (343), Temperate Desert Regime Mountains (M341).

19.2 Measurements (Tables 19.1 and 19.2)

A newborn ram weighed 4.5 kg (Sapozhnikov 1976).

19.3 Habitat

Inhabit grasslands on mountain plains, sometimes on steep slopes. Usually, avoid rocky and stony places. Canyons, very steep valley slopes or other forms of rugged terrain are important for shelter (Sapozhnikov 1976; Shakula 1989c).

Terrain Form. A combination of high and low mountains is preferred (Fedosenko and Kapitonov 1983). Banks of mountain rivers, streams and lakes; plains; hills; foothills, valleys, slopes and tops of mountains (Antipin 1947; Meklenburtsev 1948; Yanushevich et al. 1972; Sopin 1975a; Sapozhnikov 1976; Fedosenko and Kapitonov 1983; Fedosenko 1989).

Altitude (Table 19.3). 3,000–5,000 m a.s.l. (Pamirs, Tajikistan), 400–2,000 m a.s.l. (Kazakhstan), 400–2,000 m a.s.l. (Nuratau), 1,000–3,000 m

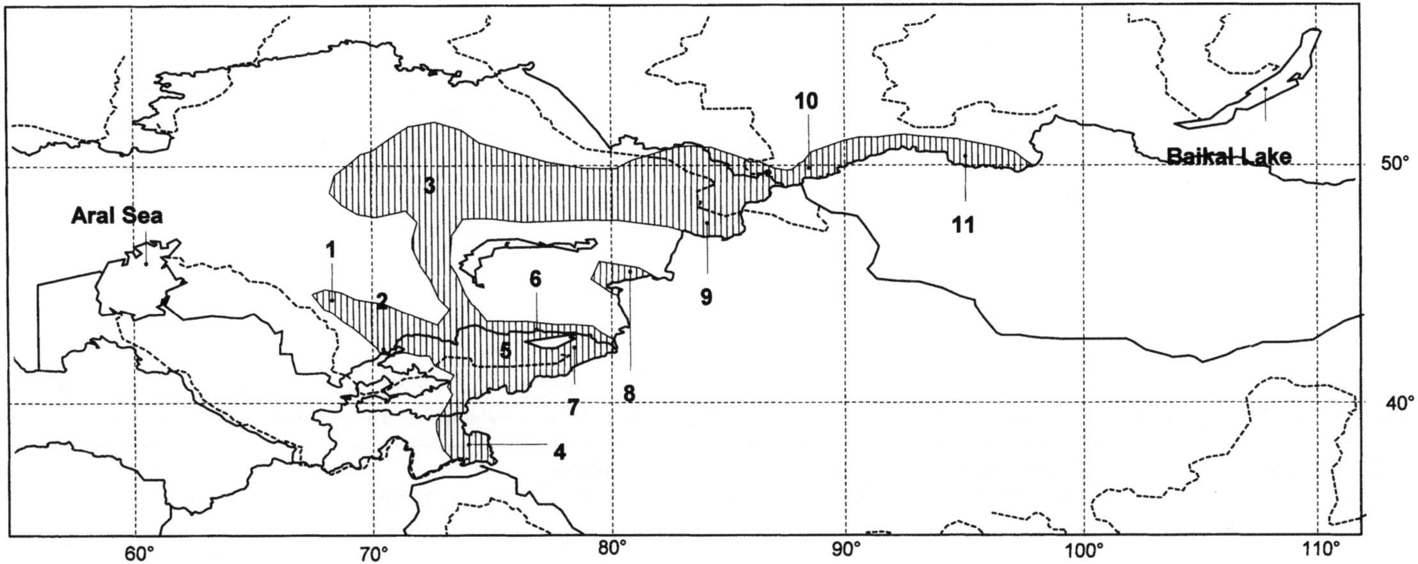

Fig. 19.2. *Ovis ammon.* Distribution map (Sapozhnikov 1976; Fedosenko and Kapitonov 1983; Heptner 1989; Smirnov 1994). *1* Karatau Range; *2* Talas Alatau Range, Aksu-Dzhabagly Nature Reserve; *3* Kazakhstan Upland; *4* Pamir Plateau, Tajikistan; *5* Tien-Shan Mountains, Kyrgyzstan; *6* Zailiyskoe Alatau Range; *7* Terskey Alatau Range; *8* Dzungarian Alatau Range; *9* Tarbagatay Mountains; *10* Altai Mountains; *11* Tuva

Table 19.1. *Ovis ammon.* Measurements

Locality[a]	Longitude (°E)	Latitude (°N)	Sex	*n*	Body mass (kg)	Head and body length (cm)	Tail length (cm)
Pamirs[1]	73–75	37–39	Male	3	100.9 (92–109.8)	163.8 (160–168)	10.3 (8.5–11.3)
			Female	2	55.3 (48.5–61.5)	143.5 (138.5–148.5)	10.8
Kyrgyzstan[2]	70–80	39–43	Male	4	82 (83–130)		
			Female	2	53.4 (47.7–59.1)		
Kazakhstan[3]	68–87	47–52	Male	3	119 (110–125)	156–163	13.0–13.1
			Female	3	57.7 (53–62)	147–150	10–12
Kazakhstan[4]	68–78	48–52	Male	58	147.6 (113–180)	167 (144–202)	
Nuratau Range[5]	66–67	41	Male[b]	4	105.5 (70–125)		
Tuva[6]	89–98	50–54	Male	4	134 (101–160)	176 (172–180)	11 (10–12)
			Female	3	90 (80–100)	170 (167–174)	10.7 (9.5–12)

[a]The superscript numbers denote the following references: 1, Sapozhnikov (1976); 2, Denisov (1935); 3, Fedosenko and Kapitonov (1983); 4, Baidavletov (1997); 5, Rakhimov (1975); 6, Sopin (1975b).
[b]In captivity.

Table 19.2. *Ovis ammon*. Measurements of males. (Berber and Kalmykov 1994; Fedosenko 2000)

Age (years)	Body mass		Head and body length		Tail length	
	n	(kg)	*n*	(cm)	*n*	(cm)
2 months	1	16.5	1	95	1	8
5	1	110	2	158 (156–160)	2	12.5 (12–13)
6	1	122	1	163	1	13
7.5	4	139 (108–146)	5	181 (165–194)	4	12.9 (11–15)
8.5	2	135.5 (125–146)	1	168	2	12 (11–13)
9.5	4	145 (135–157)	4	179 (166–199)	3	13 (10–15)
10.5	8	132 (118–146)	7	177.8 (172.2–185.5)	8	14.8 (14–15.5)
11.5	1	165	3	172.3 (165–186)	3	15.4 (14.3–17)

Table 19.3. *Ovis ammon*. Distribution of animals in the Talas Alatau Range (percent of sheep observed). (Grachev and Smirnova 1980)

	Altitude (m a.s.l.)	Winter (December–March)	Spring (April–May)	Summer (June–August)	Autumn (September–November)
Grassland					
With sparse trees	1,300–1,800	13–31	3–23	0	5–13
With sparse shrubs, *Juniperus*	1,800–2,300	39–62	43–52	2–32.4	34–47
Subalpine	2,300–2,900	7–47	24–43	40–59	48–52
Alpine	2,900–3,400	0	0.5–1	9–49	0
Glaciers	3,400–4,200	0	0	2–17	0

a.s.l. (Altai) (Yanushevich et al. 1972; Sapozhnikov 1976; Fedosenko and Kapitonov 1983; Shakula 1989c; Fedosenko et al. 1995).

Snow. Avoids snow >25–30 cm. Pressure of hooves (Altai ram) is 650 g/cm^2 (Dulkeit, cited in Nasimovich 1955).

Water. Regularly drink water and make special trips from pastures to water springs. Can live without water for some days (for example, if water shortage is due to temporary disturbance by man). Severe water shortage forces them to leave in search of a new source (Zhiryakov 1975; Fedosenko and Kapitonov 1983).

19.3.1 Biotope (Table 19.4)

A combination of grassland areas and areas without vegetation is preferred (Sapozhnikov 1976).

Table 19.4. *Ovis ammon*. Habitat[a]

Habitat type	Locality[b]					
	Kyrgyzstan, Tajikistan[1]	Kazakhstan[2]: low mountains (1); moderate mountains (2); high mountains (3)			Nuratau Range[3]	Altai[4]
Altitude (m a.s.l).	3,000–5,000	1,000	1,200–1,800	3,000–5,000	1,000–2,200	2,400–2,600
		1	2	3		
(Ecoregion)	M321, M341	343	M341	M341	M341	M251
Mature forest						
Coniferous (pine)		x				
Mixed deciduous-coniferous		x				
Broadleaf[c]				x	x	
Shrubland (sparse)[d]		x			x	
Grassland (dry)						
Shore of fresh water	x	x	x	x		x
Valleys	x	x	x	x		x
Slopes		x	x	x		x
Alpine				x		x
Tundra (alpine)						x

[a]x indicates habitat use noted, no numerical data reported.
[b]The superscript numbers denote the following references: 1, Meklenburtsev (1948); Yanushevich et al. (1972); Sapozhnikov (1976); Fedosenko et al. (1995); 2, Antipin (1947); Fedosenko and Kapitonov (1983); 3, Shakula (1989a,c); 4, Sopin (1975b); Fedosenko (1989).
[c]At the bottom of deep valleys.
[d]On slopes.

Table 19.5. *Ovis ammon.* Life forms/plant parts eaten.[a] (Fedosenko and Kapitonov 1983; Fedosenko 1989)

Life forms	Plant parts eaten (x indicates the type used)							
	Whole plant	Leaves	Stems/ twigs	Bark	Roots/ bulbs	Flowers	Fruits	Seeds
Shrubs		x	x					
Herbs	x	x	x		x	x		x
Grasses/sedges		x	x					x

[a](x indicates plant part used).

19.4 Diet and Feeding Habits

19.4.1 Diet

Herbs and grasses. For a list of plant genera eaten see Appendix 1, Table 20. Use salt licks.

19.4.2 Life Forms/Plant Parts Eaten (Table 19.5)

Consumption of Food/Day. About 10 kg of fresh green plants in summer (Sopin 1976).

Feeding Niche. In field and shrub layers, rarely below ground.

19.5 Social Organization and Behavior

19.5.1 Social Life (Table 19.6)

During spring and summer, groups of personally related females are seen with their young (<3 years). From the time of rut and through the winter, sheep form mixed groups and also groups of adult males (Sopin 1975b). Group size is about 10 (Altai) (Sobansky 1992). Gatherings sometimes include 50–90 animals (Fedosenko and Skuratov 1990). Composition of groups is constant, except during gatherings on pastures, when some groups unite. Leaders determine movement of groups through pastures and during escape.

Table 19.6. *Ovis ammon.* Group size

Locality[a]	Female groups		Male groups		Mixed groups	
	Average	Maximum	Average	Maximum	Average	Maximum
Pamirs[1]	11	35	4	15		
Kazakhstan[2]						
Kalkany	4	15	2	9	7	18
Terskey Alatau	5	15	5	18	7	28
Altai[3]	(8–14.8)	30	6.6–10.6	24		
S Siberia[4]	8	18		20		39

[a]The superscript numbers denote the following references: 1, Sapozhnikov (1976); Fedosenko et al. (1995); 2, Kapitonov and Makhmutov (1977); Fedosenko and Kapitonov (1983); 3, Fedosenko (1989); Filus (1992); 4, Sopin (1975a,b); Smirnov (1994).

19.5.2 Home Range

Prefer to stay in the same home range all year-around. There are permanent trails, and preferred resting places.

19.5.3 Mode of Movement

Although capable of digging in snow, they rarely do so (Nasimovich 1955). Dig in soil to find plant roots (Fedosenko and Skuratov 1990). Maximum speed on the plain is 60 km/h (when pursued by a car) (Yanushevich et al. 1972). Have great endurance when they run; dogs are not able to catch up (Meklenburtsev 1948). When escaping from danger they follow the leader.

19.5.4 Dispersal/Migration

Routine activities occur inside the home range, but animals daily move up and down slopes in order to feed, and escape blood-sucking insects. There are seasonal migrations along slopes to avoid deep snow (Yanushevich et al. 1972; Fedosenko and Kapitonov 1983). Also, there are movements to find water and fresh plants for food (Antipin 1947; Kapitonov 1980; Fedosenko and Kapitonov 1983). Migration distance in Altai is 10–12 to 20–25 km (Sobansky 1992).

Table 19.7. *Ovis ammon.* Feeding animals at different times of the day[a] (%)

Locality[b]	Season	Time of day (h)					
		6–9	9–12	12–15	15–18	18–21	21–22.30
Kazakhstan, high mountains[1]	Spring	80	70	50	50	75	
	Summer	75	75	50	50	70	
	Late autumn-winter	50	85	65	65	75	
Nuratau Range[2]	Spring	95	81	46			
	Summer	59	53	55		36	39
	Autumn	91	72	56		82	70
	Winter	86	85	70	100		
Pamirs[3]	Spring-summer	74		57	78		
Altai[4]	Summer	56		41		37	79
	Autumn	48		46		100	

[a]No data for 22.30–6.00.
[b]The superscript numbers denote the following references: 1, Fedosenko and Kapitonov (1983); Grachev (1988); 2, Shakula (1989b); 3, Sapozhnikov (1976); 4, Fedosenko and Skuratov (1990).

Table 19.8. *Ovis ammon.* Rhythm of feeding

Locality and season[a]		Length of feeding period (min)	Length of resting period
Kyrgyzstan, high mountains[1]			
Summer	Males	65 (60–70)	55 (20–92)
	Females	70 (60–90)	38 (31–43)
Kazakhstan, high mountains[2]			
Late winter	Females	49 (26–60)	36 (20–56)
Spring	Males	47 (15–70)	82 (55–135)
	Females	99 (65–140)	39 (32–44)
Early summer	Females	53 (30–71)	56 (25–100)

[a]The superscript numbers denote the following references: 1, Fedosenko et al. (1995); 2, Fedosenko and Kapitonov (1983).

19.5.5
Diurnal Activity (Tables 19.7 and 19.8)

Polyphasic rhythm of activity (Table 19.8). Activity is related to abundance and availability of food, and weather conditions. Visit watering places in the mornings and evenings. Feeding activity of males declines during rut, and for females it declines during the calving season (Antipin 1941; Meklenburtsev 1948).

Table 19.9. *Ovis ammon.* Breeding parameters

Locality[a]	Age at maturity (years)		Pregnancy (days)	Barrenness (%)	Number of lambs/ 100 females		Females with 2 lambs (%)
	Male	Female			Early summer	Early autumn	
Kyrgyzstan[1]	1.5–2.5	1.5–2.5	150				
Kazakhstan[2]	2.5[b]	2.5		24–33			(2–33)
Altai[3]				11	67	60	

[a]The superscript numbers denote the following references: 1, Yanushevich et al. (1972); 2, Antipin (1947); Kapitonov and Makhmutov (1977); Fedosenko and Kapitonov (1983); 3, Fedosenko (1989).
[b]Only males >4–6 years males can mate with females.

19.5.6
Shelters

Often use the same resting places for many days; they clean the area before lying down. Caverns in mountains; roof-like rocks are often used to find shelter from wind, these places are also used during calving.

19.5.7
Escape Behavior

Sheep are very alert; flight distance more than 700–800 m (Sobansky 1992). Leaders (usually old females and males, constant in a group) determine escape behavior, direction and route of escape. Females defend lambs against foxes, golden eagles and lammergeyer.

19.6
Reproduction (Table 19.9)

In the first phase of the rut, males follow each another, they battle fiercely to establish hierarchy. In the second phase, males unite with female groups and mate. Some males are present in mixed groups during rut, but mainly, males ≥8 years, copulate. A mother and her newborn spend the first week in one place; the newborn hides in some shelter; after a week it follows the mother (Fedosenko 1979).

19.6.1
Seasonal Breeding Events (Table 19.10)

Rut and calving occur on winter pastures. Lambs suckle up to the rutting season.

Table 19.10. *Ovis ammon.* Seasonal breeding events

Locality[a]	January	February	March		April		May		June	July	August	September	October		November		December	
Kyrgyzstan[1]						N	N	N							R	R	R	R
Kazakhstan[2]																		
Low mountains				N	N									R	R			
High mountains						N								R	R			
Nuratau Range[3]						N	N								R	R		
Altai[4]						N	N						R	R	R	R		
Tuva[5]							N						R					

R, Rutting season; N, newborns.
[a]The superscript numbers denote the following references: 1, Yanushevich et al. (1972); Sapozhnikov (1976); 2, Zhiryakov (1975); Fedosenko and Kapitonov (1983); 3, Shakula (1989b); 4, Fedosenko and Skuratov (1990); 5, Sopin (1975a).

Table 19.11. *Ovis ammon*. Gender and age compositions of populations

Locality and season[a]	Adult male: female ratio	Age and gender composition of population (%)			
		Adult		Yearlings	Lambs
		Males	Females		
Kyrgyzstan[1]	1.05:1	38	36	10	16
Kazakhstan[2]	1.84:1	27	49	8	16
Karatau Range[3]	1.43:1				
Aksu-Dzhabagly[4]	1.1:1				
Ermentau[5]					
In autumn	1.24:1				
In spring after bad winter	1:2.8				
Nuratau Range[b,6]		25–27	41–53		18–32
Altai[7]		27	51	10	12

[a]The superscript numbers denote the following references: 1, Fedosenko et al. (1995); 2, Fedosenko and Kapitonov (1983); 3, Pleshak (1980); 4, Kovshar and Yanushko (1965); 5, Kapitonov and Makhmutov (1977); 6, Shakula (1989c); 7, Sobansky (1992).
[b]In spring.

Maximum Length of Life. 20 years (Yanushevich et al. 1972).

19.7 Mortality

Severe climate with snow is the main mortality factor. Predation accounts for <30% of the total mortality. Mortality of lambs during the three summer months in Altai was about 12% (mostly leg injuries) (Fedosenko 1989). In the lower mountains of Kazakhstan, lamb mortality during the first year of life was 42–50% (Kapitonov and Makhmutov 1977; Fedosenko and Kapitonov 1983).

19.7.1 Predators

Predators were responsible for killing 23–33% of the sheep found dead in the Kazakhstan mountains (Fedosenko and Kapitonov 1983). Wolves are important predators in the years when animals are weakened by starvation.

19.7.2 Diseases, Helminths and Ectoparasites (Appendix 2, Tables 8 and 9)

Table 19.12. *Ovis ammon.* Age composition of males in groups during the rutting season. (Fedosenko and Kapitonov 1983)

Locality in Kazakhstan mountains	Age of males (years)				
	1.5	2.5–3	3.5–5	5.5–8	≥8
Talas Alatau Range	4	20	35	27	14
Terskey Alatau Range	15	20	15	30	20
Kalkany	17	24	32	17	10
Dzungarian Alatau Range	2	4	19	43	32

Table 19.13. *Ovis ammon.* Population density and size in typical habitats

Subspecies	Locality[a]	Year	Population density (animals/ 10 km^2)	Population size (×1,000)
O. a. nigrimontana	Karatau Range[1]	1997		0.3
O. a. karelini	Tien Shan[2]	1988–95	25–65	5–7
O. a. collium	Kazakhstan Upland[3]	1970–83	2.2 (0.3–20)	5–7
		1989–1991		10
O. a. severtzovi	Aktau[4]	1998		1–1.1
	Nuratinsky Nature Reserve (Nuratau Range)[b,5]	1983	8.7	1.2
		1987	6.5	1.0
		1990	9.7	1.5
		1993		1.9
O. a. polii	Pamirs[6]	1976	37–60	70–80
		1977	8–30	20
		1990		5
O. a. ammon	Altai[7]	1973		0.6
		1986		0.3
		1994		0.15
		1999		0.2[c]
	Tuva[8]	1994	2.0	0.2
		1999		0.15

[a]The superscript numbers denote the following references: 1,2, Fedosenko (1997); 3, Fedosenko and Kapitonov (1983); Fedosenko (1989); Berber and Kalmykov (1994); 4, Reimov and Reimov (1997); 5, Chernogaev et al. (1996); Fedosenko (1997); 6, Sapozhnikov (1976); Fedosenko (1989); Fedosenko et al. (1995); Abdusalyamov (1997); 7, Fedosenko (1989); Sobansky (1986, 1989); Fedosenko (1997); 8, Smirnov (1989b, 1994); Fedosenko (2000).

[b]In spring.

[c]In summer up to 500 because of arrival from Mongolia.

19.8 Population Structure and Density (Table 19.11 and 19.12)

Seven to 8-year-old males constitute 29% of the population in the Kazakh Upland (Berber and Kalmykov 1994).

19.8.1 Population Density and Size (Table 19.13)

There is a decline in numbers.

19.9 Relations to Man

Important hunting species for meat and hunting trophies (horns). There is trophy hunting in Pamirs and the Kazakh Upland (Berber and Kalmykov 1994; Fedosenko et al. 1995).

20 Mouflon – *Ovis aries*

Ovis aries Linnaeus, 1758.
Artiodactyla, Bovidae, Caprinae

20.1 Status and Distribution (Fig. 20.1)

20.1.1 Taxonomic Remarks

In Red List 2000, the synonym *Ovis orientalis* Gmelin, 1774 is used. In Russian literature, the name *O. ammon gmelini* Blyth 1840 is used (Heptner 1989).

20.1.2 Status

Vulnerable (VU–A2cde) (Hilton-Taylor 2000). *Ovis orientalis gmelini* (synonym of *O. aries*) is indeterminate (Groombridge 1994). There is one acclimatized population (since 1913) of hybrids (10 Corsican mouflons, *Ovis ammon musimon* Pallas 1811, and 3 domestic sheep were ancestors) in the Crimean Peninsula (Dal 1954). Hunting is banned.

20.1.3 Distribution (Fig. 20.2)

Distribution Area. Ca. 16,000 km^2.

Occurrence in Ecoregions. Prairie Regime Mountains (M252), Mediterranean Regime Mountains (M262).

20.2 Measurements (Table 20.1 and Fig. 20.3)

In Armenia, the body mass of newborn males is 2.9 kg (1.7–4.4), females 2.7 kg (1.5–3.5), male head + body length is 65.8 cm (64–70), female length is 62.8 cm (57–68) (Minasyan et al. 1986).

Fig. 20.1. Armenian mouflon

Table 20.1. *Ovis aries*. Measurements (Minasyan et al. 1986)

Locality	Longitude (°E)	Latitude (°N)	Gender	Age	*n*	Body mass (kg)	*n*	Body length (cm)	Head length (cm)	Tail length (cm)
Armenia[a]	43–47	39–41	Male	Adult	3	48.6 (40.0–56.0)	2	132.5 (129–136)	28.5 (28–29)	12.5 (12–13)
			Female	Adult	6	31.8 (30.0–34.0)	5	112.6 (105–123)	22.1 (21–23)	12.2 (11–13)

[a]In captivity.

20.3 Habitat

Stony steppe, semi-deserts.

Terrain Form. Plains; hills; plateau slopes (Crimean Peninsula); valleys and tops of mountains (Sarkisov 1950; Vereshchagin 1959; Yavruyan 1970; Dulitsky 1992).

Altitude. 400–1,000 m a.s.l. (Crimean Peninsula), 2,500–4,000 m a.s.l. (Armenia) (Vereshchagin 1959; Yavruyan 1970; Dulitsky 1992).

Snow. Snow depths of 35–60 cm in the Crimean Peninsula in 1954, 1957, 1963, 1965, 1967 were disastrous (Parshintsev 1992). In severe winters with deep snow, most animals migrate downwards (Sarkisov 1950; Yavruyan 1970).

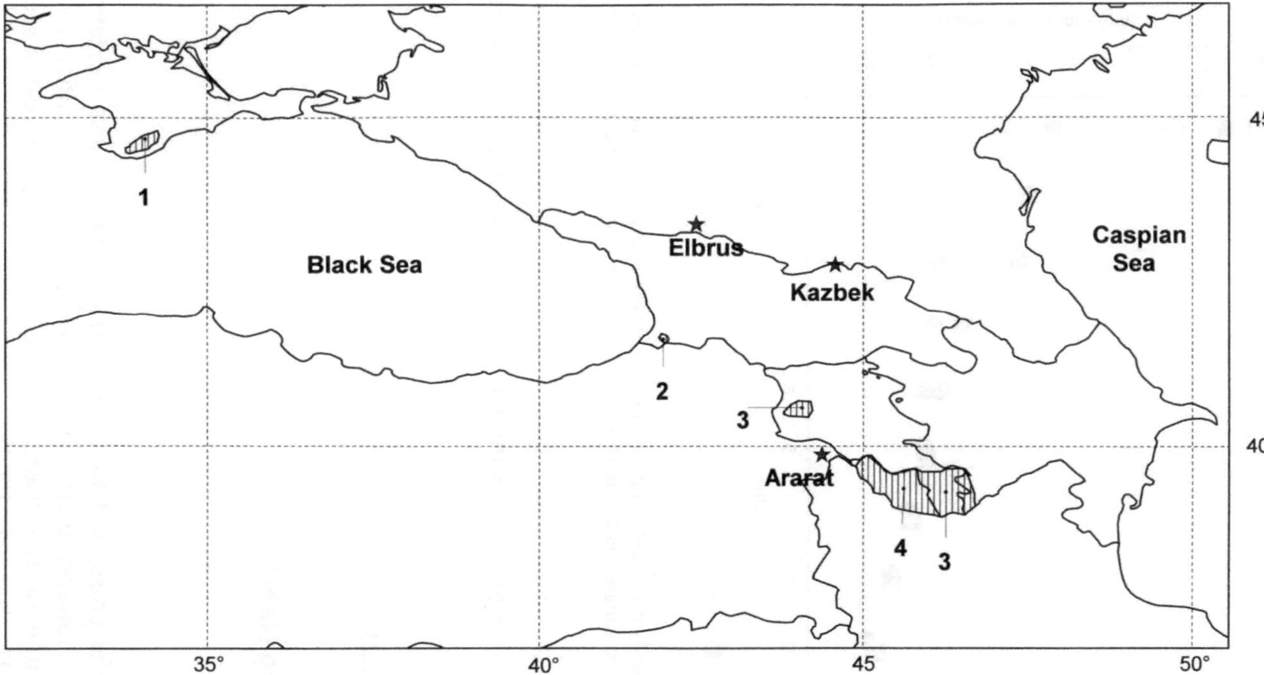

Fig. 20.2. *Ovis aries.* Distribution map (Yavruyan 1969; Arabuli 1989; Sokolov and Tembotov 1993). *1* Krymsky Nature Reserve; *2* Kedsky rayon, Georgia; *3* Armenia; *4* Nakhichevan, Azerbaijan

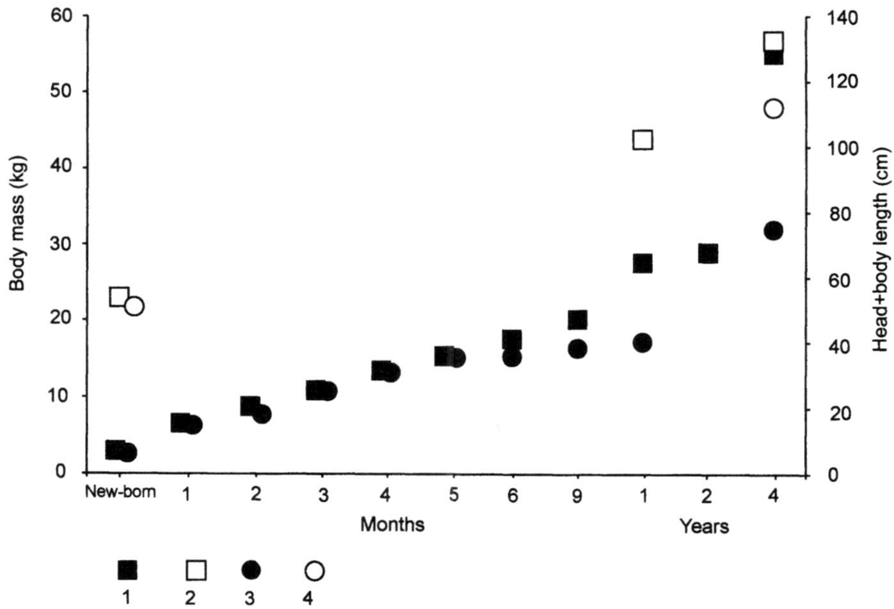

Fig. 20.3. *Ovis aries.* Body mass and length versus age for the Armenian sheep (Minasyan et al. 1986). *1* Male mass; *2* male head + body length; *3* female mass; *4* female head + body length

Water. Watering is regular; in summer twice a day (Ayrumyan and Gasparyan 1976).

20.3.1 Biotope (Table 20.2)

20.4 Diet and Feeding Habits

20.4.1 Diet

Ovis orientalis prefer grass, herbs and also shrubs, but to a lesser extent than *Capra hircus*. However, their ecological niches overlap significantly. Use salt licks and mineral-rich water. For a list of plant genera eaten see Appendix 1, Table 21 (Yavruyan 1969).

20.4.2 Life Forms/Plant Parts Eaten (Table 20.3)

Feeding niche. In field and shrub layers.

Table 20.2. *Ovis aries.* Habitat[a]

Habitat type[b]	Crimean Peninsula[1]	Armenia[2]
(Ecoregion)	M252	M262
Forest		
Coniferous	r	
Broadleaf	+++	
Clearing	+++	
Savanna[c]		x
Grassland		
Foothill		x
Slopes		x
Plateau	++	x
Alpine		x

[a]Symbols indicate percent of habitat use was >30% (+++); 20–30% (++): 10–20% (+); <10% (r); x indicates habitat use noted, no numerical data reported.
[b]The superscript numbers denote the following references: 1, Dulitsky (1992); 2, Sarkisov (1950); Vereshchagin (1959); Yavruyan (1970).
[c]Semi-desert type.

Table 20.3. *Ovis aries.* Life forms/plant parts eaten (x indicates the type used)

Life forms	Plant parts eaten							
	Whole plant	Leaves	Stems/twigs	Bark	Roots/bulbs	Flowers	Fruits	Seeds
Trees		x	x					
Shrubs		x	x					
Herbs		x	x					
Grasses/sedges		x	x					

20.5 Social Organization and Behavior

20.5.1 Social Life

During rut, herds are composed of up to 12–15 individuals (Yavruyan 1969). In the past, 200–300 animals were observed in each herd (Dinnik 1910). In the Crimean Peninsula, 4–14 animals (maximum 47) were observed (Dulitsky 1992).

20.5.2 Dispersal/Migration

In the past, when the border fences of the Soviet Union were absent, there were regular migrations from Armenia and Nachichevan (where winters are severe) to Iran, along traditional migratory routes (Sarkisov 1950; Yavruayan 1969).

20.5.3 Diurnal Activity

During summer, they are active in the morning, evening and early night.

20.5.4 Shelters

In bad weather, they prefer cavities in mountains, and stay under roof-like cliffs; on hot days, they use windy places on slopes and move to glaciers in order to escape blood-sucking insects (Yavruyan 1969).

20.6 Reproduction

Pregnancy lasts for 5 months. A female stays with her newborn in the same place for 6–9 days, after which they return to the herd. Lambs suck for about 5 months.

20.6.1 Seasonal Breeding Events

Rut is in the second part of November–the first part of December; calving is in late April–the first half of May (Dinnik 1910; Yavruyan 1969).

Maximum Length of Life. 13–15 years (Tsalkin 1951; Yavruyan 1969).

20.7 Mortality

20.7.1 Predators

There are few predators in the habitats of *Ovis aries* and they are of minor importance (Sokolov and Tembotov 1993).

Table 20.4. *Ovis aries.* Male age composition of the Crimean Peninsula population. (Dulitsky 1992)

Age (years)	1	2	3	4	5	6	7
Proportion of males (%)	11	12	15	13	12	12	25

20.7.2 Diseases, Helminths and Ectoparasites (Appendix 2, Table 8)

20.8 Population Structure and Densities (Table 20.4)

In 1990, in the Krymsky Nature Reserve, the population consisted of 18% males, 49% females and 33% lambs (Parshintsev 1992).

20.8.1 Population Density and Size

In Armenia, about 350–400 mountain sheep were estimated to be present in 1975 (Ayrumyan and Gasparyan 1976; Sokolov and Tembotov 1993), in Nakhichevan, 1,000–1,200 (Sokolov and Tembotov 1993). European mouflons were brought into the Crimea in 1913 (10 animals); by spring 1941, the population had reached 490 animals; in 1945, there were 35 animals (strong decline because of the war); in 1955, 254 animals; in 1983, 300 animals; in 1990, 296 animals (Parshintsev 1992); in 1997, 300 animals (Fedosenko 1997).

20.9 Relations to Man

Population decline is connected with hunting and competition from sheep farming (Yavruyan 1970).

21 Snow Sheep – *Ovis nivicola*

Ovis nivicola Eschscholtz, 1829.
Artiodactyla, Bovidae, Caprinae

21.1 Status and Distribution (Fig. 21.1)

21.1.1 Taxonomic Remarks

In Russia, *Ovis canadensis* was often used as a synonym for *Ovis nivicola* (Heptner 1989). Four subspecies are recognized: *O. n. nivicola* Eschscholtz, 1829 (Kamchatka Peninsula); *O. n. alleni* Matschie, 1907 (Okhotsk-Kolyma Upland); *O. n. lydekkeri* Kowarzik, 1913 (Yakutia, Okhotsk-Kolyma Upland, the Kodarsky Ridge); *O. n. borealis* Severtzov, 1873 (the Putoran Mountains).

21.1.2 Status

O. nivicola borealis is vulnerable (VU–D2), other subspecies are at lower risk – conservation-dependent (LR/cd) (Hilton-Taylor 2000). In the Russian Federation, *O. nivicola borealis* is undetermined, *O. n. lydekker* is rare (Rozhnov 2001). All subspecies are protected in the nature reserves and by hunting regulations (Revin et al. 1988; Zheleznov-Chukotsky 1994).

21.1.3 Distribution (Fig. 21.2)

Distribution Area. Ca. 3,760,000 km^2. Range of *O. nivicola borealis* occupies ca. 120,000 km^2 where only 12 localities with >100 animals were found (Larin and Sipko 1999).

Occurrence in Ecoregions. Tundra Regime Mountains (M121), Subarctic Regime Mountains (131, M132, M133, M134, M135).

Fig. 21.1. Snow sheep

21.2
Measurements (Table 21.1)

21.3
Habitat

Open terrain, or sites with sparse trees or shrubs and with rough relief. Steep rocky slopes are used for shelter. In winter, patches with thin snow cover are important. Females prefer habitats on the upper part of the slopes.

Terrain Form (Table 21.2).

Altitude (Table 21.3).

Temperature. Inhabits areas with a temperature range from –60 to +30 °C.

Snow. Snow >25–30 cm with a hardness of 0.26–0.30 g/cm^3, or >20–25 cm with a hardness of 0.24–0.26 g/cm^3 limits feeding (Zheleznov-Chukotsky 1994).

21.3.1
Biotope (Table 21.4)

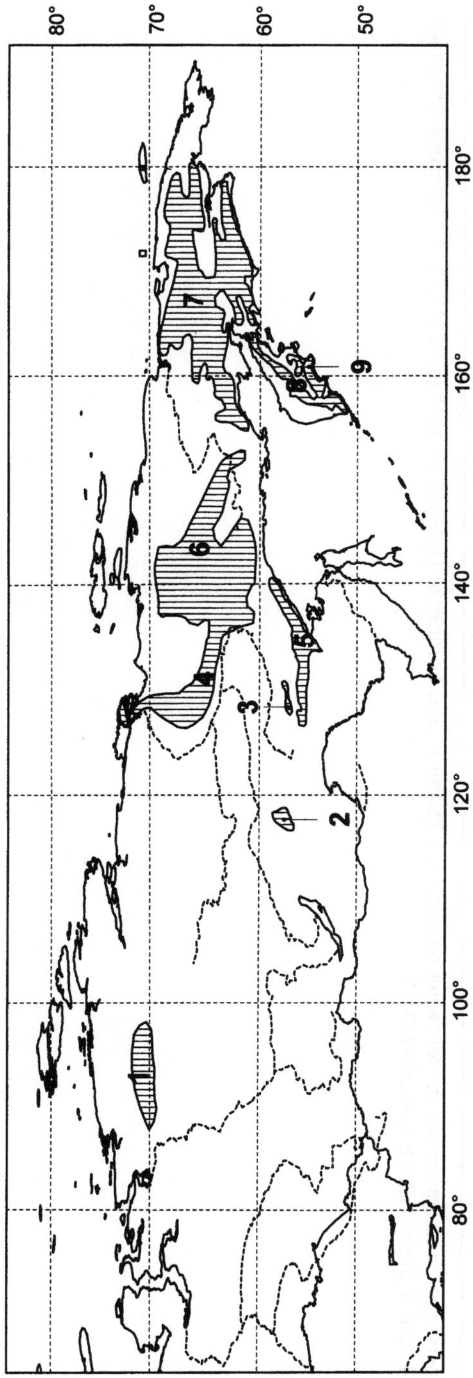

Fig. 21.2. *Dvis nivicola.* Distribution map (Revin et al. 1988). *1* Putoran Mountains; *2* Kodar Range; *3* Aldan Upland; *4* Verkhoyansk Range; *5* Stanovoy Range, Dzhugdzhur Range; *6* Chersksy Range; *7* Okhotsk-Kolyma Upland; Koryak Upland; *8* Kamchatka; *9* Kronotsky Nature Reserve

Table 21.1. *Ovis nivicola.* Measurements

Locality[a]	Longitude (°E)	Latitude (°N)	Gender	*n*	Body mass (kg)	Head and body length (cm)	Tail length (cm)
Putoran Mountains[1]	87–98	67–70	Male	2	107.8 (106–109)	152.3 (151–153)	6.8 (6.5–7)
Verkhoyansk Mountains[2]	127–137	63–69	Male	9	69.2 (56–81)	152.7 (136–161)	
			Female		48.0 (42–50)	136.4 (128–156)	
Okhotsk–Kolyma Upland[3]	170	65	Male	5	98.8 (86–105)	182.5 (178–188)	12.5 (12–13)
			Female	5	61.8 (45–68)	175.8 (169–179)	10.9 (9–13.2)
Koryak Mountains[4]	165–175	62–63	Male	8	73.1 (59.3–81)	150 (148–154)	10.3 (9.5–11.5)
			Female	9	41.5 (33–49)	133.9 (126–142)	9.6 (7.5–10)
Kronotsky Nature Reserve[5]	161	55	Male	1	99.6	178.0	
			Female	6	49.6 (45.5–62.4)	138.1 (135–145)	

[a]The superscript numbers denote the following references: 1, Michurin and Mironenko (1966); 2, Krivoshapkin (1995); 3, Zheleznov-Chukotsky (1994); 4, Chernyavsky (1962); 5, Averin (1948).

Table 21.2. *Ovis nivicola.* Use of terrain forms[a]

Terrain form	Locality[b]				
	Putoran Mountains[1]	Yakutia[2]	Okhotsk-Kolyma Upland[3]	Koryak Mountains[4]	Kamchatka[5]
(Ecoregion)	M131	M132, M134	M132	M132	M135
Seashores			+++	+	+++
Shores of fresh water		+++			++
River and lake terraces	x				
Flood plain		++			
Mountain					
Foothill	x	+	+		
Slope	x	+++	+++	+++	+++
Hill top	x	+++	++	++	+++

[a]Symbols indicate percent of habitat use was >30% (+++); 20–30% (++): 10–20% (+); <10% (r); x indicates habitat use noted, no numerical data reported.
[b]The superscript numbers denote the following references: 1, Fedosenko (1985); Larin (1990); 2, Egorov (1965); Krivoshapkin (1995); 3, Kishchinsky (1967); Zheleznov-Chukotsky (1994); 4, Chernyavsky (1962); Baskin (1968); 5, Averin (1948); Mosolov (1986).

Table 21.3. *Ovis nivicola.* Habitat use in different altitudes[a]

Altitude (m a.s.l.)	Locality[b]			
	Putoran Mountains[1]	Yakutia[2]	Okhotsk-Kolyma Upland[3]	Kamchatka[4]
(Ecoregion)	M131	M132, M134	M132	M135
0–50		+		+++
201–500		+++	+++	
501–1,000	+++[c]			+
1,001–2,000		+++	+++	+++
2,001–3,000		r		r

[a]Symbols indicate percent of habitat use was >30% (+++); 20–30% (++): 10–20% (+); <10% (r); x indicates habitat use noted, no numerical data reported.
[b]The superscript numbers denote the following references: 1, Fedosenko (1985); Larin (1990); 2, Egorov (1965); Krivoshapkin (1995); 3, Kishchinsky (1967); Zheleznov-Chukotsky (1994); 4, Chernyavsky (1962); Baskin (1968); 5, Averin (1948); Mosolov (1986).
[c]From 700 m.

Table 21.4. *Ovis nivicola*. Habitat[a]

Habitat	Locality[b]				
	Putoran Mountains[1]	Yakutia[2]	Okhotsk-Kolyma Upland[3]	Koryak Mountains[4]	Kamchatka[5]
(Ecoregion)	M131	M132, M134	M132	M132	M135
Sparse larch forest	x	x	x		
Birch forest					+
Shrubland (alpine)[c]		x	x	+	+
Grassland					
Flood plain		x			
Slope	x	x	x	+++	+++
Alpine			x	+	+++
Tundra (alpine)		x	x	r	+++
Steep shores of sea and fresh water with shrubs and grass		x	x	+	+++

[a]Symbols indicate percent of habitat use was >30% (+++); 20–30% (++): 10–20% (+); <10% (r); x indicates habitat use noted, no numerical data reported.
[b]The superscript numbers denote the following references: 1, Fedosenko (1985); Larin (1990); 2, Egorov (1965); Krivoshapkin (1995); 3, Kishchinsky (1967); Zheleznov-Chukotsky (1994); 4, Chernyavsky (1962); Baskin (1968); 5, Averin (1948); Mosolov (1986).
[c]Sparse shrubs of *Pinus pumila*.

21.4 Diet and Feeding Habits

21.4.1 Diet (Table 21.5)

Herbs, grass; in winter trees, shrubs, and lichens (Zheleznov 1980; Larin and Shelkovnikova 1985, 1988; Zheleznov-Chukotsky 1994). For a list of plant genera eaten see Appendix 1, Table 22.

21.4.2 Life Forms/Plant Parts Eaten (Table 21.6)

Feeding Behavior. Mode of feeding similar to goats; snow sheep pasture for a long time in the same place. Good diggers in the snow and ground.

Feeding Niche. In field and shrub layers.

Table 21.5. *Ovis nivicola.* Preferred seasonal food (percent of stomach content)

Life forms	Locality[a]							
	Putoran Mountains[1]		Yakutia[2]		Chukchi okrug (1), Okhotsk-Kolyma Upland (2), Koryak Mountains (3)[3]			Kamchatka[4]
	No snow	Snow	No snow	Snow	1	2	3	
Trees, shrubs	0–10	15–43	12–22	26	15	6	10	30
Twigs								
Larch twigs	0–2	0–10.0						
Leaves, needles			17–22	22	13	6	9	22
Flowers, fruits, seeds			4–8	14	2	33	7	6
Dwarf shrubs	3–25							
Herbs, grass, sedges	31–39		43–63	26	65	55	74	36
Cereals, sedges	30–50	15–55						
Roots[b]	x							x
Mosses	2	1–2	1	3	1			2
Lichens	1–2	2–58	1–2	6	4			4
Fungi			1–3	3	2			

[a]The superscript numbers denote the following references: 1, Larin and Shelkovnikova (1985, 1988); 2, Egorov (1965); 3, Zheleznov (1981); Zheleznov-Chukotsky (1994); 4, Averin (1948).
[b]*Oxytropis* (Fedosenko 1985).

Table 21.6. *Ovis nivicola.* Life forms/plant parts eaten[a] (x indicates the type used)

Life forms	Plant parts eaten							
	Whole plant	Leaves	Stems/ twigs	Bark	Roots/ bulbs	Flowers	Fruits	Seeds
Trees		x	x			x	x	x
Shrubs		x	x			x	x	
Dwarf shrubs		x	x					
Herbs		x	x			x		
Grasses/sedges		x	x		x			
Mosses	x							
Lichens	x							
Fungi	x							

[a]x indicates plant part used.

Table 21.7. *Ovis nivicola*. Group size and composition

Locality[a]	Group composition	Number of animals	
		Average	Maximum
Putoran Mountains[1]		2–7	
Yakutia[2]	Males	2–4.3	9
	Females and lambs	3.4–5.9	72
Okhotsk-Kolyma Upland[2]		4–7	
Koryak Mountains[2]	Males	4	8
	Females and lambs	3	7
Kamchatka[3]		2–10	18
	Males	2–12	
	Females with lambs	5–13	
	Mixed	9–29	
	Females in harem	4	

[a]The superscript numbers denote the following references: 1, Larin (1990); 2, Revin et al. (1988); Krivoshapkin et al. (1997); 3, Mosolov (1992).

21.5 Social Organization and Behavior

21.5.1 Social Life (Table 21.7)

Groups of females with newborns and yearlings, groups of young (2–5 year old) males, groups of adult (over 5-year-old) males, and single males over 11–12 years of age. In Kamchatka, of the observed groups, 12% consisted of single animals, 39% of 2–3 animals, 26% of 4–6 animals, 18% of 7–10 animals, and 5% of 11–12 animals (Gribkov 1969). During rut, harems consist of one male with a group of females and young animals. The male demonstrates territorial behavior, such as the marking of territory and guarding the group. The size of groups increases from summer to autumn (Chernyavsky 1967).

21.5.2 Dispersal/Migration

Daily routes are short, but on some days (especially in the case of rams during rut), they reach 15 km. Isolated, small populations are typical. During the rutting season, rams can visit neighboring groups 10–50 km away (Fedosenko 1985; Mosolov 1986). Seasonal migrations evoked by snow conditions are usual. In Yakutia, migrations from 50–60 up to 100–

120 km are known (Egorov 1965). However, in the Kronotsky Nature Reserve, they spend the entire year in the same area (Averin 1948).

21.5.3 Home Range

Use the same home range for several years and for many generations. The home range includes trails, areas of defecation, resting places, and shelters on rocks, which are usually used to escape predators (Chernyavsky 1967; Larin 1983; Mosolov 1992; Zheleznov-Chukotsky 1994).

21.5.4 Mode of Movement

Dig in snow for food. Move through rocks carefully and follow the footprints of the leader exactly.

21.5.5 Diurnal Activity

Polyphase diurnal activity; two thirds of their time is used for feeding and one third for resting. Females feed for 92–100 min per feeding period and sleep for 55–66 min per sleeping period; males feed and sleep for 20–104 and 195–206 min, respectively (Fedosenko 1985).

21.5.6 Shelters

Select comfortable places with a good view for resting. Scratch the ground before lying down, use the same resting places for many years (Zheleznov-Chukotsky 1994).

21.5.7 Escape Behavior

One third of the feeding time is used for looking around. A low level of alertness towards humans was found in areas where sheep are rarely disturbed by them; sheep sometimes stand for a long time at a short (30 m) distance from humans (Egorov 1965; Fedosenko 1985).

21.6 Reproduction (Tables 21.8 and 21.9)

Combat between males is observed before the rut. Observations have shown that male snow sheep are less aggressive during rut than males of

Table 21.8. *Ovis nivicola*. Breeding parameters

Locality[a]	Age at sexual maturity (years)		Participation in breeding (years)		Pregnancy (days)	Litter size
	Male	Female	Male	Female		
Putoran Mountains[1]				1.5–2.5	180	1
Yakutia[2]	1.5	1.5	3.5	1.5		1–2[b]
Chukchi okrug[3]		2.5				1, rare 2
Kamchatka[4]			6.5	2.5		1, rare 2

[a]The superscript numbers denote the following references: 1, Larin (1990); 2, Revin et al. (1988); 3, Zheleznov-Chukotsky (1994); 4, Fil (1978).
[b]In some areas about 17% of females have 2 lambs.

Table 21.9. *Ovis nivicola*. Age of males participating in breeding and frequency of mounted females in Putoran Mountains. (Larin 1990)

Age (years)	Participating males (%)	No. of mounted females (%)
5	25	13
6	21	14
7	16	27
8	18	30
9	11	10
10	6	3
11	3	3

other sheep species. Such relatively low levels of aggression may be a result of a need to conserve energy due to severe living conditions (Fil 1978; Fedosenko 1985). In Kamchatka, within the group, 4 (2–9) females were breeding. 1–5 males compete for the females. A male can copulate 3–9 times an hour (Fil 1978). Only 11% of 1.5-year-old females take part in breeding, the rest breed at 2.5 years of age (Larin 1990).

The females abandon the group 1.5–2 days before lambing, which usually takes place among rocks in the upper part of mountains. The lambs can stand up after 20–30 min; follow the mother after 1 h; eat green plants after 7–8 days (Larin 1989).

21.6.1 Seasonal Breeding Events (Table 21.10)

Maximum Length of Life. 18 years (Egorov 1965). Tsalkin (1951), Chernyavsky (1962), and Egorov (1965) observed that the teeth of

Table 21.10. *Ovis nivicola.* Seasonal breeding events

Locality[a]	January		Feb	March	April		May		June		July	Aug	Sept	Oct	Nov		Dec	
Putoran Mountains[1]								N	N							R	R	
Yakutia[2]	R								N	N						R	R	R
Kamchatka[3]						N	N	N	N						R	R[b]	R[b]	R

R, Rutting season; N, newborns.
[a]The superscript numbers denote the following references: 1, Larin (1990); 2, Revin et al. (1988); 3, Averin (1951).
[b]Highest rutting activity.

Table 21.11. *Ovis nivicola*. Gender and age composition of populations (%)

<table>
<tr><th>Locality[a]</th><th>Males (2–5 years)</th><th>Males (≥5 years)</th><th>Females (adult)</th><th>Yearlings</th><th>Lambs</th></tr>
<tr><td>Putoran Mountains[1]</td><td colspan="2">25</td><td>35</td><td>24</td><td>16</td></tr>
<tr><td>Putoran Mountains[2]</td><td colspan="2">27</td><td>31</td><td>15</td><td>27</td></tr>
<tr><td>Yakutia[3]</td><td>11</td><td>12</td><td>35</td><td>17</td><td>25</td></tr>
<tr><td>Yakutia[4]</td><td colspan="2">25–34</td><td>29–35</td><td>11–15</td><td>22–29</td></tr>
<tr><td>Koryak Mountains[5]</td><td>13</td><td>22</td><td>35</td><td>11</td><td>19</td></tr>
<tr><td>Okhotsk - Kolyma Upland[6]</td><td>1–7</td><td>34–65</td><td>15–37</td><td>7–11</td><td>12–19</td></tr>
<tr><td>Kamchatka[7]</td><td>4–12</td><td>17–31</td><td>34–55</td><td>3–11</td><td>7–16</td></tr>
</table>

[a]The superscript numbers denote the following references: 1, Borzhonov et al. (1979); 2, Larin (1990); 3, Egorov (1965); 4, Krivoshapkin et al. (1997); 5, Chernyavsky et al. (1990); 6, Zheleznov-Chukotsky (1994); 7, Mosolov (1992).

snow sheep wear down much slower with age than in other species of wild sheep.

21.7 Mortality

Mortality of newborns reaches 38–59%, especially in low mountains (wolf predation) (Krivoshapkin et al. 1997). Mortality of lambs belonging to young females reaches 60%, the mortality of lambs of mature females is 25% (Larin 1990). In Yakutia, mortality of lambs during the first year of life was found to be 30% (Egorov 1971), in Kamchatka (Averin 1948) and in the Koryak Mountains, (Chernyavsky 1962) it was 50%.

21.7.1 Predators

Predators do not play a significant role because of the good shelter conditions for sheep (Egorov 1965; Borzhonov et al. 1979). In the Putoran Mountains, remnants of snow sheep were found in 62% of wolf scats in summer and 4.8% in winter (Larin 1989). One third of the sheep killed by wolves were lambs, other sex and age groups were killed in equal proportion. Other predators include wolverine, brown bear, lynx, golden eagle and fox (Zheleznov 1983; Larin and Krashevsky 1989b; Zheleznov-Chukotsky 1994).

Table 21.12. *Ovis nivicola.* Population density (in typical habitats) and size

Locality[a]	Year[a]	Population density (animals/10 km^2)	Population size (×1,000)
Putoran Mountains[1]	1979	1–6	1.45
Putoran Mountains[2]	1990	0.05–5	3.6
Putoran Mountains[2]	1996		5.5
Verkhoyansk Mountains[3]	1988	1–1.5	29–32
Verkhoyansk Mountains[4]	1995	1.5–2.3, max. 20	21–24
Kodarsky[5]	1988		0.2
Aldan Upland[3]	1988	0.3	2.4–2.85
Chersky[3]	1988	0.5–1	24.5–26.5
Okhotsk–Kolyma Upland[6]	1990	0.4–9	1.5–2
Chukchi okrug[6]	1994		2.85
Koryak Mountains[6]	1988	0.5	(6–7)
Kamchatka oblast[7]	1969	3–12	9.1
Kamchatka oblast[3]	1988		12–13
Kamchatka oblast[8]	1999		3.1
Russian Federation[8]	1999		49.7

[a]The superscript numbers denote the following references: 1, Borzhonov et al. (1979); 2, Larin (1990); Larin and Sipko (1999); 3, Revin et al. (1988); 4, Krivoshapkin (1995); 5, Sopin and Ermolin (1988); 6, Zheleznov-Chukotsky (1994); 7, Gribkov (1969); 8, Sidorov (official data) (2000).

21.7.2
Diseases, Helminths and Ectoparasites (Appendix 2, Table 8)

No diseases have been reported (Revin et al. 1988). Parasites: Appendix 2, Table 8.

21.8
Population Structure and Densities (Table 21.11)

The ratio of males to females of adult sheep in the Putoran Mountains was found to be from 1 : 0.88 to 1 : 1.11 (Larin 1990), in Yakutia the male to female ratio was from 1 : 0.87 to 1 : 1.44 (Krivoshapkin et al. 1997).

21.8.1
Population Density and Size (Table 21.12)

Zheleznov-Chukotsky (1994) estimated the total number of snow sheep in N Asia to be 55,000.

21.9
Relations to Man

Important hunting animal. In Yakutia, in the 1970s, 5,000 were hunted annually. Hunted for meat, fur (for traditional local cloth), which is regarded as warmer than reindeer fur, and for horns as trophies.

22 Urial – *Ovis vignei*

Ovis vignei Blyth, 1841.
Artiodactyla, Bovidae, Caprinae

22.1 Status and Distribution (Fig. 22.1)

22.1.1 Taxonomic Remarks

Ovis vignei includes populations considered earlier as *O. vignei arkal, O. vignei bocharensis, O. vignei varentsovi* and *O. vignei* (Tsalkin 1951; Nadler et al. 1973). Three subspecies occur: *O. v. vignei* Blyth, 1840 (SW Pamirs Mountains); *O. v. bocharensis* Nasonov, 1914 (W Pamirs, Uzbekistan); *O. v. cycloceros* Hutton, 1842 (Badkhyz, Kopet-Dag Range, Bolshoy Balkhan Range, Ustyurt Plateau) (Grubb 1993).

22.1.2 Status

O. v. vignei (SW Pamir Mountains) CITES I; *O. v. bocharensis* (W Pamirs, Uzbekistan) and *O. v. cycloceros* (Badkhyz, Kopet-Dag Range, Bolshoy Balkhan Range, Ustyurt Plateau) included in CITES II (Hilton-Taylor 2000). All species are protected in nature reserves and by hunting regulations.

22.1.3 Distribution (Fig. 22.2)

Distribution Area. Ca. 183,000 km^2.

Occurrence in Ecoregions. Tropical/Subtropical Steppe Regime Mountains (M313), Tropical/Subtropical Desert Regime Mountains (M321), Temperate Desert (343).

22.2 Measurements (Table 22.1, Fig. 22.3)

Fig. 22.1. Urial

22.3 Habitat

Semi-desert and desert grasslands with stony and gypsum-covered ground, savannas on mountain slopes and plateaus, with rugged terrain (canyons, rifts, and cliffs) for shelter.

Terrain form. Seashores; shores of fresh water; hills; plateaus; foothills, valleys, slopes and tops of mountains (Baskin 1976; Sapozhnikov 1976; Savinov and Bekenov 1983).

Altitude. 0–400 m a.s.l. (Ustyurt); 1,000–2,000 m a.s.l. (Kopetdag), 400–800 m a.s.l. (Badkhyz), 400–3,000 m a.s.l. (Tajikistan) (Sapozhnikov 1976; Savinov and Bekenov 1983; Korshunov 1988b).

Snow. Inhabits areas with snow <30 cm, usually <15 cm. Use snow as water source.

Temperature. Inhabits areas with temperatures ranging from –30 °C (in winter in the mountains) to +45 °C (in shade; summer, deserts).

Water. All parts of the range are dry or extremely dry. The animals need water during summer and regularly visit watering places (Gorelov 1964; Gorbunov 1986). In desert mountains they use three sources of water: small streams, seepage of groundwater, and puddles with rainwater. Drink fresh water and can use salt water with a salt content up to 20 g/l, during dry years (Gorbunov 1986).

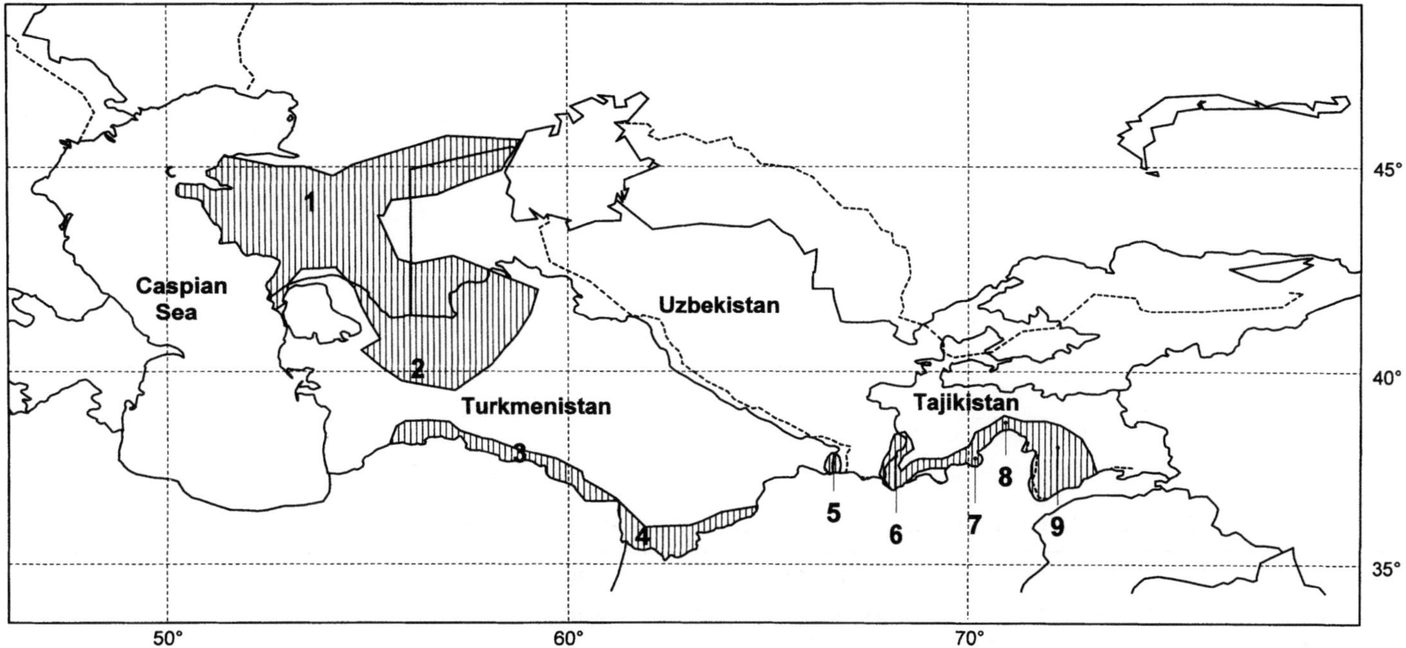

Fig. 22.2. *Ovis vignei.* Distribution map (Ishunin 1961; Ishadov 1972; Sapozhnikov 1976; Heptner 1989). *1* Ustyurt; *2* Bolshoy Balkhan Range; *3* Kopet-Dag Range; *4* Badkhyz Nature Reserve; *5* Kugitangtau Range; *6* Babatag Range; *7* Karatau Range; *8* Khazratishokh Range; *9* Pamir Plateau

Table 22.1. *Ovis vignei.* Measurements

Locality[a]	Longitude (°E)	Latitude (°N)	Gender	*n*	Body mass (kg)	Head and body length (cm)	Tail length (cm)
Ustyurt Plateau[1]	52–56	42–45	Male	5	55.0 (46.4–68.3)	146.4 (112–160)	12.4 (11–17)
	52–56	42–45	Female	8	39.2 (36.8–41.2)	125.5 (121–138)	12.2 (10–15)
Kopet-Dag[2]	57	38	Male	18	62.1 (48–79)	140.2 (132–147)	11.2 (11–12)
			Female	4	36.2 (30–41)	126.6 (121–133)	11.4 (10–12)
			Female	2	34.9 (34.5–35.2)		

[a]The superscript numbers denote the following references: 1, Savinov and Bekenov (1983); 2, Tsalkin (1951); Gritsuk (1975).

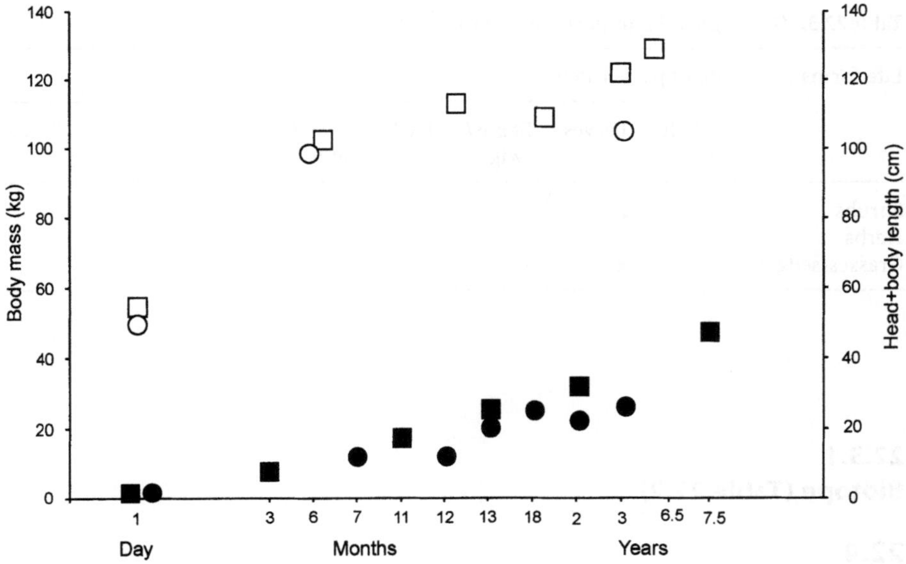

Fig. 22.3. *Ovis vignei.* Body mass and head and body length versus age (Pamirs) (Sapozhnikov 1976). *1* Male mass; *2* male head + body length; *3* female mass; *4* female head + body length

Table 22.2. *Ovis vignei.* Habitat[a]

Habitat type	Locality[b]			
	Ustyurt Plateau[1]	Kopet-Dag[2]	Badhyz[3]	Tajikistan[4]
(Ecoregion)	343	M321	M321	
Savanna[c]			x	x
Shrubland (sparse)[d]		x		x
Grassland (desert type)				
Foothill	x		x	x
Valleys (canyons)	x	x	x	x
Slopes		x		x
Plateau	x			x
Alpine		x		

[a]x indicates habitat use noted, no numerical data reported.
[b]The superscript numbers denote the following references: 1, Savinov and Bekenov (1983); 2, Baskin (1976); Korshunov (1988b); 3, Baskin (1976); 4, Sapozhnikov (1976).
[c]Sparse *Pistacia vera* on slopes, *Amygdalus bucharica* on canyon slopes.
[d]Sparse *Juniperus sabina.*

Table 22.3. *Ovis vignei.* Plant parts eaten (x indicates plant part used)

Life forms	Plant parts eaten							
	Whole plant	Leaves	Stems/ twigs	Bark	Roots/ bulbs	Flowers	Fruits	Seeds
Shrubs		x	x			x		
Herbs		x	x			x		
Grasses/sedges		x	x					

22.3.1 Biotope (Table 22.2)

22.4 Diet and Feeding Habits

22.4.1 Diet

Stems, twigs and leaves of shrubs, herbs and grasses. For a list of plant genera eaten see Appendix 1, Table 23.

22.4.2 Life Forms/Plant Parts Eaten (Table 22.3)

Feeding niche. In field and shrub layers, and rarely below ground. Dig in the snow and the ground to find food.

22.5 Social Organization and Behavior

22.5.1 Social life (Table 22.4)

From spring into the summer, there are families (a female and her lambs), groups of 2–3 personally connected adult females with their young, male groups, herds (groups with coordinated behavior, such as following after the leader), and gatherings (maximum 350 animals) in areas with plenty of food. Mixed herds (males and females of all ages) exist from October. Leaders direct the movement through pastures and during escape (Baskin 1976; Fedosenko 1984a).

Table 22.4. *Ovis vignei.* Group size in different seasons

Locality[a]	Season	No. of animals per group (observed groups during season, %)					Average group size
		1	2–5	6–10	11–15	<15	
Ustyurt Plateau[1]	Winter–spring	16	58	16	4	6	2.6
	Lambing	13	52	24	10	1	4.2 (2–6)
	Watering	22	52	12		14	5.4 (3–13)
	Rut	20	55	17		8	13.6
Badkhyz[2]	Winter–spring		42	37		21	13.7 (4–30)

[a]The superscript numbers denote the following references: 1, Savinov and Bekenov (1983); Fedosenko (1984a); 2, Baskin (1976).

22.5.2 Home Range

Within the home range there are permanent trails and distinct places for rest and shelter (in caves, under rocks). A sheep group has 2–3 heavily used areas, and animals move between them when disturbed (Baskin 1976).

22.5.3 Dispersal/Migration

During lambing, the mothers stay with their lambs within a 1–10 ha area. They need daily watering. Later, sheep are able to move 30–70 km, but during hot periods they usually keep within 20 km of water (Mambetzhumaev 1968; Gorbunov 1986).

22.5.4 Diurnal activity (Table 22.5)

Mostly active during the day. During the night, the sheep are pasturing, but they stay within the same area and do not try and escape when disturbed (Baskin 1976).

22.5.5 Shelters

Use cavities in mountains; roof-like rocks give shelter from the sun, wind and snowfall and are also used during calving. Lambs spend most of their first month in these shelters (Sapozhnikov 1976; Savinov and Bekenov 1983).

Table 22.5. *Ovis vignei*. Diurnal activity[a]

<table>
<tr><th rowspan="2">Locality[b]</th><th rowspan="2">Season</th><th colspan="6">Time of day (hours)</th></tr>
<tr><th>6–9</th><th>9–12</th><th>12–15</th><th>15–18</th><th>18–21</th><th>21–22.30</th></tr>
<tr><td>Ustyurt Plateau[1]</td><td>Spring</td><td colspan="2">x</td><td>Rest</td><td colspan="3">x</td></tr>
<tr><td></td><td>Summer</td><td>x</td><td colspan="3">Rest</td><td colspan="2">x</td></tr>
<tr><td></td><td>Autumn</td><td colspan="2">x</td><td>Rest</td><td colspan="3">x</td></tr>
<tr><td></td><td>Winter</td><td colspan="2">x</td><td>Rest</td><td colspan="3">x</td></tr>
<tr><td>Tadjikistan[2]</td><td>Spring</td><td colspan="2">56</td><td>46</td><td>Rest</td><td colspan="2">100</td></tr>
<tr><td></td><td>Summer</td><td colspan="2">79</td><td colspan="2">Rest</td><td colspan="2">100</td></tr>
</table>

[a]x, animals were observed feeding; number indicates percent of feeding animals, observations made during day light hours.
[b]The superscript numbers denote the following references: 1, Savinov and Bekenov (1983); 2, Sapozhnikov (1976).

Table 22.6. *Ovis vignei*. Breeding parameters. (Ishadov 1965; Bogolyubsky et al. 1978)

Locality	Age at sexual maturity (years)		Participation in breeding (years)		Pregnancy (days)	Litter size	Newborns per female (all females, %)		
	Male	Female	Male	Female			1	2	3
Ustyurt Plateau					160				
Kopet-Dag	2.5	1.5	4.5–5.5			1–2	40	52	8

22.6 Reproduction (Table 22.6)

In the first phase of the rutting season, males follow each other. Behavioral ceremonies establish hierarchy; battles are rare, but cruel. In the second phase of the rut, males join female groups and mate. Some males are present in the mixed groups, but mainly males, 8 years and older, copulate. Births take place in the lower part of the mountains. The mother and her newborn spend the first week in the same place and the newborn hides. After 1 week the lamb can follow the mother.

22.6.1 Seasonal Breeding Events

In Tajikistan, females lactate for only 2 months (Sapozhnikov 1976) (Table 22.7).

Table 22.7. *Ovis vignei*. Seasonal breeding events[a]

Locality[b]	Jan	Febr	Mar		Apr		May		June	July	Aug	Sept	Oct		Nov		Dec	
Ustyurt Plateau[1]				N	N			N						R	R	R	R	R
Kopet-Dag [2]					N	N										R	R	
Tajikistan [3]						N	N									R	R	

[a]R, Rut; N, newborns.
[a]The superscript numbers denote the following references: 1, Ishunin et al. (1981); Savinov and Bekenov (1983); 2, Baskin (1976); 3, Sapozhnikov (1976).

Table 22.8. *Ovis vignei.* Gender and age composition in populations (%)

Locality[a]	Age of males (years)					Age of females (years)		
	1.5	2.5–3	3.5–5	5.5–8	≥8	1.5	Adult	Lambs
Ustyurt Plateau[1,b]		28					50	22
Kopet-Dag[2]		20–21					58–59	19–22
Kopet-Dag[3]	4.8	9.5	16.3	4.8	1.9	0.9	44	17.8
Badkhyz Nature Reserve[3,c]	4.6	7.6	5.6	6.3	3.9	5.1	44.4	22.5
Badkhyz Nature Reserve[4,c]	7	24				7	34	28

[a]The superscript numbers denote the following references: 1, Savinov and Bekenov (1983); 2, Korshunov (1988a); 3, Fedosenko (2000); 4, Zhestkova (1999).
[b]In autumn.
[c]After lambing.

Table 22.9. *Ovis vignei.* Population density and size in typical habitats

Subspecies	Locality[a]	Year	Population density (animals/10 km²)	Population size (×1,000)
O.v. cycloceros	Ustyurt Plateau (total)[1]	1997		6.0
	Ustyurt Plateau (Kazakhstan)[2]	1960		7–10
		1980	0.5–80	1.7–1.8
		1998		5.5–6.5
	Ustyurt Plateau (Turkmenistan)[3]	1960		5–7
		1998		1.1–11
	Ustyurt Plateau (Karakalpakstan)[4]	1997		0.3
	C Kopet-Dag[5]	1995	1.5–31.9	3.5
	Badkhyz Nature Reserve[5]	1998		1.2
	Turkmenistan (total)[6]	1975		8–10
		1990		10–12
		2001		2.5–2.8
O. v. bocharensis	Turkmenistan[7]			
	Kugitangtau Range	1998		0.1
	Uzbekistan (total)[7]	1988		0.05
	Babatag Range	1993		0.005
	Tajikistan (total)[7]	1978		1.1–1.4
		1988		0.8–1.0
	SW Tajikistan	1961	4.8–20.7	2.6
	Surkhukh Range	1999		0.07–0.27
	Khazratishokh Range	1965	6.2–10.4	
		1999		0.5–0.6
	Karatau Range	1961	20.7	0.1
		1965	4.8	0.25
		1999		0.25–0.68

[a]The superscript numbers denote the following references: 1, Fedosenko (1997); 2, Savinov and Bekenov (1983); Plakhov (1999); 3, Fedosenko (1997); 4, Reimov and Reimov (1997); 5, Fedosenko (2002); 6, Lukarevsky (1999); Fedosenko (2002); 7, Sapozhnikov (1976); Tyshkevich (1999a); Fedosenko (2002).

Maximum Length of Life. The average for males is 6.2 years, maximum 11 years (Korshunov 1988a).

22.7 Mortality

In Kopet-Dag, 1.65 calves per female were counted just after lambing, 0.42/female in November, and 0.35/female by the next April (Ishadov 1965; Korshunov 1988a).

22.7.1 Predators

Wolf (Savinov and Bekenov 1983), leopard (Korshunov 1985), golden eagle (*Aquila chrysaetos*) and lammergeyer (*Gypaetus barbatus*) (Mambetzhumaev 1968; Sapozhnikov 1976). In Kopet-Dag, 17% of leopard scats included urial remnants (Lukarevsky 1995).

22.7.2 Diseases, Helminths and Ectoparasites (Appendix 2, Tables 8 and 9)

22.8 Population Structure and Densities (Table 22.8)

The embryo male to female ratios were 1.14:1 in Ustyurt (Savinov and Bekenov 1983), and 1:2.9 in Kopet-Dag (Korshunov 1988b).

22.8.1 Population Density and Size (Table 22.9)

22.9 Relations to Man

In the past, the urial was an important hunting species, with valuable meat and horns as trophies. Because of human persecution, the total number strongly declined. However, because of a large range, good natural conditions to escape pursuit, sparse human population, very vigilant behavior of leaders, great alertness and clever use of home range, sheep can survive even near large towns (Baskin 1976).

23 Chamois – *Rupicapra rupicapra*

Rupicapra rupicapra Linnaeus, 1758.
Artiodactyla, Bovidae, Caprinae

23.1 Status and Distribution (Fig. 23.1)

23.1.1 Taxonomic Remarks

One subspecies, *R. r. caucasica* Lydekker, 1910 (Caucasus) is found (Heptner 1989).

23.1.2 Status

Vulnerable (VU–C1) (Hilton-Taylor 2000). Only numerous in nature reserves. Protected in reserves and by hunting regulations (Sokolov and Tembotov 1993).

23.1.3 Distribution (Fig. 23.2)

Distribution Area. Ca. 81,000 km^2.

Occurrence in Ecoregions. Prairie Regime Mountains (M252), Mediterranean Regime Mountains (M262).

23.2 Measurements

In Caucasus, the body mass of males is 30–50 kg, the body mass of females is 25–42 kg, head and body length of males is 125–135 cm, while for females it is ca. 96 cm (Dinnik 1914; Sokolov 1979).

Fig. 23.1. Chamois

23.3 Habitat

Rocky mountain forests with openings, cliffs at the forest edge in mountains (Dinnik 1910; Dormidontov and Blokhin 1977; Sokolov and Tembotov 1993).

Terrain Form. Mountain slopes.

Altitude (Table 23.1). Vertical migrations occur between 1,500–2,000 m a.s.l. in W Caucasus and between 300–1,500 in E Caucasus; between 180–2,850 m a.s.l. in Zakataly Nature Reserve; between 180–1,600 (in winter) and 2,500–2,850 m a.s.l. (in summer) in Ismailinsky Nature Reserve (Duben 1985; Kuliev 1989).

23.3.1 Biotope (Table 23.2)

23.4 Feeding

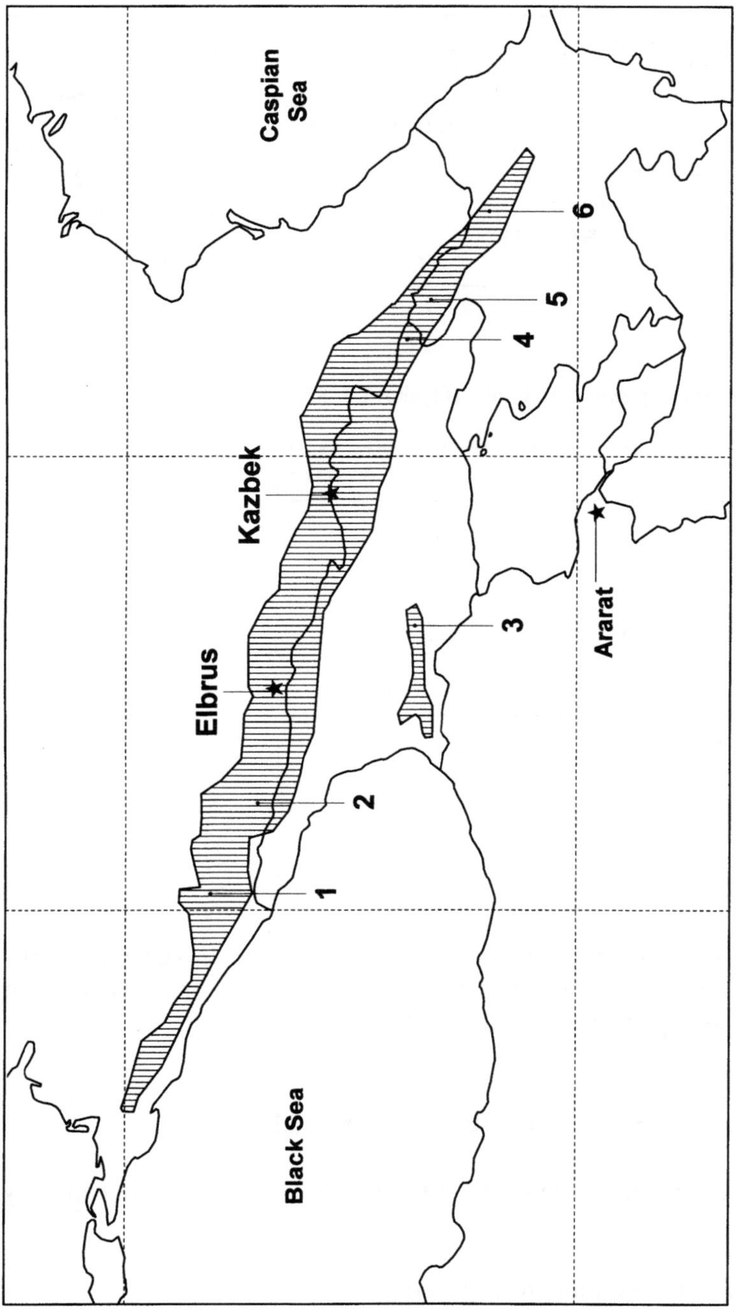

Fig. 23.2. *Rupicapra rupicapra.* Distribution map (Gineev et al. 1988; Arabuli 1989; Kuliev 1989; Sokolov and Tembotov 1993). *1* Kavkazsky Nature Reserve; *2* Teberdinsky Nature Reserve; *3* Borzhomi Nature Reserve, Georgia; *4* Lagodekhi Nature Reserve, *5* Zakataly Nature Reserve; *6* Ismailinskii Nature Reserve

Table 23.1. *Rupicapra rupicapra.* Seasonal distribution of animals (%) in different altitudes. (Duben 1985)

Altitude (m)	Vegetation zone	Season			
		December–February	March–May	June–August	September–November
300–1,600	Forest	73–91	39–91	18–32	20–27
1,600–1,900	Upper zone of forest with meadows, clearings	7–21	5–53	21–52	34–42
>1,900	Alpine zone	3–6	4–10	16–61	31–46

Table 23.2. *Rupicapra rupicapra.* Habitat.[a] (Inyakova 1957a; Duben 1985)

Habitat type	Kavkazsky
(Ecoregion)	
Mature forest	
Coniferous (sparse)	x
Broadleaf curved	x
Grassland	
Slopes with rocks	x
Alpine	x

[a]x indicates habitat use noted.

23.4.1 Diet

In winter, twigs of shrubs, dry grass and tree lichens; in summer, herbs and grass. Require salt, and frequently visit salt licks. For a list of plant genera eaten, see Appendix 1, Table 24 (Nasimovich 1949b).

23.4.2 Life Forms/Plant Parts Eaten (Tables 23.3 and 23.4)

Consumption of Food Per Day. 5.5–6 kg of fresh green plants in captivity (Duben 1985).

Feeding Niche. In field and shrub layers.

23.5 Social Organization and Behavior

Table 23.3. *Rupicapra rupicapra.* Life forms/plant parts eaten[a]

Life forms	Plant parts eaten							
	Whole plant	Leaves	Stems/ twigs	Bark	Roots/ bulbs	Flowers	Fruits	Seeds
Trees			+++					
Shrubs			+++					
Herbs		+++	+			+++		
Grasses/sedges		+++	+++					
Lichens	+							

[a]Symbols indicate frequency of use was >30% (+++); 20–30% (++): 10–20% (+); <10% (r).

Table 23.4. *Rupicapra rupicapra.* Seasonal utilization of different plant life forms.[d] (Nasimovich 1949b; Zharkov 1949; Popkova 1967)

Life forms	Winter	Spring	Summer
Trees, shrubs	+++	+++	r
Herbs			+++
Grass, sedges	++		++
Lichens	+		

[a]Symbols indicate percent of plant use was >30% (+++); 20–30% (++): 10–20% (+); <10% (r).

23.5.1 Social life

From April to October, adult males live alone, and groups consist of females, young animals and kids. Group composition is variable; after a disturbance new group members may join. During rut, harems are composed of one male and 1–4 females plus yearlings and kids; bachelor males form separate groups. After lambing the female lives with her kid for 1–2 months. Average group size in winter is 5.7, in spring 4.6, in summer 5.9–12.1, in autumn 5.9 (Nasimovich 1949b; Popkova 1967; Duben 1985).

23.5.2 Home Range

Closely attached to its home range. Moving <4–5 km per day. When heavily hunted, chamois stay in their home range, but change activity from day to night. The borders of home ranges are stable and new occupants adopt the same home range borders (Nasimovich 1949b; Duben 1985).

23.5.3
Mode of movement

Escape danger by rushing downwards on vertical slopes.

23.5.4
Dispersal/Migration

Diurnal movements of 4–5 km (maximum 20–25) in W Caucasus, 2–5 km (maximum 10–15) in E Caucasus. Seasonal shifts are short, depending on snow conditions. In winter, chamois concentrate on steep slopes (Nasimovich 1949b; Popkova 1967). For details of altitudinal migrations, see section 23.3.

23.5.5
Diurnal activity (Table 23.5)

Mostly chamois are active during the day, but in summer many animals also feed during the first part of night (Nasimovich 1949b).

23.5.6
Shelters

In summer, chamois prefer to stand in the shade of trees, lie on snow; use caverns, even if they have to crawl to reach them (Nasimovich 1949b).

23.5.7
Escape Behavior

Very alert. Good vision, hearing, and smell. After a disturbance, groups scatter in different directions. Animals raised in captivity become aggressive towards man (Inyakova 1957a,b).

Table 23.5. *Rupicapra rupicapra.* Rhythm of feeding and rest (percent of observed animals)

Time of day[a] (h)	5–11	11–17	17–20
Feeding[1]	41	37	43
Rest on snow fields[1]	19	26	3
Feeding[2]	90	34–40	

[a]The superscript numbers denote the following references: 1, Kuznetsov (1969); 2, Nasimovich (1949b).

23.6 Reproduction

The complicated behavior of males during rut includes marking, by visible and olfactory marks, as well as tending females. Combat is rare and mostly exchanged by ritual threatening or submissive poses.

Breeding parameters: age at maturity of males is 1.5 years, females 1.5 years; the first participation in breeding of males and females is 2.5 years; length of pregnancy is 160–170 days; the litter is usually composed of 1, sometimes 2 kids.

23.6.1 Seasonal Breeding Events (Table 23.6)

23.7 Mortality

Predation, poachers, and avalanches are the main mortality factors. The newborns in the population decline from summer to winter by up to 6% as a result of wolf and lynx predation.

Table 23.6. *Rupicapra rupicapra.* Seasonal breeding events

Locality[a]	Rutting period		Lambing		
	Establishing harems	Main period of sexual activity	First kids	Most kids	Last kids
Kavkazsky[1]	October	Early November–early December	Late April	May–early June	Up to mid-June
Zakataly region[2]		20 October–late November	25 April		

[a]The superscript numbers denote the following references: 1, Dinnik (1910); Sokolov and Tembotov (1993); 2, Popkova (1967).

Table 23.7. *Rupicapra rupicapra.* Age composition of populations (%)

Locality[a]	Years	Adult	Kids	Yearling
Kavkazsky[1]	1955–59	72–85	15–28	
Kavkazsky[2]	1970–81	44–56	13–29	7–14
Zakataly region[3]	1953–58		14–22	

[a]The superscript numbers denote the following references: 1, Kotov and Ryabov (1963); 2, Duben (1985); 3, Popkova (1967).

Table 23.8. *Rupicapra rupicapra.* Population density in typical habitats and size

Locality[a]	Year	Population density (animals/10 km^2)	Population size (×1,000)
N Caucasus[3]	1989	(1.2–10.5)	(3.4–3.7)
W Caucasus[1]			
Total	1890		40–50
	1905		15–20
	1924		<1
	1936		4.4
	1939		12
	1957		4.5–5.5
	1968		4.5–5
	1978		4.5–5
	1981		5.5
	1983		13–15
Kavkazsky Nature Reserve[1,6]	1957		3–4
	1960		5
	1968		5.5–6
	1972		5.5–6
	1978	3.5–131.5	4.5–5
	1981		5
	1992		2.0
	1999		2.4
Teberdinsky Nature Reserve[2]	1952		0.4
	1962		0.82
	1967		0.7
	1972		0.74
	1993	3	
	1998		0.06
The main Caucasian Ridge between Elbrus and Kazbek peaks[6]	1999		1
The main Caucasian Ridge east from Kazbek (including Dagestan)[6]	1999		0.3
Dagestan[7]	1998	0.7	0.35
Georgia[4]	1989		5
Lagodekhi Nature Reserve[5]	1952		Many
	1981		0.3
Azerbaijan[5]	1972		0.7–0.8

[a]The superscript numbers denote the following references: 1, Dinnik (1910); Zharkov (1940); Nasimovich (1941); Kotov (1960); Kotov and Ryabov (1963); Duben (1985); Trepet (1997); 2, Tarasov (1975); Bobyr (1999); 3, Gineev et al. (1988); 4, Arabuli (1989); 5, Alekperov (1966); 6, Kudaktin (pers. comm.); 7, Magomedov and Akhmedov (1999).

23.7.1 Predators

Chamois remnants were found in 12% of wolf scats and 18% of lynx scats (Teplov 1938; Kotov 1960).

23.7.2 Diseases, Helminths and Ectoparasites (Appendix 2, Table 8)

Foot-and-mouth disease, anthrax, brucellosis, necrobiosis (hoof disease), tuberculosis, kerato-conjuctivitis, coccidiosis.

23.8 Population Structure and Densities (Table 23.7)

23.8.1 Population Density and Size (Table 23.8)

By 1971–1980, population density was 3–13 animals/10 km^2 in the Kavkazsky Nature Reserve, 17 animals/10 km^2 in the Teberdinsky Nature Reserve, 1–11 animals/10 km^2 in the hunting grounds of northern Caucasus.

There is a decline in numbers everywhere, even in nature reserves.

23.9 Relations to Man

In 1940, the annual catch was 15,000 for meat (poor quality), fat and skin (Vereshchagin 1947). Nowadays, the species survive almost only in the nature reserves (Sokolov and Tembotov 1993).

24 Przewalski's Horse – *Equus caballus*

Equus caballus Linnaeus, 1758.
Perissodactyla, Equidae

24.1 Status and Distribution (Fig. 24.1)

24.1.1 Status

Extinct in the Wild (EW) (Hilton-Taylor 2000). There are now 74 horses in large enclosures of the Askania-Nova (Ukraine) (V. Gavrilenko, pers. comm.) and 14 horses in the Bukhara Ecological Center (Uzbekistan) (Flint et al. 1990b; Pereladova et al. 1999b). A free-living population (21 animals) was created in the Chernobyl area in Ukraine (Zharkikh et al. 1999).

Distribution Area. Ca. 4,268,000 km^2 (Estimated historical range) (Heptner 1989).

Fig. 24.1. Przewalski's horse

25 Mongolian Gazelle – *Procapra gutturosa*

Procapra gutturosa Pallas 1777.
Artiodactyla, Bovidae, Antilopinae

25.1 Status and Distribution (Fig. 25.1)

25.1.1 Status

Lower risk–near threatened (LR/nt) (Hilton-Taylor 2000).

25.1.2 Distribution

Since 1992, there have been 30–60 gazelles living permanently in Russian territory (Kirilyuk and Tseveenmyadag 1999). The only place in Russian territory where Mongolian gazelle visit is the Daursky Nature Reserve (Chita oblast). In January 2001, 80,000 gazelles migrated from Mongolia to Chita oblast, where they lived under strong poaching pressure (Sharov 2001).

Fig. 25.1. Mongolian gazelle

Appendix 1
Plant Genera Eaten by Ungulates (Wi, winter; Sp, spring; Su, summer; Au, autumn)

Table 1. Plant genera eaten by *Equus hemionus*[a]

Plant genera	Locality[b]				
	Badkhyz Nature Reserve[1]			Barsakelmes Island[2]	Qapshaghay Game Husbandry[3]
(Seasons)	Wi	Sp	Su		
Acanthophyllum	+++	r			++
Acroptilon				+	
Aegilops	r	++	+++		
Aeluropus	r			+++	
Agropyron				+++	
Alhagi	++	++	r	++	
Allium		+++			
Amberboa		+++			
Anabasis				+++	+++
Aphanopleura	++				
Arabis				+	
Aristida				+	
Arnebia	+++	+++	r		
Artemisia	+++			+++	
Astragalus	+++	++		+	
Atraphaxis	r			+	+
Atriplex	++			+++	
Bongardia		++			
Bromus	+++	+++	+++	+	++
Bunium		+++			
Calamagrostis				+	
Calligonum	++	++			
Caragana				+	++
Cardaria				++	
Carex	+++	+++	+++	+++	
Carthamnus			++		
Centaurea		++			
Ceratocarpus				+++	
Chorispora				+	
Convolvulus		+++			
Cousinia	+++	++	r		
Crypsis				++	
Delphinium	+++				

Table 1. *Continued*

Plant genera	Locality[b]				
	Badkhyz Nature Reserve[1]			Barsakelmes Island[2]	Qapshaghay Game Husbandry[3]
(Seasons)	Wi	Sp	Su		
Elymus				+	
Ephedra	++				
Eremopyrum	+++	++		+++	+
Eremostachys		r	+++		
Erodium		++			
Euclidium				+	
Eurotia					+++
Ferula	++	++	++		
Filipendula					++
Frankenia					++
Halocnemum	+++			++	
Haloxylon	++			+++	+
Hordeum	++	+++	++		
Isatis	+++	r			
Ixiolirion	+++				
Kochia					+
Lagonychium	++				
Lepidium					+++
Leptaleum				+++	
Limonium	+++	r		+	
Lycium				+	
Malcolmia	++	r	++		
Medicago		+++			
Mentha					+++
Nitraria	++				
Onobrychis		+++			
Papaver		++			
Phragmites				+	+++
Pistacia	++				
Poa	+++	+++	+++	+++	
Psoralea		++			
Ranunculus		+++			
Rheum				+	
Roemeria		++			
Rosa				+	
Salsola	+++	r		+	+++
Schismus		++	r		
Scorzonera				+++	
Secale		++			
Sisymbrium					+++
Sorghum			+++		
Sphenopus	r	++			
Stipa	++	r		+	+++
Tamarix	++	r	++	+	
Tanacetum	+++				

[a]Symbols indicate % of plant use was >30% (+++); 20–30% (++): 10–20% (+); <10% (r).
[b]The superscript numbers denote the following references: 1, Solomatin (1973); 2, Rashek (1967); 3, Blank and Dzhanyspaev (1990).

Table 2. Plant genera eaten by *Sus scrofa*[a]

Plant genera	Locality[b]				
	Belarus (1)[1]; Caucasus(2)[2]		Plain Kazakhstan (1)[3]; Kazakhstan Mountains (2)[4]; Sikhote-Alin Range (3)[5]		
	1	2	1	2	3
Acer	x				+
Achyrophorus	x				
Aegopodium	x			r	
Agropyron				++	
Ajuga	x				
Alchemilla		x			
Alliaria	x				
Allium		x	+	+++	
Anemone	x	x			+
Angelica	x				++
Anthriscus	x				
Apocynum			+		
Aquilegia	x				
Arctium	x				
Armeniaca				++	
Artemisia				+	
Aruncus					+
Asarum	x				
Atriplex	x				
Berberis				r	
Betula	x				
Bolboschoenus			+		
Calamagrostis	x		+++		+
Calla	x				
Calluna	x				
Caltha	x				
Campanula	x				
Cardamine	x				
Carex	x				+++
Carpinus	x				
Carum		x			
Castanea		++			
Cerastium		x			
Cerasus		+++			
Chaerophyllum	x				
Chrysosplenium	x			r	
Cirsium		x			
Colchicum		x			
Comarum	x				
Convallaria	x				+
Corydalis					+
Corylus	x	++			+++
Crataegus				+	
Crepis	x				

Table 2. *Continued*

Plant genera	Locality[b]				
	Belarus (1)[1]; Caucasus(2)[2]		Plain Kazakhstan (1)[3]; Kazakhstan Mountains (2)[4]; Sikhote-Alin Range (3)[5]		
	1	2	1	2	3
Crocus				+++	
Dactylis	x			+++	
Dryopteris		x			
Elaeagnus			+		
Epilobium		x			
Equisetum					+++
Eremopyrum				+	
Eriophorum	x				
Euonymus	x				
Euphorbia		x			
Fagus		+++			
Ferula			+++		
Festuca		x		r	
Ficaria	x				
Filipendula	x				+
Fraxinus	x				
Fritillaria	x				
Gagea				+++	
Galega		x			
Galeobdolon	x				
Galium				+	
Gentiana		x			
Geranium	x			+	
Geum	x				
Glycyrrhiza			+		
Hedysarum		x			
Heracleum	x	x		++	+
Hieracium	x			+	
Hottonia	x				
Impatiens	x	x			+
Iris	x			+	
Ixiolirion				++	
Juglans		++		+++	+++
Juniperus				++	
Lamium	x				
Laserpitium	x				
Lespedeza					+
Lilium					+++
Lonicera				r	
Malus		+++		+++	+
Melica					
Menyanthes	x				
Milium					+
Mushrooms	x			++	+

Table 2. *Continued*

Plant genera	Locality[b]				
	Belarus (1)[1]; Caucasus(2)[2]		Plain Kazakhstan (1)[3]; Kazakhstan Mountains (2)[4]; Sikhote-Alin Range (3)[5]		
	1	2	1	2	3
Nitraria			+++		
Ornithogalum		x			
Oxycoccus	x				
Padus					+
Papaver					+
Paris		x			
Pedicularis		x			
Petasites		x			+
Peucedanum	x				
Phalaris					+
Phlomis				+	
Phragmites	x		+++		+
Picea	x			+	
Picris	x				
Pinus	x				+++
Pistacia				+++	
Plantago	x				
Poa	x			++	
Polygonatum	x	x			
Polygonum	x	x		++	
Populus	x			+	
Potentilla	x				+
Primula	x			++	
Prunus		+++		+++	
Pteridium	x				
Pyrus		+++			
Quercus	x	+++			+++
Ranunculus	x				
Rheum				++	
Ribes				r	
Rosa				++	
Rubus		x	+	++	
Rumex	x	x			
Salix				r	
Sanguisorba					+
Scilla		x			
Scorzonera	x				
Senecio		x			
Silaus	x				
Silene	x				
Sonchus	x				+
Sorbus	x			r	
Spiraea					+
Stachys	x				

Table 2. *Continued*

Plant genera	Locality[b]				
	Belarus (1)[1]; Caucasus(2)[2]		Plain Kazakhstan (1)[3]; Kazakhstan Mountains (2)[4]; Sikhote-Alin Range (3)[5]		
	1	2	1	2	3
Stellaria	x				
Suaeda			+		+
Taraxacum	x			++	+
Thalictrum				r	
Tilia	x				+
Trifolium	x				
Trisetum					+
Tulipa				+++	
Typha			+++		
Ulmus	x				
Urtica	x	x			+
Vaccinium	x				
Verbascum	x				
Viscum	x				
Vitis					+

[a]Symbols indicate percent of plant use was >30% (+++); 20–30% (++): 10–20% (+); <10% (r) ; x indicates plant use noted, no numerical data reported.
[b]The superscript numbers denote the following references: 1, Lebedeva (1956); 2, Donaurov and Teplov (1938); 3, Fedosenko et al. (1984); 4, Yanushevich et al. (1972); Fedosenko and Zhiryakov (1984); 5, Bromley and Kucherenko (1983).

Table 3. Plant genera eaten by *Moschus moschiferus*[a]

Plant genera Seasons	Locality[b]				
	Yakutia[1]		Sikhote-Alin Range[2]		Transbaikal[3]
(Seasons)	Au Wi	Su Sp	Wi	Su	Wi
Abies			+++		
Acer			r	+++	
Aconitum				+	r
Agropyron				+	
Athyrium				++	
Allium	+	–			
Alnus	+		r	+	
Angelica					++
Arctous	+	–			
Atragene				+	
Betula		+		r	r
Betula nana	+++	+			

Table 3. *Continued*

Plant genera	Locality[b]				
	Yakutia[1]		Sikhote-Alin Range[2]		Transbaikal[3]
(Seasons)	Au Wi	Su Sp	Wi	Su	Wi
Bupleurum			+		
Calamagrostis			++		
Carex	+++	++			
Cereals			r	r	
Chosenia			r	+++	
Crataegus	+				
Dryopteris				++	
Eleutherococcus			+		
Epilobium				r	
Equisetum	+++	+++	r	+	
Eriophorum	+	–			
Filipendula				++	
Galium				+	
Graminea	+++	+++			
Gypsophila			+		
Juniperus	+				
Larix	+++	+	+	+	
Lathyrus				r	
Leguminosae	+	–			
Lichens on ground	+++	+++	+++	+++	+++
Lichens on trees	+++	++			
Linnaea				+	
Majanthemum				++	
Mosses	+++	+++	+	+	++
Mushrooms	+++	+++		++	r
Oxalis				r	
Padus			r	++	
Pinus		+	r		r
Populus			r	r	+
Potentilla	+++	–		r	
Pteridium				++	
Pulsatilla	+++	–			
Pyrola	+	–			
Rhododendron			++	+	++
Rosa				+	
Rubia				+	
Salix	+++	+	r	+	
Sanguisorba	+	–		r	
Spiraea	+	–		++	
Swida	+				
Taxus			+	+	
Thalictrum				r	
Tilia				+	
Vaccinium	+	+	+	+	++
Viburnum				+	
Vicia				+	
Vitis				+	

[a]Symbols indicate percent of plant use was >30% (+++); 20–30% (++): 10–20% (+); <10% (r).

[b]The superscript numbers denote the following references: 1, Egorov (1971); 2, Salmin (1972); Bromley and Kucherenko (1983); Zaitcev (1991); 3, Shvetsov et al. (1984).

Table 4. Plant genera eaten by *Cervus elaphus*[a]

Plant genera	Locality[b]											
	Belowezhskaya Forest Nature Reserve (1)[1], Kavkazsky Nature Reserve (2)[2], Kazakhstan (3)[3], Tajikistan (4)[4]							Altai (1)[5], Stolby Nature Reserve (2)[6], Far East (3)[7]				
	1		2		3		4	1	2		3	
(Seasons)	Wi	Sp, Su, Au	Wi	Su	Au, Wi	Sp, Su			Wi, Sp	Su, Au	Au, Wi	Sp, Su
Abies			++	r				+				
Acer		++	+	+								+++
Achillea					+							
Aconitum				+	+	+						
Actinidia											++	+++
Adenophora					++	++						
Adonis											+	+++
Aegopodium			r	++		++			++	++		
Aetheopappus				+								
Agasyllis				+								
Agropyron				+	++	++				++		
Agrostis				++						+		
Alchemilla				++	++							
Alectorolophus				+								
Alfredia						++						
Alhagi							+					
Allium			r	+		++						
Alnus	+	+		r					r			+
Alopecurus				++		+				+		
Alyssum				+								
Andromeda		++										
Anemone		++		+		r				+		
Angelica		+++						r	r	++		+
Anthemis				+++								
Anthoxanthum		+		++								
Anthriscus					++				++			
Anthyllis				r								
Apocynum							+++					
Aquilegia		+						+				
Aralia											+++	+++
Archangelica								r	r			
Aristella				+++								
Artemisia					++	++			r			+
Aruncus												r
Asparagus							+++					
Asperula				+								
Aster		r								r		
Astragalus						++						
Astrantia				++								
Athyrium												
Atragene					++	++				+		
Atriplex							+					
Avenastrum									++			
Berberis				+++							+	+
Bergenia								r				
Betonica				++								
Betula	r	r	+	++				+++	+		+	+
Brachypodium				+								
Briza				r						+++		

Table 4. *Continued*

Plant genera	Locality[b]											
	Belowezhskaya Forest Nature Reserve (1)[1], Kavkazsky Nature Reserve (2)[2], Kazakhstan (3)[3], Tajikistan (4)[4]							Altai (1)[5], Stolby Nature Reserve (2)[6], Far East (3)[7]				
	1		2		3		4	1	2		3	
(Seasons)	Wi	Sp, Su, Au	Wi	Su	Au, Wi	Sp, Su			Wi, Sp	Su, Au	Au, Wi	Sp, Su
Bromus										++		
Bupleurum				+++	r			+		+++		
Cacalia									r		r	+++
Calamagrostis		+++		+	r	r		+		+++	+++	+++
Calla	++						+++					
Callianthemum								r				
Calluna	+											
Caltha		+						+			+	+++
Campanula				+	+++	+++						
Caragana								+++	+++	+++		
Cardamine				r								
Carduus				+	++							
Carex		++	++	++	++	++		+++	+	+	+++	+++
Carpinus	r	r	r	+								
Carum				++						r		
Centaurea				+++		++	+					
Cephalaria				+++								
Cerastium				++								
Cerasus			r	r								
Cetraria			++	+++					++			
Chaerophyllum				+++								
Chamaeneriom		++		+								+
Chenopodium							+					
Chosenia											r	++
Cicerbita				+++								
Cinna				r							+	+++
Cirsium	+	+		++	++			+++				
Cladonia			+	+++								
Clinelymus										+		
Codonopsis					+++	+++						
Conioselinum						++						
Cornus									++			+
Coronilla				++								
Corydalis						r				r		
Corylus		++	r	++								+
Crataegus									r			++
Crepis						r		r	+++	+++		
Cuscuta							+++					
Dactylis		++		+	++	++				++		
Daphne		+++		+								
Dasiphora										r		
Delphinium				+	++					r		
Deschampsia				+						r		
Dianthus				r	r							
Doronicum								+				
Dryopteris				++								
Echinochloa							+++					
Echinopsilon					++		+					
Elaeagnus							+++					

Table 4. *Continued*

Plant genera	Locality[b]											
	Belowezhskaya Forest Nature Reserve (1)[1], Kavkazsky Nature Reserve (2)[2], Kazakhstan (3)[3], Tajikistan (4)[4]							Altai (1)[5], Stolby Nature Reserve (2)[6], Far East (3)[7]				
	1		2		3		4	1	2		3	
(Seasons)	Wi	Sp, Su, Au	Wi	Su	Au, Wi	Sp, Su			Wi, Sp	Su, Au	Au, Wi	Sp, Su
Eleutherococcus											+	++
Empetrum				++								
Epilobium				+++		++		+++		+++	+	+
Equisetum	+++				+++				+++	+++	+	+
Eremurus						++						
Erianthus							+					
Erigeron				+	r							
Euonymus	+++	+++	+++	+							+	+++
Euphorbia									+++	+++		
Eutrema					++							
Fagus				r								
Ferula					++							
Festuca	++	++	++	++		++		+		++	+++	+++
Filipendula		+							+		r	+
Fraxinus	+++	+++									+	++
Galatella						r						
Galega				+++								
Galeopsis				++								
Galium				+	r	r			++			
Genista		+										
Gentiana						++						
Geranium		+		+	++	++		+				
Geum										r		
Glycyrrhiza							+					
Gnaphalium				++								
Halimodendron							+					
Hedera			+									
Hedysarum				+++	++	++						
Helianthemum				++								
Heracleum		++		++	+++	+++		+	++		+++	+++
Hieracium		+		+++	++	++				+		
Hippophae					r							
Humulus			r		++							
Hypericum				+								
Hypochoeris										++		
Ilex				r								
Impatiens		+		+							r	++
Inula				r								
Iris		++		r	r					++		
Juncus				+								
Juniperus				+								
Knautia				+++								
Kobresia				+								
Koeleria										++		
Lamium				r						r		+
Lapsana				++								
Larix									r		r	+
Lathyrus		++				r			+++	+++	+++	+++
Laurocerasus			r	+								

Table 4. *Continued*

Plant genera	Locality[b]											
	Belowezhskaya Forest Nature Reserve (1)[1], Kavkazsky Nature Reserve (2)[2], Kazakhstan (3)[3], Tajikistan (4)[4]							Altai (1)[5], Stolby Nature Reserve (2)[6], Far East (3)[7]				
	1		2		3		4	1	2		3	
(Seasons)	Wi	Sp, Su, Au	Wi	Su	Au, Wi	Sp, Su			Wi, Sp	Su, Au	Au, Wi	Sp, Su
Leonurus					++							
Lespedeza											+++	+++
Leuzea					++			+++				
Libanotis					++	++						
Lichens											+++	
Ligularia					+							
Ligusticum				r								
Lilium										r		+
Linum				++								
Lonicera			r	r	+++			+	+++	+++		++
Lotus				++								
Maackia												r
Malus				++	r	r			+++			+
Medicago									+++	+++		
Melissitus					+	+						
Menyanthes		++						+			+++	+++
Milium				++								+
Molinia		+++										
Mushrooms											+++	+++
Nardus				++								
Neomolinia											+	++
Nepeta					+							
Ornithogallum				++								
Orobus						++						
Oxytropis								r		++		
Pachypleurum						+++						
Padus	+++		++		r				++			r
Pastinaca				+++								
Patrinia					+							
Pedicularis				+		+++						
Petasites				++							+++	+++
Phalaroides											+	+++
Phellodendron											++	+++
Philadelphus											+++	+++
Phleum				++						++		
Phlomis					+++	+++				++		
Phragmites							+++					
Picea	+	+	r		+							
Pimpinella				++								
Pinus	++	++	r						r			+
Plagiobasis					++							
Plantago				+		++				+		
Plectranthus												+
Pleurospermum								r	++	++		
Poa		r	+	++	++	++				+++	+++	+++
Polemonium					++	++						
Polygonatum				+								
Polygonum						++		+++		++		
Populus	+++	+++	+++	++	r		+	+	+++	+++	+++	+++

Table 4. *Continued*

Plant genera	Locality[b]											
	Belowezhskaya Forest Nature Reserve (1)[1], Kavkazsky Nature Reserve (2)[2], Kazakhstan (3)[3], Tajikistan (4)[4]							Altai (1)[5], Stolby Nature Reserve (2)[6], Far East (3)[7]				
	1		2		3		4	1	2		3	
(Seasons)	Wi	Sp, Su, Au	Wi	Su	Au, Wi	Sp, Su			Wi, Sp	Su, Au	Au, Wi	Sp, Su
Potamogeton											++	+++
Potentilla		r				++		r		r		
Primula				++				+++		++		
Prunus			r									
Psephellus				+++								
Pteris								r				
Pulmonaria										+		
Pulsatilla										r	+	+++
Pyrus			r	+++								
Quercus	+++	+++	++								+	++
Ranunculus						++		+		r		+
Rhamnus		++									r	+
Rhaponticum										++		
Rheum					++	++						
Rhodiola						+						
Rhododendron			+	++				r				r
Ribes		++	++	+++	+	+			++			++
Roegneria										++		
Rosa			++	+++	+	+			+	+	r	+
Rubus	++	+++	+++	++	r			r			r	++
Rumex				+	r	+++		+		r		
Sagittaria											++	+++
Salix	+++	+++	++	++	+++	+++			+++	+++		++
Salvia				r								
Sambucus		++	+	+					+		+	r
Sanguisorba						r		+++		+++	r	+
Sanicula				+								
Saussurea					++	++		+++				+
Scabiosa				+++								
Scilla				r								
Scirpus		+							+			
Scorzonera										++		
Secale				r								
Sedum					+++	+++		r				
Semenovia						++						
Senecio				++	++	++				++		
Serratula										r		
Seselopsis						++						
Schulzia								+++				
Silene				r		++						
Sisymbrium					+++							
Sium					++							
Solidago				+	r	r				+		+
Sorbaria											+	++
Sorbus	+++	+++	++	+	+++	+++		+	+++	+++	r	+
Sorghum							+++					
Spiraea					r	r		+		+	+	++
Stachyopsis					++							
Stachys				+								

Table 4. *Continued*

Plant genera	Locality[b]											
	Belowezhskaya Forest Nature Reserve (1)[1], Kavkazsky Nature Reserve (2)[2], Kazakhstan (3)[3], Tajikistan (4)[4]							Altai (1)[5], Stolby Nature Reserve (2)[6], Far East (3)[7]				
	1		2		3		4	1	2		3	
(Seasons)	Wi	Sp, Su, Au	Wi	Su	Au, Wi	Sp, Su			Wi, Sp	Su, Au	Au, Wi	Sp, Su
Staphylea			r	+								
Stellaria										+		
Stipa										+		
Streptorhamphus						++						
Suaeda							+					
Swertia				++				+				
Swida											r	++
Symphytum				+								
Syringa											+	+++
Tanacetum		r							+	+		
Taraxacum						++				+		
Taxus			++	++								++
Telekia				r								
Thalictrum					+	+						
Thamnalia			r	++						++		
Thesium				r								
Tilia	r	r									+++	+++
Tragopogon				+++	++							
Trifolium		++		++		r				+++	++	+++
Trisetum				+								
Trollius				++				+		++		
Tussilago												
Typha												++
Ulmus			++	+							++	++
Urtica	+		+	++	+++	+++				r		
Usnea			++	+++				++	+++			
Vaccinium			+++	+++					+	r	r	+
Valeriana				+		r						
Veratrum								r	+			
Verbascum				+								
Veronica				+++		r		r		++		
Viburnum		+++										+
Vicia				++		++		+		+++	+++	+++
Viscum			+++	+								
Vitis			+	++							++	+++

[a]Symbols indicate percent of plant use was >30% (+++); 20–30% (++): 10–20% (+); <10% (r).
[b]The superscript numbers denote the following references: 1, Sablina (1955); 2, Alexandrov (1968); 3, Fedosenko (1980); 4, Stroganov (1959); 5, Dmitriev (1938); 6, Zyryanov (1975); 7, Baskin (1965), Bromley and Kucherenko (1983).

Table 5. Plant genera eaten by *Cervus nippon* in the Far East.[a] (Prisyazhnyuk and Prisyazhnyuk 1974)

Plant genera (Seasons)	Wi, Sp	Su, Au
Acanthopanax	++	r
Acer	+++	r
Achillea		++
Acroptilon		++
Actinidia	+++	+
Adenocaulon		++
Adenophora		+++
Allium	+	r
Alnus	+++	r
Angelica		++
Anthriscus		+
Aquilegia		r
Aralia	+++	+
Artemisia	+	r
Aruncus		++
Aster		+++
Astilbe		+
Athyrium		r
Atriplex	+	++
Berberis	+	r
Betula	r	
Cacalia		r
Caltha	++	+
Campanula		+++
Carduus	++	+++
Carex	+++	+
Carpinus	+	
Cerasus	+	r
Cirsium		++
Corylus	+++	+
Dianthus	+++	
Eleutherococcus	++	+
Elymus	r	+
Euonymus	++	r
Festuca	r	
Fraxinus	+++	r
Galium		++
Geranium		+
Heracleum		+
Humulus		r
Hypericum		r
Impatiens	++	++
Juglans	+++	r
Lactuca		+
Lathyrus		r
Leonurus		+
Lespedeza	+	r

Table 5. *Continued*

Plant genera (Seasons)	Wi, Sp	Su, Au
Ligularia	++	++
Ligusticum		+
Lilium		++
Lonicera	r	r
Lycopus		+
Maackia		+
Malus	+	r
Melampyrum		+
Mentha		++
Milium	r	
Osmundarium	++	++
Padus	+++	r
Pedicularis		++
Phellodendron	+++	
Philadelphus	+	r
Plantago	+	++
Poa	+	r
Polemonium		+
Polygonum		+
Populus	+	r
Prenanthes		+
Pyrus	r	
Quercus	+++	+++
Rabdosia		+++
Rhamnus	r	
Rhododendron		r
Ribes	++	
Rosa	+	
Rubia	r	++
Rubus idaeus	+++	+++
Rumex		+
Salix	+	r
Sambucus	++	
Sanguisorba		++
Saussurea		++
Saxifraga	++	++
Schisandra	++	+
Scirpus	r	
Sedum		+
Serratula		++
Sonchus		+
Sonicula	+	++
Sorbaria	++	r
Syrenia	+	r
Taraxacum	+	+
Thalictrum	++	++
Tilia	+++	+
Trifolium	+	+

Table 5. *Continued*

Plant genera (Seasons)	Wi, Sp	Su, Au
Trollius		+
Ulmus	+	r
Urtica	+	+
Valeriana		r
Veratrum		r
Veronica	+++	
Viburnum	++	
Vitis	++	r

[a]Symbols indicate percent of plant use was >30% (+++); 20–30% (++): 10–20% (+); <10% (r).

Table 6. Plant genera eaten by *Alces alces*[a]

Plant genera	Locality[b]					
	Leningrad oblast, Belarus (1)[1]; CS European Russia (2)[2]; Altai (3)[3]			C Yakutia (1)[4]; Kolyma[a] Uppland (2)[5]; Amur, Sikhote-Alin' (3)[6]		
	1	2	3	1	2	3
Abies			++			r
Acer		++				++
Achillea		+				
Aconitum			+			
Acorus			+++			+++
Aegopodium	+++	++				
Agropyron	++					
Algae						+++
Alnus	r	+	+	+++		r
Amelanchier		++				
Angelica	+++	++	++			
Anthriscus			++			
Archangelica			+			
Arctophila					+	
Arctostaphylos		+		r		
Artemisia			+			
Avena		++				
Bergenia			++			
Beta		+++				
Betula	+++	+	+++	+++	+	++
Betula nana				+++	r	+++
Boletus			+			
Brassica		+++				

Table 6. *Continued*

Plant genera	Locality[b]					
	Leningrad oblast, Belarus (1)[1]; CS European Russia (2)[2]; Altai (3)[3]			C Yakutia (1)[4]; Kolyma Uppland (2)[5]; Amur, Sikhote-Alin' (3)[6]		
	1	2	3	1	2	3
Bupleurum			+++			
Cacalia					r	
Calamagrostis	++	+			r	
Calla	+++					+++
Callitriche					r	
Calluna		+++				
Caltha	+++	++	+++			+++
Caragana			++			
Carex	+++	++	+++	++	++	++
Cerasus		+				
Chamaedaphne					r	
Chamaenerion	++	+++			+	
Chelidonium		+				
Chosenia				++	+++	+
Cicuta					+	+++
Comarum		+++			+	
Convallaria	+++	+				
Convolvulus		++				
Corylus		+				r
Cotoneaster			++			
Crataegus		++		r		++
Crepis			++			
Cytisus		++				
Dactylis		++				
Dasiphora			+			
Daucus		+				
Deschampsia	++					
Dryopteris			+			
Epilobium	+++		+++	+++		+++
Equisetum	+++	+++	+	+++	+++	+
Eriophorum			++	+	+	+++
Eunonymus		+++				+
Euphorbia			+			
Filipendula	+++	+++	++	r		
Fraxinus						+
Galium			++			
Geranium		++				
Geum		+++				
Graminea	+++			+++		
Heracleum			++			
Hieracium		++				
Hypericum			+			
Iris		+	+	+	+	+
Juncus	+++					

Table 6. *Continued*

Plant genera	Locality[b]					
	Leningrad oblast, Belarus (1)[1]; CS European Russia (2)[2]; Altai (3)[3]			C Yakutia (1)[4]; Kolyma Uppland (2)[5]; Amur, Sikhote-Alin' (3)[6]		
	1	2	3	1	2	3
Juniperus	+++					
Larix			+	r		+
Lathyrus			+			
Ledum			+	r		r
Lichens				+		
Linaria		++				
Lonicera			++			++
Lysimachia			+++			
Malus			+++			
Medicago		++				
Mentha			++			
Menyanthes	+++	+++	+++	+	+	+++
Mushrooms	+++			++	r	+++
Nuphar	+++	+				+++
Orobus			+			
Padus	r	+	++			++
Paeonia			+			
Petasites		++				+++
Phalaroides	++					
Phragmites			++			
Picea		+	+			
Picris						+
Pinus	+++	+	+	+++		
Pisum		++				
Plantago		+				
Pleurospermum			+			
Poa	++				+	
Populus	+++	+++	+++	+		+++
Potamogeton			+++	r	+	+++
Prunus		+				
Pteridium			+			
Pulmonaria		++				
Pyrola					r	
Pyrus		+++				
Quercus		+++				r
Ranunculus		++	+			
Rhamnus	r	++				
Rhododendron						+++
Ribes	r	+	++		+++	+
Rosa		++	++			
Rubus		++	++		+	
Rumex		+	+		r	
Sagittaria						+++
Salix	+++	+++	+++	+++	+++	+++

Table 6. *Continued*

Plant genera	Locality[b]					
	Leningrad oblast, Belarus (1)[1]; CS European Russia (2)[2]; Altai (3)[3]			C Yakutia (1)[4]; Kolyma Uppland (2)[5]; Amur, Sikhote-Alin' (3)[6]		
	1	2	3	1	2	3
Sambucus		+				
Sanguisorba			+			
Scirpus			++			
Secale		++				
Setaria			+			
Solanum		+				
Solidago			+			
Sonchus		++	++			
Sorbus	+	+++	+++	r	r	+++
Sparganium					+	
Spiraea			+++		r	+++
Spirogyra			+++			
Stachys		++				
Swida				+		+++
Tanacetum		+				
Taraxacum		++	+			
Thalictrum			+			
Thelycrania						+++
Tilia		+				+
Trapa						+++
Trifolium		+++				
Triticum		++				
Trollius			++			
Tussilago			+			
Ulmus		+				r
Urtica		++				
Utricularia					+++	
Vaccinium	++		+	r	+	
Valeriana		+				
Verbascum		+				
Veronica		++	+	r		
Viburnum		++	++			
Vicia		++			+	
Vitis						++

[a]Symbols indicate percent of plant use was >30% (+++); 20–30% (++): 10–20% (+); <10% (r) ; x indicates plant use noted, no numerical data reported.
[b]The superscript numbers denote the following references: 1 Rusakov (1979); Kozlo (1983b); 2 Prostakov (1996); 3 Dmitriev (1938); Sobansky (1992); 4 Egorov (1965); 5 Chernyavsky and Domnich (1989).

Table 7. Plant genera eaten by *Capreolus* spp.[a]

Plant genera	Locality[b]				
	CS Russia[1]	Kazakhstan[2]		Far East[3]	
(Seasons)	Without Snow	With Snow	Without Snow	With Snow	Without Snow
Acanthopanax				+	++
Acer	+++			+++	+++
Achillea		++		++	++
Aconitum					++
Acorus	+				++
Actinidia				+	++
Adonis					+++
Aegopodium	+++	+	+		
Agropyron	+++		++		
Agrostis			+		
Allium					+
Alopecurus			+		
Amelanchier	+++				
Anemone			++		+++
Angelica		r	+		
Anthriscus			++		
Apocynum			+		
Aralia					+++
Armeniaca			+		
Artemisia	+	+++	+	+	++
Aruncus					++
Astragalus			++		
Atriplex			+		
Avena	+				
Beta	+++				
Betonica			+		
Betula		+++	++	++	++
Betula nana				+++	+++
Bidens			++		
Brassica	+++				
Bromus		++			
Cacalia				r	+
Calamagrostis		r	+	+	+++
Calla	+			++	++
Caltha					+++
Campanula	+++				
Carex	+++			r	+++
Centaurea		+	+		
Cerasus			+++		++
Chenopodium			+		
Chosenia				++	++
Chrysosplenium				r	+
Cichorium			++		
Cirsium	+++				
Convallaria	+				

Table 7. *Continued*

Plant genera	Locality[b]				
	CS Russia[1]	Kazakhstan[2]		Far East[3]	
(Seasons)	Without Snow	With Snow	Without Snow	With Snow	Without Snow
Corylus					++
Crataegus	+++			+	+
Crepis		++	++		
Cytisus	+++				
Daucus	+++				
Eleutherococcus					++
Epilobium		+	++	r	+++
Equisetum	+				
Erigeron			+		
Eriophorum					++
Euonymus	+++			+++	++
Eurotia			+		
Fagopyrum	+++				
Festuca		+++	++		
Filipendula	+		++	+	+
Fraxinus	+++			++	++
Galium	+++		+		
Gentiana			+		
Geranium	+				
Geum	+++		+		
Glechoma	+				
Glycyrrhiza			+		
Heracleum			+		+++
Hieracium	+++		++		
Hordeum	+++				
Humulus	+				
Impatiens				++	++
Iris					++
Koeleria			+		
Kochia		+	+		
Lathyrus		r	+	++	+++
Leonurus		++			
Lespedeza				+++	+++
Libanotis		++			
Ligularia			++		
Lilium					++
Lonicera			r	+	+
Lysimachia		r			
Malus	+++	++			
Medicago	+				
Melilotus	+				
Menyanthes	+				+++
Mulgedium		+			
Mushrooms	+		+		+++
Neomolinia				+	++

Table 7. *Continued*

Plant genera	Locality[b]				
	CS Russia[1]	Kazakhstan[2]		Far East[3]	
(Seasons)	Without Snow	With Snow	Without Snow	With Snow	Without Snow
Nitraria			+		
Orobus	+++		+		
Padus				++	++
Pedicularis			+		
Peucedanum			+		
Phellodendron				+++	+++
Philadelphus				+++	+++
Phleum		+	++		
Phlomis		r			
Phragmites		+	++		
Picea		+++			
Pinus (needles)			+		
Pisum	+++				
Plantago	+				
Pleurospermum		r	+		
Poa			++		
Polygonatum	+			r	+
Polygonum					+
Populus pruinosa	+++	+++			
Populus sp.			+	+	+
Populus tremula	+++	+++	+		
Potentilla	+				
Prunus	+++				
Pulmonaria	+++		++		
Pulsatilla	+++				+++
Pyrus	+++				
Quercus (acorns)	+++			+++	+++
Ranunculus			+		++
Ribes				++	++
Rosa		+++	++		
Rubus caesius			+		
Rubus sp.	+		+	++	++
Sagittaria					+++
Salix	+++	+++	+++	+++	+++
Salsola			+		
Sanguisorba	+++	r		+	++
Saussurea			+	r	+
Schisandra					+
Sonchus		r	++		+
Sorbaria				+	++
Sorbus	+++		++	+	++
Spiraea					+
Suaeda		+	+		
Swida					++
Tilia	+++			+++	+++

Table 7. *Continued*

Plant genera	Locality[b]				
	CS Russia[1]	Kazakhstan[2]		Far East[3]	
(Seasons)	Without Snow	With Snow	Without Snow	With Snow	Without Snow
Tragopogon		+			
Trifolium	+++				+++
Triticum	+++				
Tulipa			+		
Typha			+		
Ulmus				++	++
Urtica		++			
Vaccinium				++	++
Viburnum				++	++
Vicia	+++	+		++	+++
Vitis					+++

[a]Symbols indicate percent of food use was >30% (+++); 20–30% (++): 10–20% (+); <10% (r).

[b]The superscript numbers denote the following references: 1, Prostakov (1996); 2, Sludsky and Baidavletov (1984); Bromley and Kucherenko (1983).

Table 8. Plant genera eaten by *Rangifer tarandus*[a]

Plant genera	Locality[b]				
	Kola Peninsula (1)[1]; Taymyr Peninsula (2)[2]; Sayan Mountains (3)[3]			Chukchi okrug (1)[4]; Kamchatka Peninsula (2)[5]	
	1	2	3	1	2
Aconitum				x	
Aegopodium			x		
Agrostis		++		x	
Allium			x		
Alnaster			x	X	
Alnus				X	x
Alopecurus		+		X	
Andromeda	r	+			
Androsace				x	
Anemone			x	x	
Antennaria				x	
Arctagrostis				x	
Arctophila		++			
Arctous		+		x	
Arnica				x	
Artemisia				x	x
Aruncus					x
Aster				x	
Astragalus				x	
Bergenia			x		
Betula	+++		x	x	
Betula nana	r				
Betula rotundifolia			x		
Bromus				x	
Bromopsis		+			
Calamagrostis	r	+	x	x	
Callianthemum			x		
Calluna	++				
Caltha				x	
Cardamine				x	
Carex	r	+++	x	x	x
Cassiope		+			
Chamaenerion		+		x	x
Chosenia				x	
Claytonia				x	
Cobresia		+			
Conioselinum				x	
Crepis				x	
Critesion				x	
Dasiphora				x	
Delphinium				x	
Deschampsia		+		x	x
Dianthus				x	
Dracocephalum				x	

Table 8. *Continued*

Plant genera	Locality[b]				
	Kola Peninsula (1)[1]; Taymyr Peninsula (2)[2]; Sayan Mountains (3)[3]			Chukchi okrug (1)[4]; Kamchatka Peninsula (2)[5]	
	1	2	3	1	2
Dryas				x	
Dupontia				x	
Elymus				x	
Empetrum	+++	+		x	
Endocellion		++			
Epilobium			x		
Equisetum	+	++	x	x	x
Erigeron				x	
Eriophorum		+++		x	x
Erysimum				x	
Eutrema				x	
Festuca		++		x	x
Galium				x	x
Gentiana				x	
Geranium			x		x
Hedysarum		+++			
Hierochloe		++		x	
Iris					x
Juncus				x	x
Juniperus				x	
Koeleria			x		
Lagotis		+		x	
Larix		++		x	
Lathyrus				x	
Ledum	+++	+		x	
Leuzea			x		
Lonicera			x		
Lycopodium				x	
Menyanthes		++		x	
Mertensia				x	
Minuartia				x	
Moehringia				x	
Mulgedium				x	
Nardosmia				x	
Oxycoccus		+			
Oxyria				x	
Oxytropis		+++		x	
Pachypleurum		++			
Pedicularis		++		x	
Petasites		++			
Phalaris				x	
Picea	r				
Pinus	+++			x	
Pleurospermum		+			

Table 8. *Continued*

Plant genera	Locality[b]				
	Kola Peninsula (1)[1]; Taymyr Peninsula (2)[2]; Sayan Mountains (3)[3]			Chukchi okrug (1)[4]; Kamchatka Peninsula (2)[5]	
	1	2	3	1	2
Pleurozium	+++				
Poa	+++	++	x	x	
Polemonium		+			
Polygonum				x	x
Potamogeton				x	
Populus				x	
Puccinellia				x	
Pulsatilla			x		
Pyrola		+		x	
Ranunculus			x	x	
Rhodiola				x	
Rhododendron			x		
Ribes				x	
Rubus				x	x
Rumex			x	x	
Salix	+++		x	x	x
Sanguisorba			x		x
Saussurea		++		x	
Saxifraga				x	
Schulzia			x		
Senecio		+		x	
Silene				x	
Spiraea				x	
Stellaria				x	
Swida				x	
Taraxacum				x	
Thalictrum			x		
Trisetum		+			
Urtica				x	
Vaccinium	+++	+++	x	x	
Valeriana		++		x	
Veratrum				x	x
Vicia				x	

[a]Symbols indicate percent of food use was >30% (+++); 20–30% (++): 10–20% (+); <10% (r) ; x indicates food use noted, no numerical data reported.
[b]The superscript numbers denote the following references: 1, Semenov-Tien-Shansky (1948b); Makarova (1988); 2, Michurin and Vakhtina (1968); Shchelkunova (1980); Zyryanov and Kolpashchikov 1981; Larin et al. (1990); 3, Fetisov (1955); Vodopyanov 1971; Sokolov 1983; 4, Zheleznov (1990); 5, Averin (1948); Baskin (1970); Mosolov (1986).

Table 9. Lichens, mosses and mushrooms eaten by *Rangifer tarandus*.[a] (Andreev 1940; Fetisov 1955; Karev 1956; Larin and Shelkovnikova 1985; Larin et al. 1990; Kupriyanov 1990; Zheleznov 1990)

	Western Siberia	Putorany	Sayan	Chukotka
Lichens				
Alectoria		++	x	
Cetraria			x	x
Cladonia	+++	+++	x	x
Dactilina		+		x
Peltigera		+		
Stereocaulon				x
Thamnolia		+		
Usnea		+	x	
Mosses			x	
Aulacomnium		+		
Calliergon		+		
Dicranum	+++	+		x
Drepanocladus		+		
Hylocomium		+		
Polytrichum	+++	+		
Ptilidium	r	+		
Sphagnum				x
Mushrooms				
Armellaria				x
Boletus		x	x	x
Lactarius			x	x
Russula			x	x

[a]Symbols indicate percent of food use was >30% (+++); 20–30% (++): 10–20% (+); <10% (r) ; x indicates food use noted, no numerical data reported.

Table 10. Plant genera eaten by *Gazella subgutturosa*[a]

Plant genera	Locality[b]				
	Ustyurt Plateau (1)[1]; Kyzylkum Desert (2)[2]; Barsakelmes Island (3)[3]; E Kazakhstan; Tajikistan (4)[4]				
	1	2	3		4
(Seasons)			Wi, Au	Sp, Sum	
Aellenia					x
Aeluropus			++	+++	
Agropyron			++	+++	
Alhagi		x			+
Allium					x
Alyssum		x			
Ammodendron		x			
Anabasis	x	x	++	+++	
Aristida	x		++	+++	x
Artemisia	x	x			++
Astragalus		x	++	+++	x
Atraphaxis	x	x	+++	+++	+
Atriplex			++	+++	
Bromus	x	x	++	+++	
Calligonum	+++	x	++	+	+
Capparis			+	+	x
Caragana	x		++	+	+
Ceratocarpus			++	+	
Carex	x	x	+++	+++	x
Chorispora					x
Convolvulus	+++				
Elaeagnus					x
Ephedra	+x		+	+++	
Eremopyrum		x	++	++	
Eurotia	x		++	++	+++
Frankenia			+++	+++	
Garmala	+++				
Girgensohnia					x
Halimodendron					+++
Halostachys	x				
Haloxylon	x	x	+	+++	++
Kochia	x	x	++	++	+++
Leptaleum	+++				
Limonium	x		+++	+++	
Nanophyton					+
Nitraria				++	+
Onobrychis					x
Phragmites			++	+	x
Pistacia					x
Poa	+		+++	+++	x
Populus					x
Rheum	x				

Table 10. *Continued*

Plant genera	Locality[b]				
	Ustyurt Plateau (1)[1]; Kyzylkum Desert (2)[2]; Barsakelmes Island (3)[3]; E Kazakhstan; Tajikistan (4)[4]				
	1	2	3		4
(Seasons)			Wi, Au	Sp, Sum	
Salsola	x	x	++	+++	+++
Senecio		x			
Sewerzowia					x
Smirnowia	x				
Tamarix		x	+	+++	x
Vulpia					x
Zygophyllum					x

[a]Symbols indicate percent of food use was >30% (+++); 20–30% (++): 10–20% (+); <10% (r) ; x indicates food use noted, no numerical data reported.
[b]The superscript numbers denote the following references: 1, Mambetzhumaev (1970); 2, Sludsky (1977); 3, Zhevnerov (1984); 4, Sludsky (1977); Sokov (1993).

Table 11. Plant genera eaten by *Saiga tatarica.*[a] (Sludsky 1955; Lebedeva 1959; Fadeev and Sludsky 1983)

Achillea	++
Agropyron	+++
Allium	+
Alyssum	+++
Anabasis	++
Arabidopsis	+
Arenaria	+
Artemisia	+++
Astragalus	+++
Atriplex	+++
Bromus	+++
Cancrinia	+
Capsella	++
Ceratocephala	+
Chorispora	+
Cirsium	++
Convolvulus	+
Descurainia	+++
Dodartia	+
Elymus	+
Ephedra	+++
Eremopyrum	+++
Euclidium	+
Eurotia	+
Ferula	+
Festuca	+++
Galium	+
Glycyrrhisa	+++
Goebelia	+
Hordeum	+
Hulthemia	+
Inula	++
Kochia	+++
Koeleria	+
Lappula	+
Lepidium	++
Limonium	++
Medicago	+
Melandrium	+
Myosurus	+
Nanophyton	+
Panicum	+
Poa	+++
Polygonum	+
Potentilla	+
Pulicaria	+
Rheum	+++
Rubia	+
Rumex	++
Salsola	+++
Sonchus	++
Spiraea	+
Stipa	+
Tanacetum	++
Taraxacum	+
Triticum	+
Tulipa	+++

[a]Symbols indicate peracent of food use was >30% (+++); 20–30% (++): 10–20% (+); <10% (r).

Table 12. Plant and mushroom genera eaten by *Bison bonasus*.[a] (Alexandrov 1958; Kalugin 1968; Gębczynska et al. 1991; Kazmin 1992)

Plant genera	Caucasus[b]		Belovezhskaya Forest Nature Reserve[c]	
(Seasons)	W	Sp, Su, Au	W	Sp, Su, Au
Abies	+++	++		
Acer	+++	+++	r	+++
Actaea			r	+
Aegopodium		++	r	++
Agasyllis		r		
Agropyron	++	++		
Agrostis		++		
Ajuga			+	+++
Alchemilla		+		
Allium		+		
Alnus	+++	+++	+	++
Alopecurus		+		
Anemone		++	r	++
Angelica		+		
Antennaria				
Anthemis		+		
Arrhenatherum		++		
Artemisia			r	
Aruncus		+		
Asarum				+
Asperula		+		++
Astragalus		++		
Astrantia		+		
Avenastrum		r		
Berberis	++	++		
Beta			+++	++
Betonica		+		
Betula			+++	+++
Bidens		++		
Boletus		r		
Brachypodium	+++	+++		
Brassica				+
Bromus	++	++		
Bunias		++		
Calamagrostis		++		
Calamintha		+		
Caltha		+		
Calluna			+	+
Campanula		r		
Capsella				++
Cardamine		+		+
Carduus		++		
Carex	++	++		
Carpinus	++	++	+++	+++
Centaurea		++		
Cephalaria		+++		

Table 12. *Continued*

Plant genera	Caucasus[b]		Belovezhskaya Forest Nature Reserve[c]	
(Seasons)	W	Sp, Su, Au	W	Sp, Su, Au
Cerastium		++		
Cerasus	+	+		
Chaerophyllum		++		
Chamaenerion		+		
Chenopodium				+
Chrysosplenium			r	
Cicerbita		+		
Circaea				+
Cirsium		++	r	++
Convallaria		+		
Coronilla		++		
Corylus	++	++	r	+++
Crataegus		+		
Crepis		r		
Dactylis	+++			
Daphne	+	+		
Dentaria				++
Deschampsia	+	+		
Dipsacus		++		
Doronicum		++		
Dryopteris		r		
Epilobium		r		+
Euonymus	+++	+++	r	
Festuca	++	++		
Ficaria				+
Filipendula		++	+	+++
Fragaria				++
Frangula		r	r	++
Fraxinus	+	+	++	
Galega		+++		
Galeobdolon			+	++
Galium		+		+
Genista				+
Geranium		+		+
Geum		+	r	++
Glyceria		r		
Hedysarum		+++		
Hepatica			r	+
Heracleum		+++		
Hesperis		++		
Humulus		+		
Hypericum		+		
Ilex	r			
Impatiens		+		
Juncus	r	r		
Knautia		+		
Lactuca		++		

Table 12. *Continued*

Plant genera	Caucasus[b]		Belovezhskaya Forest Nature Reserve[c]	
(Seasons)	W	Sp, Su, Au	W	Sp, Su, Au
Lamium		+		
Lathraea				+
Lathyrus		+++	r	+
Ledum				+
Leontodon			r	+
Ligusticum		++		
Lilium		++		
Linum		+		
Lonicera		r		
Lotus				+
Lysimachia		+		
Majanthemum				++
Malus	++	++		
Mentha		r		
Mercurialis				+
Milium	++	++		
Molinia		+		
Mycelis				+
Nardus		r		
Origanum		+		
Orobus		+		
Oxalis			++	+++
Padus		++		
Paris				+
Petasites		+++		
Philadelphus		r		
Phleum		+++		
Picea			+++	+++
Pimpinella		r		
Pinus			+	+++
Plantago		r	r	+++
Poa	++	++		
Polygonatum		++		
Populus tremula	+++	+++	+++	+
Potentilla		r		+
Primula		+	r	
Prunus	+	+		
Pulmonaria		r		
Pyrola				++
Pyrus	++	++		
Quercus			+++	+++
Ranunculus		+	+	+
Ribes		++		
Rosa	++	++		
Rubus	+	+		
Rubus caesius	+++	+++		
Rumex		r	+	++

Table 12. *Continued*

Plant genera	Caucasus[b]		Belovezhskaya Forest Nature Reserve[c]	
(Seasons)	W	Sp, Su, Au	W	Sp, Su, Au
Salix		+++	r	
Salvia		++		
Sambucus		r		
Scirpus		++		
Scrophularia		++		
Secale		++		
Senecio		+		
Serratula		++		
Silene		+		
Solanum			+	+++
Solidago		++		
Sorbus		++		+
Stachys		++	+	+
Stellaria		r	r	+++
Symphytum		++		
Tanacetum		+		
Taxus		+		
Telekia		r		
Thalictrum		+		
Tilia			+	+++
Tragopogon		r		
Trifolium		+++		
Tussilago		++		
Ulmus	+++	+++	+	
Urtica			r	++
Usnea	++			
Vaccinium		++	+	+++
Valeriana		r		+
Veratrum		++		
Veronica		+		+
Viburnum		r	r	
Vicia		++	r	
Viola		+		
Viscum	+++	+++		

[a]Symbols indicate percent of food use was >30% (+++); 20–30% (++): 10–20% (+); <10% (r).
[b]Field observations.
[c]Stomach content.

Table 13. Plant genera eaten by *Capra caucasica*.[a] (Kotov 1968)

Plant genera	Kavkazsky Nature Reserve
Abies	++
Acer	+
Aconitum	+
Agrostis	++
Alchemilla	+++
Alopecurus	+++
Androsace	+
Anemone	++
Anthemis	++
Arenaria	++
Aster	++
Astragalus	++
Astrantia	+
Asyneuma	++
Athyrium	+
Betonica	+
Betula	++
Bromus	+++
Bupleurum	+
Calamagrostis	++
Caltha	+
Campanula	++
Cardamine	+
Carex	++
Carum	+
Centaurea	+++
Cephalaria	+
Cerastium	+
Cetraria	++
Chaerophyllum	++
Chamaenerion	+
Cicerbita	+
Cirsium	+++
Cobresia	++
Colpodium	++
Coronilla	+
Cystopteris	+
Daphne	+
Delphinium	+
Deschampsia	+++
Doronicum	++
Draba	+++
Empetrum	+
Erigeron	+
Euphorbia	+
Festuca	+++
Fritillaria	+
Galium	+
Gentiana	+
Geranium	+

Table 13. *Continued*

Plant genera	Kavkazsky Nature Reserve
Geum	+
Gymnadenia	+
Gypsophila	+
Hedysarum	++
Helictotrichon	+
Hieracium	++
Hypericum	+
Iris	+
Juniperus	+
Jurinea	+
Knautia	+
Koeleria	+
Lactarius	++
Lamium	+
Licopodium	+
Ligularia	++
Lloydia	+
Lonicera	+
Luzula	+
Macrotomia	+
Milium	++
Mosses	+
Muscari	+
Myosotis	+
Onobrychis	++
Orchis	+
Orobus	++
Oxalis	+
Oxytropis	++
Pastinaca	+
Pedicularis	+
Petasites	+
Phleum	+++
Picea	++
Pinus	++
Plantago	+
Poa	++
Polygala	++
Polygonum	+
Polystichum	+
Populus	+
Potentilla	+
Primula	+
Psephellus	+
Pulsatilla	+
Ranunculus	+
Rhododendron	++
Rhynchocorys	+
Rosa	+
Rumex	+

Table 13. *Continued*

Plant genera	Kavkazsky Nature Reserve
Russula	+
Salix	+
Salvia	+
Sanguisorba	+
Saxifraga	+
Scabiosa	+
Senecio	++
Sibbaldia	+
Silene	+
Solidago	++
Swertia	+
Taraxacum	+
Trifolium	++
Trollius	++
Usnea	+
Vaccinium	++
Valeriana	+
Veratrum	+
Verbascum	+
Veronica	++
Viola	+

[a]Symbols indicate percent of food use was >30% (+++); 20–30% (++): 10–20% (+); <10% (r) of field observations, stomach contents also examined.

Table 14. Plant genera eaten by *Capra cylindricornis*.[a] (Tochiev and Batkhiev 1980; Sokolov and Tembotov 1993)

Species	Caucasus
Betula	+++
Campanula	+
Carex	+++
Euonymus	+++
Fabaceae	+++
Festuca	+++
Filipendula	+++
Lactarius	+
Pinus (needles)	+++
Plantago	+
Primula	+
Rosa	+++
Rubus	+++
Russula	+
Salix	+++
Vaccinium	+++

[a]Symbols indicate percent of food use was >30% (+++); 20–30% (++): 10–20% (+); <10% (r) of observations.

Table 15. Plant genera eaten by *Capra falconeri*[a]

Plant genera	Locality[b]	
	Tajikistan[1]	Uzbekistan[2]
Acer	+++	x
Agropyron		x
Amygdalus	++	x
Artemisia	+	x
Astragalus	++	
Bromus		x
Carex		x
Cerasus	+	x
Colutea	+	
Cotoneaster	+	
Crataegus	+	
Dianthus	+	
Ephedra		x
Eremurus		x
Ferula	+	x
Hordeum	+	x
Juniperus		x
Lonicera		x
Oxytropis	+	
Pistacia	+++	
Poa	+	x
Prangos	+	x
Pyrus		x
Rheum		x
Rosa		x
Rumex	+	
Salsola	+	
Scaligeria	+	
Scrophularia	+	
Ungernia	++	
Veronica		x
Vicia	+	x
Ziziphora		x

[a]Symbols indicate percent of observed food use was >30% (+++); 20–30% (++): 10–20% (+); <10% (r) ; x indicates food use noted, no numerical data reported.
[b]The superscript numbers denote the following references: 1, Sapozhnikov (1975); Sokov (1993); 2, Ishunin (1961).

Table 16. Plant genera eaten by *Capra hircus*.[a] (Gasparyan 1964)

Plant genera	Armenia
Achillea	++
Allium	++
Alyssum	+
Amygdalus	++
Asparagus	++
Bellevalia	+++
Centaurea	+
Cerasus	++
Chenopodium	++
Corydalis	++
Ephedra	+
Eryngium	++
Erysimum	++
Euphorbia	+
Eurotia	++
Gagea	++
Geranium	+
Gladiolus	++
Kochia	+
Lappula	+
Onobrychis	++
Papaver	+++
Prangos	+
Pyrethrum	++
Rosa	++
Rubia	++
Scilla	++
Scorzonera	++
Silene	++
Sonchus	+
Stachys	+
Thalictrum	+++
Thlaspi	+
Tragopogon	++
Zozimia	++

[a]Symbols indicate percent of observed food use was >30% (+++); 20–30% (++): 10–20% (+); <10% (r).

Table 17. Plant genera eaten by *Capra sibirica*[a]

Plant genera	Tajikistan[1]	Zailiyskoe Alatau Range (1)[2], Dzungarian Alatau Range (2)[2]		Altai[3]
		1	2	
Aegopodium			++	
Agropyron		+++		
Agrostis		+		
Alchemilla		+++	+++	
Alfredia		++		
Allium	x		+++	+++
Anemone				x
Anthriscus		+		
Antonina			+	
Artemisia	x	+++	+++	+++
Aster			+	
Astragalus		+++		
Atragene		+++		
Atraphaxis		+	+++	
Berberis		+	+	
Betula				+++
Bromus			+	
Cachrys			+	
Calamagrostis	x			
Campanula		+++		+++
Caragana				x
Carex		+++	+	
Centaurea		+	+	
Cerastium	x		+	
Chamaenerion		+++		
Cobresia	x		+++	
Codonopsis		+++		
Conioselinum		+	+	
Corydalis		+		
Cotoneaster	x	+++		
Cousinia	x			
Crepis		+++		
Dactylis		+++		
Delphinium	x			
Dictamnus			+	
Dracocephalum	x			
Echinops		++		
Elymus			+	
Ephedra	x	+++	+++	
Equisetum		+++		
Eremurus			+	
Erysimum			+	
Euonymus		+++		
Euphorbia			+++	
Ferula			+++	
Festuca	x		+++	
Galatella			+	

Table 17. *Continued*

Plant genera	Tajikistan[1]	Zailiyskoe Alatau Range (1)[2], Dzungarian Alatau Range (2)[2]		Altai[3]
		1	2	
Geranium	x	+	++	
Goniolimon			+	
Hedysarum		++	+++	
Heliotropium			+	
Hieracium		+++	+++	
Hypericum		+	+	
Iris			+	
Juncus		+		
Juniperus	x			
Kochia			+	
Koeleria		+++		
Lappula			+++	
Leonurus			+++	
Lepidium		++		
Libanotis		+++		
Ligularia		++		
Ligusticum		+++		
Lonicera		+		+++
Medicago			+	
Myosotis			+	
Origanum		++		
Oxytropis	x			
Pachypleurum			+++	
Padus				x
Patrinia			+++	
Pedicularis		+		
Phleum		+++	+++	
Phlomis			+++	
Phragmites		+		
Picea		+++		
Piptatherum	x			
Poa	x		+++	
Polemonium			+	+++
Polygonum		+++	+++	
Puccinellia	x			
Rheum			+	
Rhodiola			+++	
Rosa		+++	+++	+++
Rubus		++	++	
Rumex		+++		
Salix	x	++	+	+++
Saussurea			++	
Sedum		+++	+++	
Semenovia		+++		
Senecio		+++		
Silene		+		
Sisymbrium			+++	

Table 17. *Continued*

Plant genera	Tajikistan[1]	Zailiyskoe Alatau Range (1)[2], Dzungarian Alatau Range (2)[2]		Altai[3]
		1	2	
Sorbus		++		+++
Spiraea		+++	+++	+++
Stellaria			+	
Stilpnophleum	x			
Stipa	x		+++	
Thalictrum		+	++	
Trifolium			+	
Tulipa			++	
Tussilago		+		
Urtica		++		
Verbascum		+		
Vicia		++	+++	

[a]Symbols indicate percent of observed food use was >30% (+++); 20–30% (++): 10–20% (+); <10% (r) ; x indicates food use noted, no numerical data reported.
[b]The superscript numbers denote the following references: 1 Sapozhnikov (1984); 2 Fedosenko and Savinov (1983); 3 Sobansky (1992).

Table 18. Plant genera eaten by *Naemorhaedus caudatus*.[a] (Veinger 1963; Bromley and Kucherenko 1983)

Plant genera	Season			
	Wi	Sp	Su	Au
Acanthopanax		++	++	
Acer	r	+		
Adenophora			++	++
Allium		++	+++	
Aralia		r		
Angelica			+	r
Artemisia	+	+	+	+
Aruncus			+++	
Atriplex	++	++		
Bupleurum			+	
Calamagrostis		r		
Cacalia		+		
Carex	r	+++	r	
Carpinus		r		
Corylus		++		
Crataegus			+++	
Cystopteris		+		

Table 18. *Continued*

Plant genera	Season			
	Wi	Sp	Su	Au
Dianthus			r	
Eleutherococcus	r		++	
Elymus	++	+		
Euonymus	++	++	++	++
Festuca		+		
Galium			++	
Geranium		+	+	+
Gypsophila			++	+
Hierochloe		r	+	
Impatiens		+++	+++	+++
Iris		++	++	
Koeleria			+	
Lespedeza	+++	r	++	+++
Ligularia		+	+	+
Maackia	+++			
Malus			+	
Pedicularis		++	+++	
Plantago		+	+	
Poa			r	
Polygonum	+	+	+	
Pulsatilla			+	+
Quercus	+++	++		++
Rabdosia		+++	+++	
Rhododendron	++			
Rubus	+++	+++	+++	+++
Rubia		r	r	
Rumex		+++	+++	
Salix		+		
Sambucus		+++	+++	+++
Sanguisorba	r		r	r
Sanicula			r	
Saussurea			++	+
Serratula	++	++	++	
Solidago			+	
Spodiopogon			+++	
Thalictrum			+	+
Thermopsis			++	
Tilia	+++	r	+++	r
Trifolium	+++	+++	+++	+++
Vicia		+++	+++	
Vitis	+++	r	+	+++

[a]Symbols indicate % of observed food use was >30% (+++); 20–30% (++): 10–20% (+); <10% (r).

Table 19. Plant genera eaten by *Ovibos moschatus*.[a] (Rapota 1981; Chernyavsky et al. 1981; Zheleznov 1990)

Plant genera	Season	
	Au, Wi	Sp, Su
Alopecurus	x	x
Arctagrostis		x
Arctophilla	x	x
Artemisia		x
Astragalus	x	x
Calamagrostis		x
Carex	x	x
Cetraria	x	
Cladonia	x	
Deschampsia	x	x
Dryas	x	
Dupontia	x	x
Equisetum		x
Eriophorum	x	x
Festuca		x
Hedysarum		x
Hierochloe		x
Juncus		x
Koeleria	x	x
Lloydia		x
Nardosmia		x
Novosieversia		x
Oxyria	x	x
Oxytropis		x
Papaver		x
Parrya		x
Poa		x
Polemonium		x
Polygonum		x
Rumex	x	x
Salix	x	x
Saussurea		x
Saxifraga		x
Valeriana		x

[a]x indicates food use noted, no numerical data reported.

Table 20. Plant genera eaten by *Ovis ammon*[a]

Plant genera	Locality[b]		
	Kazakhstan[1]	Pamirs[2]	Altai[3]
Agropyron			+++
Allium	+++	++	
Alopecurus	++		
Artemisia	+++	++	+++
Aster	+++		++
Astragalus	++	++	+++
Avenastrum	+++		
Aulacospermum			++
Bromus		+++	
Carex	++	++	
Ceratocarpus		++	
Chamaenerion	++		
Cobresia			++
Cotoneaster	+++		
Dactylis	+++		
Deschampsia			+++
Dracocephalum			++
Elymus			+++
Erophila		++	
Festuca	+++		+++
Gentiana	+++		+++
Gypsophila	++		
Hedysarum	+++		+++
Helictotrichon			++
Hieracium	++		
Hordeum		++	
Kochia	++		
Koeleria	++		
Lagotis			+++
Leymus			+++
Ligularia			++
Libanotis	++		
Lonicera	+++		
Medicago	+++	+++	
Onobrychis	+++		
Onosma	++		
Orostachys	+++		
Oxytropis	+++		+++
Pachypleurum			++
Papaver			++
Patrinia	++		
Pedicularis	+++		+++
Phleum	++		
Polygonum	++	++	+++
Poa			+++
Potentilla	++	++	
Puccinellia			+++
Pulsatilla	++		+++

Table 20. *Continued*

Plant genera	Locality[b]		
	Kazakhstan[1]	Pamirs[2]	Altai[3]
Ranunculus		++	++
Ribes	+++		
Rosa	++		
Rumex			+++
Saussurea			+++
Sedum	+++		
Spiraea	+++		
Stipa	++		
Trifolium	+++		
Vicia	++	++	

[a]Symbols indicate percent of observed food use was >30% (+++); 20–30% (++): 10–20% (+); <10% (r).
[b]The superscript numbers denote the following references: 1, Fedosenko and Kapitonov (1983); Fedosenko et al. (1995); 2, Sapozhnikov (1984); 3, Dmitriev (1938); Sobansky (1992).

Table 21. Plant genera eaten by *Ovis aries*[a]

Plant genera	Season	
	Wi, Au	Sp, Su
Allium		x
Amygdalus	x	x
Asparagus		x
Bellevalia		x
Berberis	x	
Carex		x
Cerasus	x	x
Chenopodium		x
Corydalis		x
Eryngium		x
Erysimum		x
Euphorbia		x
Gladiolus		x
Juniperus	x	
Karkas	x	
Onobrychis		x
Papaver		x
Rosa	x	x
Salix	x	
Thalictrum		x

[a]x indicates food use noted, no numerical data reported. Yavruyan (1969).

Table 22. Plant genera eaten by *Ovis nivicola*[a]

Plant genera	Locality[b]			
	Putoran Mountains[1]	Yakutiya[2]	Okhotsk-Kolyma Upland[3]	Kamchatka Peninsula[4]
Agropyron		x		
Allium	x			
Alnus		x	r	
Alopecurus			+	
Andromeda		x		
Androsace				++
Anemone				+++
Antennaria				++
Arabis		x		
Arctagrostis		x	+	+++
Arctophila		+++		
Arctous			+++	x
Arenaria		x		
Artemisia		x	+++	++
Aster		x	+++	
Astragalus		x	+++	+++
Beckwithia			+	
Betula		+	++	
Bromopsis			+++	+++
Calamagrostis		x	+++	
Calluna			+	
Caragana		x		
Cardamine			+	
Carex		x	+++	++
Cassiope		x	+	++
Castilleja			+	++
Cerastium		x	+	
Chamaedaphne			+	
Chamaenerion		x	+++	
Chrysoplenium			+	
Claytonia			+++	
Delphinium			+	
Diapensia			+	
Draba		x		
Dracocephalum			+	
Dryas		x	+++	++
Dryopteris			+	x
Empetrum			+	
Ephedra		x		
Equisetum		x	+++	
Erigeron			+	
Eriophorum		x	x	
Eritrichium			+	
Festuca			+++	
Galium			+	
Gentiana			+	
Harrimanella				x

Table 22. *Continued*

Plant genera	Locality[b]			
	Putoran Mountains[1]	Yakutiya[2]	Okhotsk-Kolyma Upland[3]	Kamchatka Peninsula[4]
Hedysarum		x	+++	
Helictotrichon		x		
Hierochloe		x	+	
Juncus			+	+++
Juniperus		x		
Koeleria		x		
Lagotis			+	
Larix		x		
Ledum			x	
Leymus		x	+++	
Lloydia				++
Loiseleuria			+	+
Luzula			+	
Lycopodium				x
Minuartia				+
Myosotis			+	
Nardosmia			+++	
Neuroloma				++
Novosieversia			+	
Oxyria			+++	
Oxytropis		x	+++	++
Parrya				+
Pedicularis		x	+++	+++
Pentaphylloides			+	
Phlojodicarpus		x		
Phyllodoce			+	x
Pinus		x	r	x
Poa		x	+++	+++
Polemonium		xc	+	++
Polygonum		x	+++	
Populus		x		
Potentilla		x	+	
Pulsatilla		x	+	
Pyrola			+	
Rhodiola			x	+
Rhododendron			+	++
Ribes		x	+++	
Rosa		x		
Rubus			x	
Rumex		x	+++	
Salix		xx	+++	+++
Sanguisorba		x		
Saussurea			+	+
Saxifraga			+++	++
Sedum			+++	
Senecio			+	
Sorbaria			+++	

Table 22. *Continued*

Plant genera	Locality[b]			
	Putoran Mountains[1]	Yakutiya[2]	Okhotsk-Kolyma Upland[3]	Kamchatka Peninsula[4]
Spiraea			+++	++
Stellaria			+	
Taraxacum		x	+	
Thalictrum				+
Vaccinium		x	+++	
Valeriana			+	x
Veronica				++
Viola			x	+

[a]Symbols indicate percent of observed food use was >30% (+++); 20–30% (++): 10–20% (+); <10% (r) ; x indicates food use noted, no numerical data reported.
[b]The superscript numbers denote the following references: 1, Fedosenko (1985); Larin and Shelkovnikova (1988); 2, Egorov (1965); Revin et al. (1988); 3, Zheleznov (1990); Zheleznov-Chukotsky (1994); 4, Averin (1948); Mosolov (1986).

Table 23. Plant genera eaten by *Ovis vignei*[a]

Plant genera	Locality[b]	
	Ustyurt[1]	Tajikistan[2]
Agropyron	++	
Alhagi		+
Allium	++	
Amaranthus	++	
Amygdalus		+++
Artemisia	++	+
Astragalus	+++	+
Atraphaxis	++	
Atriplex	++	
Bromus		+++
Calligonum	++	
Caragana	+++	+++
Carex	++	
Cenolophium	++	
Centaurea		+++
Cerasus		+
Cousinia	++	+
Crataegus		+
Delphinium		+
Ephedra	++	+
Euphorbia		+++
Ferula		+
Garhadiolus	++	
Haplophyllum		+

Table 23. *Continued*

Plant genera	Locality[b]	
	Ustyurt[1]	Tajikistan[2]
Hypericum		+
Jurinea	++	
Koelpinia		+++
Lepidium	++	
Medicago		+++
Onobrychis		+++
Peganum	++	
Phragmites	++	
Pistacia		+++
Poa	++	+
Potentilla		+++
Rosa		+
Rumex		+++
Salsola	++	+++
Saussurea	++	
Scaligeria		+
Schumannia	++	
Seseli	++	
Suasda	++	
Syrenia	++	
Tragopogon	++	
Trigonella		+++
Zosima	++	

[a]Symbols indicate % of observed food use was >30% (+++); 20–30% (++): 10–20% (+); <10% (r).
[b]The superscript numbers denote the following references: 1, Fedosenko (1984a), 2, Sapozhnikov (1984).

Table 24. Plant genera eaten by *Rupicapra rupicapra*.[a] (Nasimovich 1949b; Duben 1985)

Plant genera	Kavkazsky Nature Reserve
Abies	r
Acer	+
Aegopodium	r
Agrostis	+
Alchemilla	+++
Alnus	r
Alopecurus	+
Anemone	+
Anthemis	++
Anthoxanthum	++
Aquilegia	+
Astragalus	r
Berberis	r

Table 24. *Continued*

Plant genera	Kavkazsky Nature Reserve
Betonica	++
Betula	+
Bupleurum	r
Calamagrostis	++
Campanula	+
Cardamine	+
Carex	+
Carpinus	r
Carum	+
Castanea	++
Centaurea	++
Cephalaria	++
Cerasus	r
Cerinthe	r
Cetraria	++
Chaerophyllum	+++
Chamaenerion	++
Chamaesciadium	+
Cicerbita	+++
Cirsium	r
Colpodium	++
Corydalis	r
Dactylis	r
Deschampsia	+++
Dianthus	+
Erigeron	r
Fagus	++
Festuca	+++
Fragaria	+
Gentiana	++
Geranium	+
Grossularia	+
Hedera	r
Hedysarum	++
Heracleum	+
Hieracium	++
Inula	+
Juniperus	r
Jurinea	+++
Knautia	+
Kobresia	+
Lactuca	+
Lapsana	+
Laurocerasus	++
Leontodon	++
Luzula	r
Macrotomia	r
Malus	+
Milium	+++
Muscari	+

Table 24. *Continued*

Plant genera	Kavkazsky Nature Reserve
Myosotis	+
Nardus	r
Onobrychis	+
Oxytropis	+++
Pedicularis	r
Petasites	r
Phleum	++
Picea	r
Pinus	r
Plantago	+++
Poa	++
Polygala	r
Polygonum	+
Populus	++
Potentilla	r
Prunus	+
Psephellus	+
Pulmonaria	r
Pulsatilla	+
Pyrus	++
Quercus	++
Ranunculus	r
Rhododendron	++
Ribes	++
Rosa	r
Rubus	+++
Rumex	++
Ruscus	+++
Salix	+++
Scabiosa	+
Scrophularia	+
Secale	++
Sedum	r
Senecio	++
Sibbaldia	+
Silene	++
Sisymbrium	r
Sorbus	+
Taraxacum	+
Thymus	r
Tragopogon	++
Trifolium	++
Trollius	+
Usnea	+
Vaccinium	++
Valeriana	++
Veronica	r
Vicia	+
Viola	r
Viscum	+++

[a]Symbols indicate percent of observed food use was >30% (+++); 20–30% (++): 10–20% (+); <10% (r).

Appendix 2
Helminths and Ectoparasites

Table 1. Helminths and ectoparasites of *Equus hemionus*. (Rashek 1964; Solomatin 1973; Bannikov 1981; Ivashkin and Dvoynos 1984; Rastegaev 1992)

Helminths	Ectoparasites
Alfortia	*Bovicola*
Anoplocephala	
Gasterophilus	
Craterostomum	*Rhinoestrus*
Cylicocyclus	
Cylicodontophorus	
Delafondia	
Dictyocaulus	
Gyalocephalus	
Habronema	
Oxyuris	
Parascaris	
Petroviema	
Poteriostomum	
Probstmayria	
Schultzitrichonema	
Setaria	
Trichonema	
Triodontophorus	

Table 2. Helminths and ectoparasites of *Sus scrofa*. (Rukhlyadev 1964; Egorov 1994, 1995; Gorelov 1994; Govorka et al. 1988)

Helminths	Ectoparasites
Alaria	*Boophilus*
Alveococcus	*Dermacentor*
Ascaris	*Haematopinus*
Ascorops	*Hyaloma*
Bourgelatia	*Rhipicephalus*
Capillaria	
Chabertia	
Coenurus	
Cysticercus	
Dicrocoelium	
Diphyllobotrium	
Echinochasmus	
Echinococcus	
Eupariphium	
Fasciola	
Gastrodiscoides	
Globocephalus	
Gnathostoma	
Gongylonema	
Hyostrongylus	
Macracanthorhynchus	
Metacercaria	
Metastrongylus	
Oesophagostomum	
Paracoenogonimus	
Physaloptera	
Physocephalus	
Rabditata	
Setaria	
Simondsia	
Songylonema	
Sparganum	
Spirocerca	
Spirometra	
Strigea	
Strongyloides	
Taenia	
Trichinella	
Trichocephalus	
Trichostrongulus	

Table 3. Helminths and ectoparasites of *Moschus moschiferus*. (Egorov 1971; Bromley and Kucherenko 1983)

Helminths	Ectoparasites
Armocappilaria	*Cerathophylla*
Cestodes	*Cordylodia*
Cysticercus	*Ixodes*
Dicrocoelium	*Trichodectes*
Dyctyocaulon	*Trombidiidae*
Eimeria	*Xanthocalliphorinae*
Moniezia	
Moniza	
Nematodirella	
Nematodirus	
Pneumocaulus	
Protostrongylus	
Sarcosporidia	
Setaria	
Taenia	
Thysaniezia	
Trematoda	
Trichocephalus	

Table 4. Helminths of *Cervidae*.[a] (Gubanov 1964; Rukhlyadev 1964; Pyatkov and Pryadko 1971; Yanushevich et al. 1972; Prisyazhnyuk and Pigolkin 1974; Pryadko 1976; Shostak and Vasilyuk 1976; Litvinov 1981; Lunitsyn 1988; Prostakov 1996; Egorov 1994)

	C. elaphus	*C. nippon*	*A. alces*	*C.* spp.	*R. tarandus*
Acanthospiculum	x		x	x	
Achworthius	x	x	x	x	
Alcefilarai					x
Alveococcus					x
Aonchotheca	x		x	x	
Ascaris	x		x	x	x
Ascarops	x		x	x	x
Ashworthius	x		x	x	
Avitellina			x	x	x
Bicaulus	x		x	x	
Bunostomum	x	x	x	x	
Capillaria	x	x	x	x	x
Capreocaulus	x		x	x	
Capreolagia	x			x	
Cephenemyja	x		x	x	
Chabertia	x		x	x	
Cooperia	x	x	x	x	x
Cotilophorum	x				
Cysticercus	x	x	x		
Dicrocoelium	x	x	x	x	x
Dictyocaulus	x	x	x	x	x

Table 4. *Continued*

	C. elaphus	*C. nippon*	*A. alces*	*C.* spp.	R. tarandus
Echinococcus	x	x	x	x	x
Elaeophora			x		
Elaphostrongylus	x	x	x	x	x
Eurytrema	x	x	x	x	
Fasciola	x	x	x	x	x
Fascioloides					x
Fischoederius	x	x	x	x	x
Globocephalus	x		x	x	
Gongylonema	x	x	x	x	
Haemonchus			x	x	
Liorchis		x			x
Marshallagia			x	x	x
Mecistocirrus					x
Metacercaria	x		x	x	
Metastrongylus	x		x	x	
Moniezia	x	x	x	x	x
Muellerius				x	
Multiceps	x		x	x	x
Nematodirella	x	x	x	x	x
Nematodirus	x	x	x	x	x
Oesophagostomum	x	x	x	x	x
Ogmocotyle	x	x			
Onchocerca	x	x	x		x
Orloffia			x	x	
Ostertagia	x	x	x	x	x
Parabronema	x	x	x	x	x
Parafasciolopsis	x	x	x	x	
Parafilaria	x				
Paramphistomatis					x
Paramphistomum	x	x	x	x	
Parelaphostrongylus	x				x
Pneumocaulus		x			
Pygarginema		x	x	x	
Rinadia	x	x	x	x	
Schulzinema	x	x	x	x	x
Setaria	x	x	x	x	x
Skrjabinagia	x	x	x	x	x
Skrjabinema	x		x	x	x
Sparganum	x		x	x	
Spiculocaulus				x	
Spiculopteragia	x	x	x	x	x
Strongyloides	x		x	x	
Taenia	x	x	x	x	x
Taeniarhynchus	x	x	x	x	x
Teladorsagia					x
Thelaria	x		x		
Thysaniezia	x			x	x
Trichinella	x		x	x	x
Trichocephalus	x	x	x	x	x
Trichostrongylus	x	x	x	x	x
Wardecmansia	x	x	x	x	x
Varestrongylus	x	x	x		

[a]x indicates occurence of parasite noted.

Table 5. Ectoparasites of *Cervidae*.[a] (Grunin 1957, 1962; Gomoyunova 1976; Bromley and Kucherenko 1983; Lunitsyn 1988; Raabe and Solopov 1992; Prostakov 1996)

Genera	*C. elapus*	*C. nippon*	*A. alces*	*C.* spp.	*R. tarandus*
Aedes	x		x	x	x
Allantotrichus		x			
Anopheles	x		x	x	
Booponus	x	x	x	x	
Cephenemyia	x		x	x	
Chrysops	x	x	x	x	x
Culex	x	x	x	x	x
Culicoides		x			
Demodex		x			
Dermacentor	x	x	x	x	
Haemaphysalis	x		x	x	
Haematobia	x		x	x	
Haematopota	x	x	x	x	x
Heleidae	x		x	x	
Hybomitra					x
Hypoderma	x	x	x	x	x
Ixodes	x	x	x	x	
Lipoptena	x	x	x	x	
Oedemagena					x
Oestrus	x		x	x	
Pharingomyia	x	x			x
Simulidae	x		x	x	
Simulium	x	x			x
Stomoxyscalutrans	x		x	x	
Tabanus	x	x	x	x	x
Trichodectes	x		x	x	
Vermipsylla	x	x			

[a]x indicates occurence of parasite noted.

Table 6. Helminths and ectoparasites of Antilopinae.[a] (Boev et al. 1962, 1963; Grunin 1962; Bannikov et al. 1961; Petrov 1983, 1985; Fadeev and Sludsky 1983; Nazarova and Prisyazhnyuk 1986; Berkinbaev 1992)

Helminths			Ectoparasites		
G. subgutturosa (1), *S. tatarica* (2)			*G. subgutturosa* (1), *S. tatarica* (2)		
Genera	1	2	Genera	1	2
Alveococcus		x	*Argas*		x
Avitellina		x	*Damalinia*	x	
Chabertia		x	*Dermacentor*		x
Coenurus	x		*Hyalomma*	x	x
Cysticercus	x		*Linognathus*	x	
Echinococcus		x	*Pallasiomyia*		x
Eimeria	x	x	*Pavlovskiata*	x	
Fasciola	x		*Przhevalskiana*	x	
Haemonchus		x	*Rhipicephalus*	x	x
Marshallagia	x	x			
Moniezia	x	x			
Moniliformis		x			
Multiceps	x	x			
Nematodirella	x	x			
Nematodirus	x	x			
Oesophagostomum		x			
Ostertagiella	x	x			
Parafilaria		x			
Setaria	x				
Skrjabinema	x				
Taenia	x				
Trichocephalus	x				

[a]x indicates occurence of parasite noted.

Table 7. Helminths and ectoparasites of *Bison bonasus*. (Zablotsky 1957; Kozlov and Nazarova 1979; Nasimovich 1989)

Helminths	Ectoparasites
Bunostomum	*Aedes*
Capillaria	*Anopheles*
Chabertia	*Booponus*
Coenurus	*Cephenomyja*
Cooperia	*Culex*
Dicrocoelium	*Dermacentor*
Dictyocaulus	*Haemaphysalis*
Fasciola	*Haematobia*
Gongylonema	Heleidae
Haemonchus	*Hypoderma*
Liorchis	*Ixodes*
Moniezia	*Lipoptena*
Nematodirella	*Oestrus*
Nematodirus	Simulidae
Neoascaris	*Stomoxyscalutrans*
Oesophagostomum	*Tabanus*
Ostertagia	*Trichodectes*
Paramphistomum	
Setaria	
Spiculopteragia	
Thelazia	
Trichocephalus	
Trichostrongylus	

Table 8. Helminths of Caprinae.[a] (Grigoryan 1949; Asadov 1960; Boev et al. 1962, 1963; Bromley 1964; Gubanov 1964; Rukhlyadev 1964; Ovsyukova 1966; Mambetzhumaev 1968; Zheleznov 1980; Fedosenko and Kapitonov 1983; Dobrynin et al. 1992; Voloshina and Solomkina 1992)

	C. caucasica, C. cylindricornis	*C. sibirica*	*N. caudatus*	*O. ammon*	*O. aries*	*O. nivicola*	*O. vignei*	*Rupicapra rupicapra*
Alveococcus				x				
Avitellina				x				
Bunostomum	x				x		x	
Camelostrongylus							x	
Capillaria	x			x	x		x	
Capreolagia					x			
Chabertia		x		x	x			
Coenurus	x							
Cooperia					x			
Cysticercus	x			x	x			x
Cystocaulus	x	x		x	x			x
Dicrocoelium		x		x	x			
Dictyocaulus	x			x	x		x	
Diphyllobotrium			x					
Echinococcus				x	x		x	x
Eimeria		x		x				
Fasciola				x				
Gelanocaulus	x							
Gongylonema					x		x	
Haemonchus	x			x	x			x
Marshallagia	x	x		x	x	x	x	x
Moniezia	x	x	x	x	x	x	x	x
Moniliformis				x				
Muellerius	x				x			
Multiceps		x		x				
Nematodirella				x		x	x	
Nematodirus	x	x	x	x	x	x	x	x
Neostrongylus	x	x			x			x
Oesophagostomum	x			x	x		x	
Ostertagia	x	x	x	x	x	x	x	x
Ostertagiella		x		x			x	
Parabronema				x			x	
Parafilaria				x				
Protostrongilus		x	x	x	x	x	x	
Rinadia					x			
Sarcocystis				x				
Setaria				x				
Skrjabinagia				x				
Skrjabinema		x		x	x	x	x	
Skrjabinodera				x				
Spiculocaulus		x		x		x		x
Spiculopteragia					x			
Stilesia						x		
Strongyloides				x				
Taenia		x		x				
Teladorsagia	x						x	
Thelazia				x				
Thysaniezia				x	x		x	
Toxoplasma				x				
Trichocephalus	x		x	x	x	x	x	x
Trichostrongylus	x			x	x		x	x

[a]x indicates occurence of parasite noted.

Table 9. Ectoparasites of Caprinae.[a] (Myslenkov and Voloshina 1989; Fedosenko and Savinov, 1983; Govorka et al. 1988; Voloshina and Solomkina 1992)

	Capra sibirica	*N. caudatus*	*Ovis ammon*	*Ovis vignei*
Cruvellia			x	
Dermacentor	x	x	x	
Haemaphysalis	x	x		x
Hyalomma	x		x	x
Ixodes		x	x	
Linognathoides	x			
Liporosia			x	
Melophagus	x		x	
Oestrus	x		x	
Ornithodorus				x
Rhinoestrus			x	
Rhipicephalus	x		x	x
Vermipsylla	x		x	
Xenopsylla	x			

[a]x indicates occurence of parasite noted.

References

Abaturov VD (1984) Mammals as a component of the ecosystem. Nauka, Moscow (in Russian)

Abaturov VD, Smirnov KA (1992) Moose browsing impact on reforestation in Central European USSR. Alces [Suppl] 1:213

Abaturov VD, Kholodova MV, Subbotin AE (1982) Intensity of consuming and ability to digest forage by saiga (*Saiga tatarica*) (in Russian with English summary). Zool Zh 61(12):1870–1881

Abdurakhmanov MG (1973) Ecology, conservation and use of goats of Dagestan. Thesis. Dagestansky gosudarstvennyi universitet, Makhachkala (in Russian)

Abdurakhmanov MG (1977) Caucasian tur. In: Kaletsky AA (ed) Redkie zhivotnye SSSR. Kopytnye zveri. Lesnaya Promyshlennost, Moscow, pp 186–200 (in Russian)

Abdusalyamov IA (1997) Status and conservation of Pamirs mountain sheep *Ovis ammon polii* in Tadjikistan. In: Sokolov VE (ed) Redkie vidy mlekopitayushchikh Rossii i sopredelnykh stran. Institut Problem Ecologii i Evolutsii, Moscow, p 4 (in Russian)

Abramov KG (1963a) Ungulates of the Far East and their hunting. Primorsckoe knizhnoe izdatelstvo, Vladivostok (in Russian)

Abramov KG (1963b) Notes on moose in Amur area. Voprosy geografii Dalnego Vostoka 5:95–98 (in Russian)

Adlerberg GP (1950) Preliminary review of Russian and Mongolian boars. Dok Akad Nauk SSSR 70:91–99 (in Russian with English summary)

Adolf TA (1957) Dynamics of range and numbers of saiga. Okhrana Prirody i Zapovednoe Delo v SSSR 2:125–139 (in Russian)

Afanasev AV, Sludsky AA (1947) Data on mammals and birds of Central Kazakhstan. Izvest Akad Nauk Kazakhskoi SSR, Ser Biol 36(6):48–64 (in Russian)

Akaevsky AI (1939) Anatomy of reindeer. Trudy Nauchno-isledovatelskogo instituta polyarnogo zemledeliya, zhivotnovodstva i promyslovogo khozyaistva, series Olenevodstvo 5:1–327 (in Russian)

Akhmedov EG (1995) Structure-functional organization of Dagestan goat populations (*Capra cylindricornis* Blyth) in eastern Caucasus. Dissertation (ms). Dagestansky nauchnyi centr, Moscow (in Russian)

Akhmedov EG (1999) The bezoar goat population structure in the optimum stations. In: Aristov AA (ed) Redkie vidy mlekopitayushchikh Rossii i sopredelnykh territoriy. Teriologicheskoe Obshchestvo, Moscow, pp 5–10 (in Russian with English summary)

Akhmedov EG, Magomedov MRD (1995) Dependence of age-sex dynamics of Dagestan *Capra cylindricornis* on habitat conditions and population density. Zool Zh 74(7):109–118 (in Russian with English summary)

Alekperov KM (1966) Mammals of the south-west Azerbajian. Academy of Sciences of Azerbaijan SSR, Baku (in Russian)

Alekperov KM, Kuliev SM (1987) Some data on wild boar and roe deer ecology in Kuba-Khachmasskoi zone of Azerbadijan. In: Bolshakov VN (ed) Ekologiya i okhrana

gornykh vidov mlekopitayushchikh. Institut Ekologii i Morphologii Zhivotnykh, Moscow, pp 8–9 (in Russian)
Alexandrov VN (1958) Study of natural diet of bison in the Kavkazsky Nature Reserve. Trudy Kavkazskogo Gosudarstvennogo Zapovednika 4:37–55 (in Russian)
Alexandrov VN (1968) Ecology of Caucasian deer. Trudy Kavkazskogo Gosudarstvennogo Zapovednika 10:95–200 (in Russian)
Alexandrova IV, Krasovsky, LI (1957) Observations on moose summer feeding in the Prioksko-Terrasny Nature Reserve. Trudy Prioksko-Terrasnogo gosudarstvennogo zapovednika 1:157–166 (in Russian)
Alexandrova VD (1937) Winter diet of tame reindeer on Novaya Zemlya, archipelago. Sovetskoe Olenevodstvo 9:127–139 (in Russian)
Alexandrova VD (1940) Food characteristics of the northern plants. Trudy Nauchno-isledovatelskogo instituta polyarnogo zemledeliya, zhivotnovodstva i promyslovogo khozyaistva, Series Olenevodstvo 11:1–96 (in Russian)
Andreev VN (1940) Pastures and pasture rotation in reindeer husbandry. In: Gulchak YF, Troshin DM (eds) Voprosy olenevodstva Krainego Severa. VASKHNIL publications, Moscow, pp 18–39 (in Russian)
Andreev VN (1968) Problems of rational utilisation and improvement of reindeer pastures. Problemy Severa 13:76–87 (in Russian)
Annenkov BP (1989) Boar in the Alakul basin. In: Sablina TB (ed) Ekologiya, morphologiya, ispolzovanie i okhrana dikikh kopytnykh, vol 2. Nauka, Moscow, pp 191–192 (in Russian)
Annenkov BP (1992) Modern distribution, numbers and conservation of rare mammals in Dzungarian Alatau Range. In: Fedosenko AK (ed) Okhrana i izuchenye redkikh i ischezayushchikh zhivotnykh v zapovednikakh. Tsentralnaya nauchno-issledovatelskaya laboratoriya okhotnichego khozyaistva i zapovednikov, Moscow, pp 69–74 (in Russian)
Antipin VM (1941) Ungulates. In: Afanasev AV (ed) Mlekopitayushchie Kazakhstana, vol 3. Kazakhskoe obedinennoe gosudarstvennoe izdatelstvo, Alma-Ata, pp 1–107 (in Russian)
Antipin VM (1947) Ecology, origin and distribution of wild sheep in Kazakhstan. Izvest Akad Nauk Kazakhskoi SSR, Ser Zool 6:3–22 (in Russian)
Antonez NV (1993) On peculiarities of parameters of wild boar in the Khopyorsky Nature Reserve. In: Noskov GA (ed) Vid i ego productivnost v areale. Hidrometeo, St Petersburg, pp 35–37 (in Russian)
Antonov AL (1999) Status of ungulate populations in pine-broadleaved forests of Amur area under human influence. Thesis. Institut vodnykh I ekologicheskikh problem, Vladivostok (in Russian)
Arabuli AB (1963) On feeding of the European roe deer (*Capreolus capreolus* L.) on the Tsiv-Gombor ridge. Dok Gruzinskoi Akad Nauk 30(4):467–474 (in Russian with English summary)
Arabuli AB (1966) Roe deer of Eastern Georgia. Thesis. Tbilissky gosudarstvennyi universitet, Tbilisi (in Russian)
Arabuli AB (1967) Study of roe deer (*Capreolus capreolus*) ecology in Kakhetia Caucasus. Trudy Instituta zoologii Akademii Nauk Gruzinskoi SSR, Materiali k Faune Gruzii 2:166–199 (in Russian)
Arabuli AB (1975) Wolf impact on numbers of Caucasian deer in the Borzhomi Nature Reserve. In: Sokolov VE (ed) Kopytnye fauny SSSR, vol 1. Nauka, Moscow, pp 189–190 (in Russian)
Arabuli AB (1989) Modern status of rare ungulate species of Georgia. In: Sokolov VE (ed) Kopytnye fauny SSSR, vol 1. Nauka, Moscow, pp 28–29 (in Russian)
Asadov SM (1960) Helminths of ruminant animals of USSR and its ecological-geographical analysis. Izdatelstvo Akademii nauk Azerbajanskoi SSR, Baku (in Russian)
Aspisov DI (1930) Moose in Tatarskoi republic. Materiali Po Izucheniju i Okhrane Pamjatnikov Prirody TSSR 1:15–34 (in Russian)
Aspisov DI (1956) Distribution and numbers of moose in Tatarstan republic. Uchenye Zapiski Kazanskogo Gosudarstvennogo Universiteta 115(8):231–239 (in Russian)

Averin YV (1948) Land mammals of the eastern Kamchatka. Trudy Kronotskogo Gosudarstvennogo Zapovednika 1:1–223 (in Russian)

Averin YV (1951) *Ovis nivicola* at Kamchatka. Byulleten Moskovskogo Obshchestva Ispytatelei Prirody, Otdelenie Biologicheskoe 56(1):15–26 (in Russian with English summary)

Avramchik MN (1939) Winter feeding of reindeer on Yamal North. Trudy Nauchno-issledovatelskogo instituta polyarnogo zemledeliya, zhivotnovodstva i promyslovogo khozyaistva, series Olenevodstvo 4:47–67 (in Russian)

Ayrumyan VA, Gasparyan KM (1976) Rare ungulates and predatory mammals of Armenia. In: Sokolov VE (ed) Redkie vidy mlekopitayushchikh fauny SSSR i ikh okhrana. Nauka, Moscow, pp 35–42 (in Russian)

Ayzin BM (1969) Game animals of Kyrgyzstan. Ilym, Frunze (in Russian)

Azarov VI (1976) Wild reindeer migrations and changes of wintering places in the Konda basin. Nauchno-issledovatelsky byulleten Naucho-issledovatelskogo instituta Krainego Severa 12/13:3–5 (in Russian)

Azarov VI, Klimov YP (1990) Wild boar of Tumen oblast. In: Evsikov VI (ed) Resursy zhivotnogo mira Sibiri. Okhotnichi zhivotnye i ptitsy. Nauka, Novosibirsk, pp 187–189 (in Russian)

Baidavletov RZ (1984) Musk deer – (*Moschus moschiferus*). Moose – (*Alces alces*). In: Gvozdev EV, Strautman EI (eds) Mlekopitayushchie Kazakhstana, vol 3(4). Nauka, Alma-Ata, pp 88–144 (in Russian)

Baidavletov RZ (1988) Ecology and behaviour of moose in Western Altai. In: Bekenov AB (ed) Ekologiya i povedenie mlekopitayushchikh v Kazakhstane. Nauka, Alma-Ata, pp 7–35 (in Russian)

Baidavletov RZ (1997) Exterior of Kazakhstan mountain sheep. In: Sokolov VE (ed) Redkie vidy mlekopitayushchikh Rossii i sopredelnykh stran. Institut Problem Ecologii i Evolutsii, Moscow, pp 9 (in Russian)

Baidavletov RZ, Murzov VN (1982) Aerial counts of ungulates in Western and Northern Kazakhstan. In: Gvozdev EV (ed) Zhivotnyi mir Kazakhstana i problemy ego okhrany. Nauka, Alma-ata, pp 13–15 (in Russian)

Bailey RG (1989) Explanatory supplement to ecoregions map of the continents. Environ Conserv 16:307–309

Bailey RG (1998). Ecoregions. The ecosystem geography of the oceans and continents. Springer, Berlin Heidelberg New York

Bakhmutov VA, Azarov VI (1981) Distribution, numbers and migrations of reindeer in the north of the Tyumen oblast. In: Dobrinsky LN, Sosin VF (eds) Chislennost i raspredelenie nazemnykh pozvonochnykh Yamala i prilegayushchikh territoriy. Uralsky Nauchnyi Centr, Sverdlovsk, pp 19–26 (in Russian)

Balĉauskas LP (1988) Spatial distribution of ungulates in the anthropogenic landscape of Lithuania. Dissertation (ms). Vilnius (in Russian)

Balĉauskas LP (1990) Trail net as reflection of animals movements and use of the inhabited territory. In: Bluzma P (ed) Mammals in the cultural landscape of Lithuania. Mokslas, Vilnius, pp 131–145 (in Russian with English summary)

Baleiŝis R (1990) Spatial migrations of moose. In: Bluzma P (ed) Mammals in the cultural landscape of Lithuania. Mokslas, Vilnius, pp 146–150 (in Russian with English summary)

Baleiŝis R, Bluzma P, Balĉiauskas L (2002) Lietuvos kano piniai -verys. Akstis, Vilnius (in Lithuanian with English summary)

Bannikov AG (1972) The numbers of the Bukharsky deer have increased. Priroda 10:110 (in Russian)

Bannikov AG (ed) (1978) Red data book of the USSR. Lesnaya promyshlennost, Moscow (in Russian)

Bannikov AG (1979) Gazelle in Azerbaijan, hopes for survival. Priroda 10:109–110 (in Russian)

Bannikov AG (1981) Kulan. Lesnaya Promyshlennost, Moscow (in Russian)

Bannikov AG (1989) Biology and economic importance of kulan, wild boar, saiga, goitered gazelle, mongolian gazelle. In: Heptner VG, Naumov NP (eds) Mammals of the Soviet Union, vol 1. Ungulates. Model Press Pvt, Leiden, pp 45–83, 608–617, 636–642, 662–672, 1025–1036 (in Russian)

Bannikov AG, Lebedeva LS (1956) Role of red deer in the forest of the Belovezhskaya Forest. Byulleten Moskovskogo obshchestva ispytatelei prirody, otdelenie biologicheskoe 61(4):75–80 (in Russian with English summary)

Bannikov AG, Pivovarova EP (1983) Red deer in the USSR. In: Belov AD (ed) Biologicheskie osnovy ispolzovaniya i okhrany okhotnichikh zhivotnykh. Moskovskaya Veterinarnaya Akademiya, Moscow, pp 34–40 (in Russian)

Bannikov AG, Pivovarova EP, Fandeev AA (1983) Principles of rational use of some species of ungulates in the USSR. Sbornik Trudov Moskovskoi Veterinarnoi Akademii 125:88–99 (in Russian)

Bannikov AG, Prisyazhnyuk VE (1977) Aboriginal sika deer. Priroda 7:96–106 (in Russian)

Bannikov AG, Teplov VP (1964) Changes of numbers and density of moose population in RSFSR. In: Bannikov AG (ed) Biologiya i promysel losya, vol 1. Rosselkhozizdat, Moscow, pp 5–12 (in Russian)

Bannikov AG, Zhirnov LV, Lebedeva LS, Fandeev AA (1961) Biology of saiga. Selkhozizdat, Moscow (in Russian)

Bannikov AG, Treus VD, Lobanov NV (1976) Przewalsky horse and turkmenian kulan in Askaniya-Nova. Sbornik Trudov Moskovskoi Veterinarnoi Akademii 84:83–90 (in Russian)

Bannikov AG, Ustinov SK, Lobanov PN (1978) Musk deer in the USSR. Sbornik Trudov Moskovskoi Veterinarnoi Akademii 97:6–36 (in Russian)

Barabash-Nikiforov II (1957) New data on intensive settling of some animal species during recent decade. Zool Zh 35(2):303–310 (in Russian with English summary)

Barancheev LM (1962) Massive unregular roe deer migrations in Amur oblast. In: Pokrovsky VS (ed) Migratsii zhivotnykh, vol 3. Nauka, Moscow, pp 26–36 (in Russian)

Baratashvili T (1980) Wild ungulates of the Borhomi Nature Reserve. In: Sokolov VE (ed) Kopytnye fauny SSSR. Nauka, Moscow, pp 62–63 (in Russian)

Bashkirov I (1939) Problems of restoration of Caucasian European bison. Nauchno-metodicheskie zapiski Komiteta po zapovednikam 5:138–145 (in Russian)

Baskin LM (1965) Feeding of red deer. Zool Zh 44(3):465–469 (in Russian with English summary)

Baskin LM (1968) Distribution of some mammals on the Kamchatka peninsula. Nauchnye Doklady vysshei shkoly. Biologicheskie Nauki 1:11–12 (in Russian)

Baskin LM (1970) Reindeer. Ecology and behaviour. Nauka, Moscow (in Russian)

Baskin LM (1974) Management of ungulate herds in relation to domestication. IUCN Publ New Ser 24(2):530–541

Baskin LM (1976) Behaviour of ungulates. Nauka, Moscow (in Russian)

Baskin LM (1979) Ecology and behaviour of European bison. In: Sokolov VE (ed) Zubr. Morphologiya, sistematika, evolutsiya, ekologiya. Nauka, Moscow, pp 442–470 (in Russian)

Baskin LM (1983) The causes of reindeer mortality. Acta Zool Fenn 175:133–134

Baskin LM (1986) Differences in the ecology and behaviour of reindeer populations in the USSR. Rangifer 1:333–340

Baskin LM (1989) Herding. Reindeer husbandry in the Soviet Union. In: Hudson RJ, Drew KR, Baskin LM (eds) Wildlife Production Systems. Cambridge University Press, Cambridge, pp 187–206

Baskin LM (1990) Population dynamics of reindeer. Rangifer 3:151–156

Baskin L (1992) Reindeer husbandry in the Soviet Union. In: Renecker L, Hudson J (eds) Wildlife production: conservation and sustainable development. University of Alaska, Fairbanks, pp 218–226

Baskin LM (1994) Population ecology of the moose in the Russian southern taiga. Alces 30:51–55

Baskin LM (1998) Moose conservation in ecosystems of Eastern Europe. Alces 34(2):395–407

Baskin LM, Skogland T (1997) Direction of escape in reindeer. Rangifer 17(1):37–40
Batkhiev AM (1980) Distribution and numbers of *Capra hircus* in Chechen-Ingush Republic. In: Tembotov AK (ed) Fauna, ekologiya i okhrana zhivotnykh Severnogo Kavkaza. Checheno-Ingushskoe Knizhnoe Izdatelstvo, Nalchik, pp 46–56 (in Russian)
Batkhiev AM (1989) *Capra hircus* in mountains of Chechen-Ingush Republic (ecological analysis of geographical distribution). In: Anisimov VI (ed) Priroda i khozyaistvo Checheno-Ingushskoi ASSR, vol 5. Checheno-Ingushskoe Knizhnoe Izdatelstvo, Groznyi, pp 103–109 (in Russian)
Bekenov AB (1988) Analysis of mammalian desert fauna in Kazakhstan. In: Bekenov AB (ed) Ekologyia i povedenie mlekopitayushchikh Kazakhstana. Nauka, Alma-Ata, pp 166–183 (in Russian)
Bekenov AB (1999) Resources of game mammals in Kazakhstan and their use. In: Orlov VN (ed) VI sezd Teriologicheskogo obshchestva. Teriologicheskoe Obshchestvo, Moscow, pp 23 (in Russian)
Bekenov AB, Grachev YA (1998) Kazakhstan. In: Sokolov VE, Zhirnov LV (eds) Saiga. Filogeniya, sistematika, ekologiya, okhrana i ispolzovanie. Nauka, Moscow, pp 105–115 (in Russian)
Bekenov AB, Grachev YA, Baidavletov RZ, Ismagulov E (1990) Actual state of mammal populations in Western Kazakhstan and perspectives of their use and protection. In: Bekenov AB (ed) Okhrana, ispolsovanie I vosproizvodstvo rastitelnykh, zhivotnykh i pochvennykh resursov Zapadnogo Kazakhstana. Orenburgskoe Knizhnoe Izdatelstvo, Chapaev, pp 71–76 (in Russian)
Bekenov AB, Baidavletov RZ, Grachev YA (1992) Wild ungulate resourses in Kazakhstan and perspectives of their rational use. Vestnik Akademii Nauk Respubliki Kazakhstan 7:19–25 (in Russian)
Belonogov AI (1977) Morphological indicators of ovaries and oviducts of maral females. Thesis. Barnaulsky pedagogichesky institut, Barnaul (in Russian)
Belov SN (1990) Density and population structure of wild reindeer in the Ob-Irtysh area. In: Evsikov VI (ed) Resursy zhivotnogo mira Sibiri. Okhotniche-promyslovye zveri i ptitsy. Nauka, Novosibirsk, pp 191–212 (in Russian)
Berber A, Kalmykov B (1994) Arkhar in Karaganda oblast. Okhota i Okhotniche Khozyaistvo 11:12–14
Berkinbaev O (1992) Study of saiga diseases in Kazakhstan. In: Khakhin GV (ed) Bolezni I parazity dikikh zhivotnykh. Vsesoyuznyi nauchno-issledovatelsky institut okhrany prirody i zapovednogo dela, Moscow, pp 32–41 (in Russian)
Bgatov VI, Panichev AM, Sobansky GG (1988) Animal salt licks in Siberian mountains. Byulleten Moskovskogo Obshchestva Ispytatelei Prirody, Otdelenie Biologicheskoe 93(2):42–53 (in Russian with English summary)
Bibikov DI, Zhirnov LV (1975) Role of wolf and dog in regulation of saiga numbers. In: Sokolov VE (ed) Kopytnye fauny SSSR. Nauka, Moscow, pp 190–191 (in Russian)
Blank DA (1990) Dzheyran - *Gazella subgutturosa* Guld, 1780. In: Kovshar AF (ed) Redkie zhivotnye pustyn. Institut Zoologii Akademii Nauk Kazakhskoi SSR, Alma-Ata, pp 56–79 (in Russian)
Blank DA, Dzhanyspaev A (1990) Kulan - *Equus hemionus onager* Boddaert, 1785. In: Kovshar AF (ed) Redkie zhivotnye pustyn. Institut Zoologii Akademii Nauk Kazakhskoi SSR, Alma-Ata, pp 80–93 (in Russian)
Blank DA, Kovshar AF (1986) Numbers of gazelles in deserts of the south-eastern Kazakhstan. In: Flint VE (ed) Okhrana i perspektivy restavratsii chislennosti dzheiranov v SSSR. Vsesoyuznyi nauchno-issledovatelsky institut okhrany prirody i zapovednogo dela, Moscow, pp 8–12 (in Russian)
Bliznyuk AI (1977) Saiga of the north-west Caspian region and perspectives of its use. In: Zykov KD (ed) Okhotniche khozyaistvo i zapovednoe delo. Rosselkhozizdat, Moscow, pp 60–68 (in Russian)
Bliznyuk AI (1980) Use of pastures by the saiga population of the north-west Caspian region. In: Safonov VG (ed) Khozyastvennaya deyatelnost i ohotnichya fauna. Vsesoyuznyi

Naychno-Issledovatelsky Institut Okhotnichego Khozyaistva i Zverovodstva, Kirov, pp 68–79 (in Russian)
Bludnik LV, Danilov PI, Markovsky VA, Heykura K (1989) Daily and seasonal displacements of forest reindeer in Karelia republic (1986–1988). In: Danilov PI (ed) Lesnoy severnyi olen. Institut Biologii, Petrozavodsk, pp 47–54 (in Russian)
Bluzma PP (1975) Roe deer of Lithuania (ecological-morphological characteristics). Thesis. Institut Evolutsionnoi Morphologii i Ekologii Zhivotnykh, Moscow (in Russian)
Bluzma P (1990) Living conditions and status of mammal populations in Lithuania. In: Bluzma P (ed) Mammals in the cultural landscape of Lithuania. Mokslas, Vilnius, pp 4–78 (in Russian with English summary)
Bluzma P, Baleišis R (1990) European roe deer in agrocenosises of Lithuania. In: Bluzma P (ed) Mammals in the cultural landscape of Lithuania. Mokslas, Vilnius, pp 138–145 (in Russian with English summary)
Bluzma P, Baleišis R (1992) Population density and winter distribution of moose in different landscapes of Lithuania. Alces [Suppl] 1:215
Bobyr GY (1999) Rare species of ungulates and carnivores in Karachayevo-Cherkessiya. In: Aristov AA (ed) Redkie vidy mlekopitayushchikh Rossii i sopredelnykh territoriy. Teriologicheskoe Obshchestvo, Moscow, pp 30–39 (in Russian with English summary)
Boev SN, Sokolov IB, Panin VY (1962, 1963) Helminths of ungulate mammals of Kazakhstan, vol 1/2. Izdatelstvo Akademii Nauk Kazakhskoi SSR, Alma-Ata (in Russian)
Bogatyrev AE (1982) Moose in the Kostroma region. Okhota i Okhotniche Khozyaistvo 5:8–10 (in Russian)
Bogolyubsky SN, Kurnosov KM, Petrishchev BI (1978) Embryo development of the wild sheep in Turkmenistan (*Ovis ammon cycloceros*). In: Bogolyubsky SN (ed) Ekologo-morphologicheskie cherty dikikh predkov domashnikh ovets. Nauka, Moscow, pp 47–57 (in Russian)
Bogomolova EM, Kurochkin YA (1984) Delivery of moose female: behaviour of female and new-born. Zool Zh 63(11):1713–1724 (in Russian with English summary)
Borisov BP, Gibet LA, Gubar JP, Kukushkin MA, Mosheva TS, Nazarov AA, Naumova AA, Novikov BV, Petrashov VV, Paponov VA, Tomilova TP (1992) Status of hunting grounds and numbers of main wild animals in RSFSR. Tsentralnaya nauchno-issledovatelskaya laboratoriya okhotnichego khozyaistva i zapovednikov, Moscow (in Russian)
Borodina MN, Borodin LP, Tereshkin IS, Shtarev YF (1970) Mammals of Mordovsky Nature Reserve. Trudy Mordovskogo Gosudarstvennogo Zapovednika 5:5–60 (in Russian)
Borozdin EK (1963) Spermatogenesis of reindeer. Trudy Nauchno-isledovatelskogo instituta polyarnogo zemledeliya, zhivotnovodstva i promyslovogo khozyaistva, series Olenevodstvo 12:129–133 (in Russian)
Borzhonov BB, Dorogov VF, Zyryanov VA, Katsarsky OP, Kiselev VA, Kolpashchikov LA, Pavlov BM (1979) *Ovis nivicola* at the Putoran Mountains. In: Zabrodin VA (ed) Problemy okhrany i ekonomicheskogo ispolzovaniya resursov dikikh zhivotnykh Eniseiskogo Severa. Vaskhnil, Novosibirsk, pp 44–56 (in Russian)
Bromley GF (1956) Ecology of wild sika deer in Primorsky kray. In: Yurgenson PB (ed) Sbornik materialov po izucheniyu mlekopitayushchikh v gosudarstvennykh zapovednikakh. Izdatelstvo Ministerstva Selskogo Khozyaistva SSSR, Moscow, pp 148–215 (in Russian)
Bromley GF (1959) Index of how different ungulates move through snow. Soobshcheniya Dalnevostochnogo Filiala Sibirskogo Otdeleniya Akademii Nauk SSSR 11:129–131 (in Russian)
Bromley GF (1964) Ussuri wild boar. Nauka, Moscow (in Russian)
Bromley GF, Kucherenko SP (1983) Ungulates of the southern part of the Far East. Nauka, Moscow (in Russian)
Bukreeva OM (2002) European population of saiga (*Saiga tatarica* L.) and factors of its status. Thesis. Rossiysky gosudarstvennyi agrarnyi zaochnyi universitet, Balashikha (in Russian)

Bunevich AN (1999) Reproductive parameters of European bison population in Belovezhskaya Forest. In: Aristov AA (ed) Redkie vidy mlekopitayushchikh Rossii i sopredelnykh territoriy. Teriologicheskoe Obshchestvo, Moscow, pp 72–82 (in Russian with English summary)

Bunevich AN, Kochko FP (1988) Dynamics of numbers and population structure of European bison in the Belovezhskaya Forest. In: Sokolov VE (ed) Populyatsionnye issledovaniya zhivotnykh v zapovednikakh. Nauka, Moscow, pp 96–114 (in Russian)

Chegorka PT (1984) On ecology of ungulates. In: Averin YV (ed) Priroda zapovednika "Kodry". Shtinitsa, Kishinev, pp 75–87 (in Russian)

Chermnykh NA, Roshchevsky MP, Moiseenko NA, Ovsov AS, Mochalov NN (1989) Special physiological adaptations of reindeer. In: Danilov PI (ed) Lesnoy severnyi olen. Institut Biologii, Petrozavodsk, pp 96–101 (in Russian)

Chernogaev EA, Kayumov BK, Savich OV, Pogrebnyuk AD, Aromov B (1996) Modern status and numbers of animals in the Uzbekistan Nature Reserves. Trudy Zapovednikov Uzbekistana 1:23–34 (in Russian)

Chernov VY (1986) Numbers and distribution of gazelle in the Sarykamysh depression. In: Flint VE (ed) Okhrana i perspektivy restavratsii chislennosti dzheiranov v SSSR. Vsesoyuznyi nauchno-issledovatelsky institut okhrany prirody i zapovednogo dela, Moscow, pp 28–31 (in Russian)

Chernyavsky FB (1962) On reproduction and growth of *Ovis nivicola*. Zool Zh 41(10):1556–1566 (in Russian with English summary)

Chernyavsky FB (1967) On ecology and economic significance of snow sheep. In: Uspensky SM (ed) Problemy Severa. Nauka, Moscow, pp 128–141 (in Russian)

Chernyavsky FB, Domnich VI (1989) Moose of the north-east Siberia. Nauka, Moscow (in Russian)

Chernyavsky FB, Zheleznov NK (1982) Moose distribution and systematics at Siberian North-East. Byulleten Moskovskogo Obshchestva Ispytatelei Prirody Otdelenie Biologicheskoe 87(5):25–32 (in Russian with English summary)

Chernyavsky FB, Zheleznov NK, Domnich VI (1980) Moose of the north-east Siberia. In: Sokolov VE (ed) Kopytnye fauny SSSR. Nauka, Moscow, pp 217–219 (in Russian)

Chernyavsky FB, Vovchenko VE, Domnich VI (1981) Acclimatization of musk ox (*Ovibos moshchatus* Zimmermannn 1778). In: Krivosheev VG (ed) Ekologiya mlekopitayushchikh i ptits ostrova Wrangelya. Institut biologii problem Severa, Vladivostok, pp 123–136 (in Russian)

Chernyavsky FB, Aksenov VV, Vovchenko VE (1990) Snow sheep (*Ovis nivicola*) on the northern shores of the Okhotsk sea. Zool Zh 69(3):114–119 (in Russian with English summary)

Chernyshev VI (1958) Tugai deer in Tadzhikistan: distribution, ecology and future of range increases. Trudy Instituta Zoologii i Parazitologii Akademii Nauk Tadzhikskoi SSR 89:149–163 (in Russian)

Chervonnyi VV (1975) Moose distribution, density and dynamics of numbers in European part of RSFSR. Trudy Okskogko Gosudarstvennogo Zapovednika 11:255–279 (in Russian)

Chervonnyi VV (1980) Peculiarities of boar breeding in the Oksky Nature Reserve. In: Sokolov VE (ed) Kopytnye fauny SSSR. Nauka, Moscow, pp 215–216 (in Russian)

Chervonnyi VV (1990) Peculiarities of structure and home range of wild boar in winter (the Oksky Nature Reserve). In: Amirkhanov AM (ed) Biologicheskie osnovy ucheta chislennosti okhotnichikh zhivotnykh. Tsentralnaya nauchno-issledovatelskaya laboratoriya okhotnichego khozyaistva i zapovednikov, Moscow, pp 103–120 (in Russian)

Dal SK (1944) *Vertebrata* of Saraibulag ridge. Zoologichesky sbornik Akademii nauk Armyanskoi SSR 3:5–46 (in Russian)

Dal SK (1949) Description of *Vertebrata* of Aiondzorsky ridge. Zoologichesky sbornik Akademii Nauk Armyanskoi SSR 6:5–97 (in Russian)

Dal SK (1951) Data on biology, distribution and numerical relations in herds of *Capra hircus* at Urts ridge. Izvestiya Akademii Nauk Armyanskoi SSR Biologiya 4(1):33–40 (in Russian)

Dal SK (1954) Animal world of Armenia. Akademiya Nauk Armyanskoi SSR Erevan (in Russian)
Dal SK, Gusev VM, Bednyi SN (1956) Impact of winter 1953/1954 on state, numbers and distribution of saiga in western Caspian region. Trudy Protivochumnogo Instituta Kavkaza I Zakavkaziya 1:430–440 (in Russian)
Dal SK, Gusev VM, Bednyi SN (1958) On ecology and reproduction of saiga. Zool Zh 3:447–456 (in Russian with English summary)
Danilkin AA (1992) Range. Population structure. Behaviour. In: Sokolov VE (ed) Evropeiskaya i sibirskaya kosylya. Nauka, Moscow, pp 60–85, 160–246 (in Russian)
Danilkin AA (1996) Behavioral ecology of Siberian and European roe deer. Chapman and Hall, London
Danilkin AA (1998) Moose in Russia. Okhota i Okhotniche Khozyaistvo 4:4–6 (in Russian)
Danilkin AA (1999) Deer (*Cervidae*). GEOS, Moscow (in Russian)
Danilkin AA (2002) Pigs (*Suidae*). GEOS, Moscow (in Russian)
Danilkin AA, Darman YA, Minaev AN (1993) Homing of migratory Siberian roe deer. Doklady Akademii Nauk SSSR 332(5):664–667 (in Russian with English summary)
Danilkin AA, Markov GG, Shchtubbe K, Struchkov AJ (1992) Morphometric analysis. In: Sokolov VE (ed) Evropeiskaya i sibirskaya kosylya. Nauka, Moscow, pp 25–43 (in Russian)
Danilkin AA, Darman YA, Minaev AN, Sempere A (1995) Social organisation, behaviour and ecological parameters of migratory populations of Siberian roe deer (*Capreolus pygargus* Pall.). Izvestiya Akademii Nauk Ser Biol 1:48–61 (in Russian with English summary)
Danilov DN (1949) Moose numbers and distribution in USSR. Okhrana Prirody 7:60–70 (in Russian)
Danilov DN (1950) Numbers and distribution of moose in Gorkiy oblast. Zool Zh 29(5):466–472 (in Russian with English summary)
Danilov DN (1951) Enlarging of moose range on the southern European part of USSR. Okhrana Prirody 13:93–101 (in Russian)
Danilov PI (1987) Population dynamics of moose in USSR (Literature survey, 1970–1983). Swedish Wildlife Research "Viltrevy" 1(2) [Suppl]:503–523
Darman YA (1982) Influence of roe deer on tree-shrub vegetation. Lesnoe Khozyaistvo 8:55–57 (in Russian)
Darman YA (1986) The biology of roe deer in the Khingan Nature Rezerve. Thesis. Institut Evolutsionnoi Morphologii i Ekologii Zhivotnykh, Moscow (in Russian)
Darman YA (1990) Mammals of the Khingan Nature Reserve. Blagoveshchenskoe Knizhnoe Izdatelstvo, Blagoveshchensk (in Russian)
Davydov AF (1960a) Physiological adaptations and behaviour of reindeer under influence of environmental heat. In: Kostin AP (ed) Regulyatsiya obmena tepla i drugikh funktsiy u selskokhozyaistvennykh zhivotnykh v usloviyakh vysokikh temperatur. Sovetskaya Kuban, Krasnodar, pp 71–74 (in Russian)
Davydov AF (1960b) Study of moving activity and pasture rhythm of reindeer. Thesis. Institut phisiologii AN SSSR, Leningrad (in Russian)
Demchuk VV, Omelkovich YA, Sologor EA (1997) Dynamics of European bison number in Volyn oblast (Ukraine). In: Sokolov VE (ed) Redkie vidy mlekopitayushchikh Rossii i sopredelnykh stran. Institut Problem Ecologii i Evolutsii, Moscow, p 27 (in Russian)
Demeuov D (1971) Kazakhstan saiga and their use. Thesis. Institut Zoologii, Alma-ata (in Russian)
Denisov VF (1935) Data on the variability of goats and meat productivity of argali sheep. In: Lus YY (ed) Vidovaya gibridizatsiya domashnikh s dikimi vidami. Izdatelstvo Akademii Nauk SSSR, Moscow, Leningrad, pp 81–114 (in Russian)
Deryagina MA (1972) Intra-herd relationship of European bison, American bison and their hybrids. Zool Zh 51(3):429–439 (in Russian with English summary)
Dinnik NY (1910) Mammals of the Caucasus, vol 1. Zapiski Zakavkazskogo Otdeleniya Imperatorskogo Russkogo Geographicheskogo Obshchestva 27(1):1–246 (in Russian)

Dinnik NY (1914) Mammals of the Caucasus, vol 2. Zapiski Zakavkazskogo Otdeleniya Imperatorskogo Russkogo Geographicheskogo Obshchestva 28(1):247–536 (in Russian)

Ditsevich VN (1988) The role of water plants in summer diet of wild ungulates. Problemy Ekologii Pribaikalya 4:92–96 (in Russian)

Dmitriev VV (1938) Ungulates of the Altaisky Nature Reserve and surroundings (east Altai and west Sayan). Trudy Altaiskogo Gosudarstvennogo Zapovednika 1:171–262 (in Russian)

Dobrynin MI, Babaev YB, Korshunov BM (1992) Fauna of helminths of the Turkmenian mountain sheep. Izvestiya Akademii Nauk Turkmenskoi SSR 2:45–52 (in Russian)

Domnich VI (1986) Moose (*Alces alces Buturlini*) of the extreme north-east of Siberia. Dissertation (ms). Institute of Evolutionary Animal Morphology and Ecology, Moscow (in Russian)

Donaurov SS, Teplov VP (1938) Wild boar in the Kavkazsky Nature Reserve. Trudy Kavkazskogo Gosudarstvennogo Zapovednika 1:191–225 (in Russian)

Dormidontov RV (1969) How many sika deer in the USSR? Okhota i Okhotniche Khozyaistvo 3:18–19 (in Russian)

Dormidontov RV (1977) Sika deer. In: Kaletsky AA (ed) Redkie zhivotnye SSSR. Kopytnye zveri. Lesnaya Promyshlennost, Moscow, pp 10–27 (in Russian)

Dormidontov RV, Blokhin YY (1977) Chamois. Goat. In: Kaletsky AA (ed) Redkie zhivotnye SSSR. Kopytnye zveri. Lesnaya Promyshlennost, Moscow, pp 140–152, 153–163 (in Russian)

Doszhanov TN (1980) Blood-sucking flies in Kazakhstan (*Diptera, Hippoboscidae*). Alma-Ata, Nauka (in Russian)

Drew KR, Bai Q, Fadeev EV (1989) Deer farming in Asia. In: Hudson RJ, Drew KR, Baskin LM (eds) Wildlife production systems. Cambridge Univ Press, Cambridge, pp 334–345

Druri IV (1949) Wild reindeer of Soviet arctic and subarctic. Trudy Arcticheskogo Instituta 200:1–79 (in Russian)

Duben AV (1985) The chamois number and population structure in relationship with some ecological factors. In: Kurazhkovsky JN (ed) Ekologicheskie issledovaniya v Kavkazskom biosfernom zapovednike. Izdatelstvo Rostovskogo Universiteta, Rostov-na-Donu, pp 31–63 (in Russian)

Dulitsky AI (1992) Some aspects of the European mouflon population structure in the Crimea. Vestnik Zoologii 3:50–56 (in Russian with English summary)

Dulitsky AI, Pereladova OB (1980) Composition and dynamics of mating deer groups at the peak of rut in the Crimea. In: Sokolov VE (ed) Kopytnye fauny SSSR. Nauka, Moscow, pp 288–289 (in Russian)

Dulkeit GD (1959) Distribution and migrations of mammals in montane taiga in Eastern Sayan depending on snow cover. Trudy Biologicheskogo Instituta Sibirskogo Otdeleniya Akademii Nauk SSSR 5:101–110 (in Russian)

Dulkeit GD (1964) Game animals and methods of evaluation of hunting ground production in the Altai-Sayan mountain taiga. Trudy Gosudarstvennogo Zapovednika "Stolby" 4:1–351 (in Russian)

Dunin VF, Kozlo PG (1992) Moose in Belarus. Ecology and forest impact. Navuka i Tekhnika, Minsk (in Russian)

Dunin VF, Voronetsky NN, Tishkevich VE, Eliashevich NV, Piskuniv VS, Pareiko OA, Kireenko KM (1999) Moose and roe deer. Belyi veter, Mozyr (in Russian)

Durov VV (1973) Distribution, migrations and numbers of boar in Krasnodasky kray. Nauchnye Trudy Kubanskogo Gosudarstvennogo Universiteta 163:99–136 (in Russian)

Durov VV (1980) Share of hunted animals and age structure of population of boar at the western Caucasus. In: Sokolov VE (ed) Kopytnye fauny SSSR. Nauka, Moscow, pp 239–240 (in Russian)

Durov VV (1986) Social structure of Caucasian wild boar population. In: Sokolov VE (ed) IV Meeting of All-Union Theriological Society. Nauka, Moscow, pp 199–200 (in Russian)

Durov VV (1987) Boar of the western Caucasus. Thesis. Institut Evolutsionnoi Morphologii i Ekologii Zhivotnykh, Moscow (in Russian)

Durov VV, Alexandrov VN (1968) Size and weight of boars in the Kavkazsky Nature Reserve. Trudy Kavkazskogo Gosudarstvennogo Zapovednika 10:294–301 (in Russian)
Dvornikov MG (1984) Ecology and biocenothic role of ungulates in the Ilmen Nature Reserve. Thesis. Uralsky institut ekologii rasteniy i zhivotnykh, Sverdlovsk (in Russian)
Dvornikov MG, Dvornikova NP (1985) Spatial distribution of herbivore mammals in the Ilmen Nature Reserve. Uralsky nauchnyi centr, Sverdlovsk (in Russian)
Dvornikov MG, Dvornikova NP (1989) On impact of boar digging activity on herbs in forest biocenosis in the Ilmen Nature Reserve. In: Sablina TB (ed) Ekologiya, morphologiya, ispolzozvanie i okhrana dikikh kopytnykh, vol 2. Nauka, Moscow, p 200 (in Russian)
Dvoryadkin AV, Shvets VG (1976) Wild reindeer of the South of the Far East. Nauchno-issledovatelsky byulleten Naucho-issledovatelskogo instituta Krainego Severa 12–13:6–9 (in Russian)
Dyakova GA (1989) Some exterior features of boar of the Uman-Voronezh population. In: Sablina TB (ed) Ekologiya, morphologiya, ispolzozvanie i okhrana dikikh kopytnykh, vol 2. Nauka, Moscow, p 201 (in Russian)
Egorov AN (1994) Helminthofauna of wild ungulates in the north-west of Moscow oblast. In: Tsvetkova VM (ed) Aktualnye problemy infectsionnykh i invasionnykh boleznei zhivotnykh. Moskovskaya akademiya veterinarnoi meditsiny i biotekhnilogii, Moscow, pp 41–49 (in Russian)
Egorov AN (1995) Helminths of ungulates in the game husbandry "Zavidovo" and regulation of their numbers. Thesis. Institut Evolutsionnoi Morphologii i Ekologii Zhivotnykh, Moscow (in Russian)
Egorov O (1953) Information on breeding and development of Siberian goat. Byulleten Moskovskogo Obshchestva Ispytatelei Prirody Otdelenie Biologicheskoe 58(3):3–13 (in Russian with English summary)
Egorov OV (1955) Ecology of Siberian mountain goat. Trudy Zoologicheskogo Instituta 17:7–134 (in Russian with English summary)
Egorov OV (1965) Wild ungulates of Yakutia. Nauka, Moscow (in Russian)
Egorov OV (1971) *Artiodactyla* – Ungulata. In: Tavrovsky VA (ed) Mlekopitayushchie Yakutii. Nauka, Moscow, pp 517–608 (in Russian)
Egorov OV, Egorova GN (1967) Peculiarities of distribution of wild reindeer in summer pastures in the Yakutian tundra. In: Popov MV (ed) Lyubite i okhranyaite prirodu Yakutii. Yakutskoe Knizhnoe Izdatelstvo, Yakutsk, pp 300–308 (in Russian)
Egorov OV, Krechmar AV (1967) Resources and prognosis of utilization of wild reindeer in Novosibirskie Islands. In: Popov MV (ed) Lyubite i okhranyaite prirodu Yakutii. Yakutskoe Knizhnoe Izdatelstvo, Yakutsk, pp 296–299 (in Russian)
Egorov OV, Krivosheev VG (1965) On role of horsetail in the feeding of land vertebrates in the Yakutian taiga. In: Krivosheev VG (ed) Priroda Yakutii i ee okhrana. Yakutskoe Knizhnoe Izdatelstvo, Yakutsk, pp 21–30 (in Russian)
Egorov OV, Popov MV (1984) Resources of wild reindeer in Yakutia. In: Syroechkovsky EE (ed) Wild reindeer of the Soviet Union. Oxonian Press, New Delhi, pp 139–153
Ekvtimishvili ZS (1952) Vertical distribution of some ungulate species on the southern slope of the Caucasus range. Soobshcheniya Akadeii Nauk Gruzinskoi SSR 13(6):476–485 (in Russian)
Ekvtimishvili ZS (1954) Data on breeding of beard goat. Trudy Instituta Zoologii Akademii Nauk Gruzinskoi SSR 13:93–97 (in Russian)
Fadeev EV (1975a) Territorial structure of *Sus scrofa* range in the European part of the RSFSR. In: Dobrovolsky GV (ed) Priroda i selskokhosiaistvennoe rayonirovanie SSSR. Moskovsky Gosudarstvennyi Universitet, Moscow, pp 105–107 (in Russian)
Fadeev EV (1980a) Modern status of wild boar in the Northern European area. In: Sokolov VE (ed) Kopytnye fauny SSSR. Nauka, Moscow, pp 49–50 (in Russian)
Fadeev EV (1985) Wild reindeer of European taiga. Vestnik Moskovskogo Universiteta, Ser 16. Biologiya 4:28–32 (in Russian with English summary)

Fadeev EV (1988) Distribution of wild boar population in the Vyatka-Kama region in the second part of the 20th century. Vestnik Moskovskogo Universiteta, Ser 16. Biologiya 2:22–26 (in Russian with English summary)

Fadeev EV (1990) On restoration of ungulate fauna in forests of Middle Don. Vestnik Moskovskogo Universiteta, Ser 16. Biologiya 1:47–55 (in Russian with English summary)

Fadeev EV (1993) Territorial structure of the *Sus scrofa* population in the Saratov Volga region. Vestnik Moskovskogo Universiteta, Ser 16. Biologiya 3:30–36 (in Russian with English summary)

Fadeev EV (1994) Spatial structure of roe deer population in the Volga part of the contact zone of European and Siberian roe deer. Vestnik Moskovskogo Universiteta, Ser 16. Biologiya 1:40–49 (in Russian with English summary)

Fadeev VA (1975b) Dynamics of the saiga numbers in Kazakhstan and Ustyurt Plateau. In: Sokolov VE (ed) Kopytnye fauny SSSR. Nauka, Moscow, pp 136–139 (in Russian)

Fadeev VA (1980b) Results of saiga tagging in Kazakhstan. In: Sokolov VE (ed) Kopytnye fauny SSSR. Nauka, Moscow, pp 210–212 (in Russian)

Fadeev VA, Ivanov AI (1988) Numbers of saiga in Kazakhstan depending on food resources. In: Bekenov AB (ed) Ekologiya i povedenie mlekopitayushchikh v Kazakhstane. Nauka, Alma-Ata, pp 37–47 (in Russian)

Fadeev VA, Sludsky AA (1982) Saiga in Kazakhstan. Nauka, Alma-Ata (in Russian)

Fadeev VA, Sludsky AA (1983) Saiga. In: Gvozdev EV, Kapitonov VI (eds) Mlekopitayushchie Kazakhstana, vol 3(3). Nauka, Alma-Ata, pp 56–91 (in Russian)

Fedosenko AK (1979) Behaviour of mountain sheep (*Ovis ammon*) during the reproductive period in the north of Tien-Shan and Dzungarian Ala-Tau. Zool Zh 43(6):903–911 (in Russian with English summary)

Fedosenko AK (1980) Maral. Nauka, Alma-Ata (in Russian)

Fedosenko AK (1982) Snow leopard. In: Gvozdev EV, Strautman EI (eds) Mlekopitayushchie Kazakhstana, vol 3(2). Nauka, Alma-Ata, pp 222–239 (in Russian)

Fedosenko AK (1984a) On ecology and behaviour of Ustyurt mountain sheep. Byulleten Moskovskogo Obshchestva Ispytatelei Prirody Otdelenie Biologicheskoe 89(2):23–32 (in Russian with English summary)

Fedosenko AK (1984b) Red deer. In: Gvozdev EV, Strautman EI (eds) Mlekopitayushchie Kazakhstana, vol 3(4). Nauka, Alma-Ata, pp 6–76 (in Russian)

Fedosenko AK (1985) Ecology of snow sheep in the Putoran Mountains (*Ovis nivicola borealis*). Zool Zh 64(1):107–116 (in Russian with English summary)

Fedosenko AK (1986) Relations of large predators and ungulates in mountains of the south-east of Kazakhstan and the south of Siberia. In: Fedosenko AK (ed) Role of large predators in biocenosis of Nature Reserves. Tsentralnaya nauchno-issledovatelskaya laboratoriya okhotnichego khozyaistva i zapovednikov, Moscow, pp 4–21 (in Russian)

Fedosenko AK (1989) Data on ecology of Altai mountain sheep in the south-east of Altai. In: Amirkhanov AM (ed) Ekologicheskie issledovaniya v zapovednikakh yuzhnoi Sibiri. Tsentralnaya nauchno-issledovatelskaya laboratoriya okhotnichego khozyaistva i zapovednikov, Moscow, pp 17–28 (in Russian)

Fedosenko AK (1997) Status of wild sheep populations in the south of SNG. In: Sokolov VE (ed) Redkie vidy mlekopitayushchikh Rossii i sopredelnykh stran. Institut Problem Ecologii i Evolutsii, Moscow, 96 pp (in Russian)

Fedosenko AK (2000) Argali in Russia and neighboring countries (population status, ecology, behaviour, conservation and use). Tsentrokhotkontrol, Moscow (in Russian)

Fedosenko AK (2002) Urial. Tsentrokhotkontrol, Moscow (in Russian)

Fedosenko AK, Kapitonov VI (1983) Mountain sheep (*Ovis ammon* Linn. 1758). In: Gvozdev EV, Kapitonov VI (eds) Mlekopitayushchie Kazakhstana, vol 3(3). Nauka, Alma-Ata, pp 144–214 (in Russian)

Fedosenko AK, Savinov EF (1983) Siberian goat. In: Gvozdev EV, Kapitonov VI (eds) Mlekopitayushchie Kazakhstana, vol 3(3). Nauka, Alma-Ata, pp 92–142 (in Russian)

Fedosenko AK, Skuratov NI (1990) On daily activity and feeding of argali in the south-east of Altai. Bulletin Moskovskogo Obshchestva Ispytatelei Prirody Series Biology 95(2): 11–19 (in Russian with English summary)
Fedosenko AK, Sludsky AA (1981) Wolf. In: Gvozdev EV, Strautman EI (eds) Mlekopitayushchie Kazakhstana 3(1). Nauka, Alma-Ata, pp 8–56 (in Russian)
Fedosenko AK, Zhiryakov VA (1984) Wild boar: ecology and behaviour in mountains of the south and south-eastern Kazakhstan. Byulleten Moskovskogo Obshchestva Ispytatelei Prirody Otdelenie Biologicheskoe 84(4):36–44 (in Russian with English summary)
Fedosenko AK, Zhiryakov VA, Sludsky AA (1984) Boar. In: Gvozdev EV, Strautman EI (eds) Mlekopitayushchie Kazakhstana, vol 3(4). Nauka, Alma-Ata, pp 146–188 (in Russian)
Fedosenko AK, Veinberg PI, Valdez R (1995) On ecology of the Pamirs mountain sheep in Kirgiziya. Byulleten Moskovskogo Obshchestva Ispytatelei Prirody Otdelenie Biologicheskoe 100(1):34–41 (in Russian with English summary)
Fertikov VI, Sitsko AV, Novikov BV (1983) Status and use of reindeer. In: Zabrodin VA (ed) Dikii Severnyi olen. Tsentralnaya nauchno-issledovatelskaya laboratoriya okhotnichego khozyaistva i zapovednikov, Moscow, pp 5–16 (in Russian)
Fetisov AS (1953) Roe deer in the eastern Siberia. Irkutsky Gosudarstvennyi Universitet, Irkutsk (in Russian)
Fetisov AS (1955) Distribution and habitats of feeding and protection of the Sayan reindeer in Selenga Dauria. Doklady Akademii Nauk SSSR 104(4):412–417 (in Russian with English summary)
Fil VI (1970) Peculiarities of distribution of mountain goats on winter grounds in the Zailiyskoe Alatau Range. Nauchno-Tekhnicheskaya Informatsiya (Hunting, Furs, Game) 28:21–25 (in Russian)
Fil VI (1973) Ecology of wild reindeer in the south of Kamchatka. Kraevedcheskie Zapiski Kamchatskogo Oblastnogo Kraevedcheskogo Muzeya 4:179–185 (in Russian)
Fil VI (1976) Some peculiarities of ecology and use of wild reindeer in the south of the Kamchatka Peninsula. Nauchno-issledovatelsky byulleten Naucho-issledovatelskogo instituta Krainego Severa 12/13:99–102 (in Russian)
Fil VI (1978) Rutting period of *Ovis nivicola* at Kamchatka. Ekologiya 3:98–101 (in Russian with English summary)
Filatov D (1910) Summer and winter travels in the north-west of Caucasus in 1909 to study the Caucasian bison. Ezhegodnik Zoologicheskogo Muzeya Akademii Nauk 15(4):1–47 (in Russian)
Filatov D (1912) Caucasian bison. Zapiski imperatorskoi akademii nauk, otdelenie phisiko-matematicheskoe, series 7, 30(8):1–40 (in Russian)
Filimonov AN (1979) On possible selection role of wolf in saiga during calving season. In: Sokolov VE (ed) Ekologicheskie osnovy ohrany i ratsionalnogo ispolzovaniya khishchnykh mlekopitayushchikh. Nauka, Moscow, pp 142–143 (in Russian)
Filonov KP (1983a) Variations in mortality factors in wild ungulate populations. Ekologiya 2:57–64 (in Russian with English summary)
Filonov KP (1983b) Moose. Lesnaya promyshlennost, Moscow (in Russian)
Filonov KP (1989) Ungulates and large predators in Nature Reserve territories. Nauka, Moscow (in Russian)
Filonov KP (1993) Evaluation of *Cervidae* populations status. Nauka, Moscow (in Russian)
Filonov KP, Kaletskaya ML (1988) Population studies of moose in the Darwinsky Nature Reserve. In: Sokolov VE (ed) Populyatsionnye issledovaniya zhivotnykh v zapovednikakh. Nauka, Moscow, pp 63–64 (in Russian)
Filus IA (1992) Data on ecology and behaviour of argali in the Altaisky Nature Reserve. In: Fedosenko AK (ed) Okhrana i izuchenye redkikh i ischezayushchikh zhivotnykh v zapovednikakh. Tsentralnaya nauchno-issledovatelskaya laboratoriya okhotnichego khozyaistva i zapovednikov, Moscow, pp 44–56 (in Russian)

Flerov KK (1935) Ungulates of Tadzhikistan. In: Pavlovsky EN (ed) Zveri Tadzhikistana i ikh znachenie dlya cheloveka. Izdatelstvo Akademii Nauk SSSR, Moscow-Leningrad, pp 93–130 (in Russian)

Flerov KK (1952) Musk deer and deer. In: Pavlovsky EN (ed) Fauna SSSR, new series 55, Mlekopitayushchie, vol 1(2). Izdatelstvo Akademii Nauk SSSR, Moscow-Leningrad, pp 1–255 (in Russian)

Flint VE, Prisyazhnyuk VE (1986) Status of populations, conservation and perspectives of restoration of gazelle in USSR. In: Flint VE (ed) Okhrana i perspektivy restavratsii chislennosti dzheiranov v SSSR. Vsesoyuznyi nauchno-issledovatelsky institut okhrany prirody i zapovednogo dela, Moscow, pp 3–8 (in Russian)

Flint VE, Pereladova OB, Mirutenko MV (1990a) Bukhara deer. In: Sokolov VE (ed) Redkie i ischesayushchie vidy mlekopitayushchikh SSSR. Nauka, Moscow, pp 65–81 (in Russian)

Flint VE, Pereladova OB, Mirutenko MV, Zhirnov LV (1990b) Experimental programme for establishment of free populations of Przewalsky horse in Kazakhstan - Middle Asia. In: Sokolov VE (ed) Redkie i ischesayushchie vidy mlekopitayushchikh SSSR. Nauka, Moscow, pp 98–113 (in Russian)

Flint VE, Prisyazhnyuk VE, Belousova AV, Bozhansky AT, Vishnevskaya TY, Panyutin KK, Puzachenko AY, Shilin NI (1991) A list of rare and endangered species (subspecies) of vertebrates which are recommended to be included in the Red Book of SNG. In: Ilyashenko VY, Mazin LN (eds) Izuchenie redkikh zhivotnykh v RSFSR. Tsentralnaya nauchno-issledovatelskaya laboratoriya okhotnichego khozyaistva i zapovednikov, Moscow, pp 67–89 (in Russian)

Formozov AN (1946) Snow cover as an environmental factor, its role in life of factor animals and birds in the USSR. Izdatelstvo Akademii Nauk SSSR, Moscow (in Russian)

Galako VA, Matveev AS, Vorontsov VN, Bakunin VA (1994) Coniferous plantations and moose. Uralskoe otdelenie RAN, Ekaterinburg (in Russian)

Galuso IG, Golosov VI, Gorbunova ZI (1963) *Toxoplasma gondii* received from saiga. In: Boev SN (ed) Parasity selskokhozyaistvennykh zhivotnykh Kazakhstana. Izdatelstvo Akademii Nauk Kazakhskoi SSR, Alma-Ata, pp 1–206 (in Russian)

Gambaryan PP (1974) How mammals run. Anatomical adaptations. Wiley, New York

Garross VJ, Mikheeva RV (1992) Moose management in Latvia. Alces [Suppl] 1:224

Gasparyan KM (1964) On feeding regime of *Capra hircus* at Urts (Saraibulag) ridge. Izvestiya Akademii Nauk Armyanskoi SSR. Biologiya 17(2):85–100 (in Russian)

Gasparyan KM (1975) Ecology of *Capra hircus*. Zoologichesky Sbornik Instituta Zoologii AN Armyanskoi SSR 16:78–106 (in Russian)

Gasparyan KM, Mezhlumyan SK (1969) On past and modern distribution of *Capra hircus* in Armenia. In: Dal SK (ed) Yubileinaya sessiya po faune Armyanskoi SSR posvyashchennaya 25 godovshchine Akademii nauk Armyanskoi SSR. Istitut Zoologii Akademii Nauk Armyanskoi SSR, Erevan, pp 24–28 (in Russian)

Gatikh VS (1975) Distribution and numbers of wild ungulates in the Pripyat Nature Reserve. In: Sokolov VE (ed) Kopytnye fauny SSSR. Nauka, Moscow, pp 73–74 (in Russian)

Gębczynska Z, Gębczynskii M, Martynowicz E (1991) Food eaten by the free-living European bison in Białowieza Forest. Acta Theriologica 36(3–4):307–313

Geller MK, Borzhonov BB (1984) Migrations and seasonal distribution of wild reindeer of the Taymyr population. In: Syroechkovsky EE (ed) Wild Reindeer of the Soviet Union. Oxonian Press, New Delhi, pp 71–79

Geller MX, Pavlov BM (1972) Autumn migrations of the Taymyr population of reindeer (*Rangifer tarandus*). Zool Zh 51(9):1381–1386 (in Russian with English summary)

Gern V (1891) Zoogeographic remarks for the Akmolinsky uezd. Russkoe geographicheskoe obshchestvo, Semipalatinsk (in Russian)

Gineev AM, Abdurakhmanov MG, Spasskaya TX (1988) Modern numbers and distribution of some rare and endangered mammals of the north of Caucasus. In: Khokhlov AN (ed) Resursy zhivotnogo mira Severnogo Kavkaza. Stavropolsky pedagogichesky institut, Stavropol, pp 42–49 (in Russian)

Glushkov VM (1980) Moose ecology in hunting grounds of the Vyatka basseng. Thesis. Vsesoyuznyi Naychno-issledovatelsky institut okhotnichego khozyaistva i zverovodstva, Kirov (in Russian)

Glushkov VM (1985) Resources of moose in USSR and their use. In: Grakov NN (ed) Resursy okhotnichego khozyaistva i ikh prognozis. Vsesoyuznyi Naychno-issledovatelsky institut okhotnichego khozyaistva i zverovodstva, Kirov, pp 116–128 (in Russian)

Glushkov VM, Piminov VN, Ponomarev VP (1989) Winter mortality and resources for hunting of wild ungulates. In: Glushkov VM (ed) Upravlenie populyatsiyami dikikh kopytnykh. Vsesoyuznyi Naychno-issledovatelsky institut okhotnichego khozyaistva i zverovodstva, Kirov, pp 81–93 (in Russian)

Golovnev VA (1940) Time of calves separation from mothers. Trudy Nauchno-isledovatelskogo instituta polyarnogo zemledeliya, zhivotnovodstva i promyslovogo khozyaistva, series Reindeer Husbandry 14:25–35 (in Russian)

Gomoyunova NP (1976) Biology of warble flies of reindeer. Nauka, Novosibirsk (in Russian)

Gorbunov AV (1986) Watering places of ungulates and mammalian predators in the northwest of Turkmenistan. Byulleten Moskovskogo Obshchestva Ispytatelei Prirody Otdelenie Biologicheskoe 91(1):17–24 (in Russian with English summary)

Gorbunov AV, Zarkhidze VA (1986) Long term fluctuations of numbers, modern distribution and perspectives of restoration of the gazelle population in Western Turkmenistan. In: Flint VE (ed) Okhrana i perspektivy restavratsii chislennosti dzheiranov v SSSR. Vsesoyuznyi nauchno-issledovatelsky institut okhrany prirody i zapovednogo dela, Moscow, pp 22–32 (in Russian)

Gorbunov VA (1939) Duration of reindeer pregnancy. Trudy Nauchno-isledovatelskogo instituta polyarnogo zemledeliya, zhivotnovodstva i promyslovogo khozyaistva, series Olenevodstvo 5:135–142 (in Russian)

Gordiyuk NM (1980) Mortality factors of moose and maral in the Bashkir Nature Reserve. In: Sokolov VE (ed) Kopytnye fauny SSSR. Nauka, Moscow, pp 140–141 (in Russian)

Gordiyuk NM (1982) Fertility of ungulates in southern Urals. In: Zabrodin VA (ed) Promyslovye zveri RSFSR. Tsentralnaya nauchno-issledovatelskaya laboratoriya okhotnichego khozyaistva i zapovednikov, Moscow, pp 216–232 (in Russian)

Gordiyuk NM (1985) Winter ecology of *Cervus elaphus sibiricus* in the south of Urals. Zool Zh 64(1):98–106 (in Russian with English summary)

Gorelov IT (1994) Biology of Siberian wild boar. Institut tsitologii i genetiki, Novosibirsk (in Russian)

Gorelov YK (1964) Watering places of mountain sheep in the Badkhyz Nature Reserve. Byulleten Moskovskogo Obshchestva Ispytatelei Prirody Otdelenie Biologicheskoe 69(2):61–70 (in Russian with English summary)

Gorelov YK (1972) Breeding of gazelle (*Gazella subgutturosa* Gueldenstaedt) and problem of restoration of its numbers in Badkhyz (the south-eastern Turkmenistan). In: Kolosova LD, Lukyanova IV (eds) Teriologiya, vol 1. Nauka, Novosibirsk, pp 420–424 (in Russian)

Gorelov YK (1978) Some peculiarities of the Badkhyz ungulate ecology and the problem of their conservation. Dissertation (ms). Institut Evolutsionnoy Morphologii i Ekologii Zhivotnykh, Moscow (in Russian)

Gorelov YK, Ishanov I (1999) State of ungulates in the Badkhyz in 1998. In: Orlov VN (ed) VI sezd Teriologicheskogo obshchestva. Teriologicheskoe Obshchestvo, Moscow, pp 60 (in Russian)

Gorshkov PK (1988) Ecology of wild boar in Tatarstan. In: Popov VA (ed) Ekologia, okhrana I vosproizvodstvo zhivotnykh Srednego Povolzhja. Kazansky Pedagogichesky Institut, Kazan, pp 126–133 (in Russian)

Gorshkov PK (1989) Ecology and biocenotic role of boar in Tatarstan republic. In: Sablina TB (ed) Ekologiya, morphologiya, ispolzovanie i okhrana dikikh kopytnykh, vol 2. Nauka, Moscow, pp 195–196 (in Russian)

Govorka J, Maklakova LP, Mitukh J, Pelgunov AN, Rykovsky AS, Semenova MK, Sonin MD, Erkhardova-Kotrla B, Yurashek B (1988) Helminths of wild ungulates of Eastern Europe. Nauka, Moscow (in Russian)

Grachev YA (1988) On daily activity of ungulates in Talas Alatau range. Trudy Instituta Zoologii Akademii Nauk Kazakhskoi SSR 44:48–56 (in Russian)

Grachev YA, Bekenov AB (1993) Actual status of the saiga population in Kazakhstan. Zoologicheskie issledovaniya v Kazakhstane 2:165–89 (in Russian)

Grachev YA, Smirnova ED (1980) Vertical distribution and migrations of ungulates in Talas Alatau range. In: Sokolov VE (ed) Kopytnye fauny SSSR. Nauka, Moscow, pp 82–83 (in Russian)

Grachev YA, Eszhanov B, Shaymardanov RT (1999) Results of aerial censuses of saiga in Kazakhstan. In: Orlov VN (ed) VI sezd Teriologicheskogo obshchestva. Teriologicheskoe Obshchestvo, Moscow, 62 pp (in Russia)

Gribkov PF (1969) Status of the *Ovis nivicola* population in the Central Kamchatka. Zool Zh 48(11):1716–1721 (in Russian with English summary)

Grigoryan GA (1949) Fauna of parasitic worms of wild ungulates of Armenia and their role in distribution of helminths among domestic sheep and goats. Thesis. Gelan, Moscow (in Russian)

Gritsuk VG (1975) Growth of heart, lungs and kidneys of wild ancestors of domestic sheep. Trudy Moskovskogo Obshchestva Ispytatelei Prirody 53:97–105 (in Russian with English summary)

Groombridge B (ed) (1994) IUCN Red List of Threatened Animals (1993). IUCN Gland, Cambridge

Grubb P (1993) Order *Perissodactyla*. Order *Artiodactyla*. In: Wilson DE, Reeder DAM (eds) Mammal species of the world. A taxonomic and geographic reference. Smithsonian Institution Press, Washington, pp 369–372, 377–414

Grunin KY (1957) Nose-cavity gadflies (*Oestridae*). In: Shtakelberg AA (ed) Fauna SSSR, New series 82, Insects, *Diptera*, vol 19(3). Izdatelstvo Akademii nauk SSSR, Moscow-Leningrad, pp 1–145 (in Russian)

Grunin KY (1962) Underskin gadflies, Hipodermatidae. In: Pavlovsky EN (ed) Fauna SSSR, new series 82, Insects, *Diptera*, vol 19(4). Izdatelstvo Akademii Nauk SSSR, Moscow-Leningrad, pp 1–238 (in Russian)

Gubanov NM (1964) Helminths of the economical important mammals in Yakutia. Nauka, Moscow (in Russian)

Gulchak FY (1954) Reindeer husbandry. Selkhozgiz, Moscow (in Russian)

Gusarov IV (1999) Establishing European bison populations in conditions of northern Russia. Thesis. Vologodskaya Gos. Molochnokhozyaistvennaya Akademiya, Vologda (in Russian)

Gusev AA (1989) Animals in nature reserves. Tsentralno-Chernozemnoye Izdatelstvo, Voronezh (in Russian)

Heptner VG (1956) Fauna of vertebrates in Badkhyz, the southern Turkmenistan. Izdatelstvo Akademii Nauk Turkmenskoi SSR, Ashkhabad (in Russian)

Heptner VG (1989) Taxonomy. Geographic distribution. Geographic variation. In: Heptner VG, Naumov NP (eds) Mammals of the Soviet Union, vol 1. Model Press, Leiden, pp 19–45, 101–112, 131–141, 159–190, 243–280, 307–363, 430–463, 557–573, 704–713, 723–734, 756–768, 774–798, 816–838, 859–870, 881–929, 954–969, 1011–1025, 1037–1058

Heptner VG, Formozov AN (1941) Mammals of Dagestan. Trudy Zoologicheskogo Muzeya Moskovskogo Gosudarstvennogo Universiteta 6:1–74 (in Russian with English summary)

Heptner VG, Naumov NP (eds) (1961) Mammals of the Soviet Union, vol 1. Vysshaya Shkola, Moscow (in Russian)

Heptner VG, Naumov NP (eds) (1989) Mammals of the Soviet Union, vol 1. Model Press, Leiden

Heptner VG, Sludsky AA (1972) Mammals of the Soviet Union (Predators), vol 2(2). Vysshaya Shkola, Moscow (in Russian)

Hildebrand M (1988) Analysis of vertebrate structure. Wiley, New York
Hilton-Taylor C (ed) (2000) 2000 IUCN red list of threatened species. IUCN, Gland, Switzerland and Cambridge, UK
Holisova V, Obrtel P, Kozhina I, Danilkin AA (1992) Feeding. In: Sokolov VE (ed) Evropeiskaya i sibirskaya kosulya. Nauka, Moscow, pp 124–139 (in Russian)
Igoshina KN, Florovskaya EF (1939) Utilisation of pastures and reindeer pasturing in Northern Urals. Trudy Nauchno-isledovatelskogo instituta polyarnogo zemledeliya, zhivotnovodstva i promyslovogo khozyaistva, series Reindeer Husbandry 8:1–162 (in Russian)
Inyakova AP (1957a) A numbers and distribution of game animals in the Teberdinsky Nature Reserve. Trudy Teberdinskogo Gosudarstvennogo Zapovednika 1:249–257 (in Russian)
Inyakova AP (1957b) Breeding of Caucasian goat (*Capra caucasica* Guld.) in captivity. Trudy Teberdinskogo Gosudarstvennogo Zapovednika 1:269–270 (in Russian)
Ishadov NI (1965) Ecological-geographical features of some ungulates in Turkmenistan. Thesis. Institut Zoologii i Parazitologii AN Turkmenskoi SSR, Ashkhabad (in Russian)
Ishadov NI (1972) Distribution and recent status of animals in the north-west of Turkmenistan. In: Kolosova LD, Lukyanova IV (eds) Teriologiya. Nauka, Novosibirsk, pp 417–419 (in Russian)
Ishunin GI (1961) Mammals (predators and ungulates). In: Niyazov R (ed) Fauna of Uzbekskoi SSR, vol 3. Akademiya Nauk Uzbekskoi SSR, Tashkent, pp 132–152 (in Russian)
Ishunin GI (1987) Game animals of Uzbekistan. Mekhnat, Tashkent (in Russian)
Ishunin GI, Bekenov A, Savinov EF (1981) Recent distribution and protection of Ustyurt mountain sheep. Byulleten Moskovskogo Obshchestva Ispytatelei Prirody Otdelenie Biologicheskoe 89(3):17–22 (in Russian with English summary)
Ivashkin VM, Dvoynos GM (1984) Handbook of horse helminths. Naukova Dumka, Kiev (in Russian)
Ivanova GI (1975) Peculiarities of boar reproduction near Moscow. In: Sokolov VE (ed) Kopytnye fauny SSSR. Nauka, Moscow, pp 95–99 (in Russian)
Ivanova GI (1980) Boar numbers fluctuations in the European part of the USSR. In: Sokolov VE (ed) Kopytnye fauny SSSR. Nauka, Moscow, pp 161–162 (in Russian)
Ivanova GI (1985) Ecological and biological features concerned use of the wild boar population in European part of RSFSR. In: Ivanova GI (ed) Upravlenie populyatsiyami dikikh kopytnykh. Tsentralnaya nauchno-issledovatelskaya laboratoriya okhotnichego khozyaistva i zapovednikov, Moscow, pp 133–147 (in Russian)
Ivanter EV (1968) Numbers biology and forest impact of moose in Karelia. Uchenye Zapiski Petrozavodskogo Universiteta 16(3):203–221 (in Russian)
Jêdrzejewska B, Jêdrzejewski W, Bunevich AN, Miłkowski L, Krasiñski Z (1997) Factors shaping population densities and increase rates of ungulates in Białowieża Primeval Forest (Poland and Belarus) in the 19th and 20th centuries. Acta Theriol 42(4):399–451
Kaletskaya ML (1961) Ecology and use of moose on shores of the Rybinskogo reservoir. Trudy Darvinskogo Gosudarstvennogo Zapovednika 7:175–246 (in Russian)
Kaletsky AA (1967) Vegetable food of moose calves in summer and autumn. In: Bannikov AG (ed) Biologiya i promysel losya, vol 2. Rosselkhozizdat, Moscow, pp 113–165 (in Russian)
Kalugin SG (1968) Restoration of European bison in the north-west of the Caucasus. Trudy Kavkazskogo Gosudarstvennogo Zapovednika 10:1–112 (in Russian)
Kanshina ZA, Shibanov VV (1976) Behaviour of maral and izyubr in rutting period. In: Baskin LM (ed) Group behaviour of animals. Nauka, Moscow (in Russian)
Kapitonov VI (1980) Impact of summer drought of 1974–1975 on ungulates of Kazakh Upland. In: Sokolov VE (ed) Kopytnye fauny SSSR. Nauka, Moscow, pp 163–164 (in Russian)
Kapitonov VI, Makhmutov SM (1977) On distribution and ecology of mountain sheep at the northern limits of its range in Kazakh Uppland. In: Sludsky AA (ed) Redkie i ischesayushchie zhivotnye i ptitsy Kazakhstana. Nauka, Alma-Ata, pp 7–21 (in Russian)

Kaplanov LG (1935) Biology and hunting of moose in the Demyanka basin. In: Yurgenson PB (ed) Los i ego promysel. Glavpushnina, Moscow, pp 103–124 (in Russian)
Kaplanov LG (1948) Tiger. Deer. Moose. Moscovskoye Obshchestvo Ispytatelei Prirody, Moscow (in Russian)
Karelov AM, Bulavkin VI (1976) Problems of exploitation of wild reindeer in Evenky okrug. Nauchno-Issledovatelsky Byulleten Naucho-Issledovatelskogo Instituta Krainego Severa 12(3):23–25 (in Russian)
Karev GI (1956) Reindeer food and pastures. Selkhozgiz, Moscow (in Russian)
Kartsov GP (1903) Belovezhskaya Forest. Marks AF, St Petersburg (in Russian)
Kataev GD (1999) Monitoring of natural processes in the Lapland Nature Reserve. Archives of the Lapland Nature Reserve, vol 35 (ms). Monchegorsk, pp 170–173 (in Russian)
Katsarsky OP (1989) Status of musk ox population on Taymyr. In: Pavlov BM (ed) Mlekopitayushchie i ptitsy Severa Srednei Sibiri. Naucho-Issledovatelsky Institut Krainego Severa, Novosibirsk, pp 74–78 (in Russian)
Kazmin VD (1986) Reindeer (*Rangifer tarandus* L.) on the Wrangel Island. In: Ovsyannikov NG (ed) Zhivotnyi mir ostrova Wrangelya. Dalnevostochnyi Nauchnyi Tsentr Akademii Nauk SSSR, Vladivostok, pp 114–126 (in Russian)
Kazmin VD (1992) Plant food resources and their utilization by wild bison of the Central Caucasus. Dissertation. Institute of Evolutionary Animal Morphology and Ecology, Moscow (in Russian)
Kazmin VD, Arbuzova MV, Zembatova AA (1992) Grass vegetation in summer grazing of European bison on pastures of the Central Caucasus. In: Fedosenko AK (ed) Okhrana i izuchenye redkikh i ischezayushchikh zhivotnykh v zapovednikakh. Tsentralnaya Nauchno-Issledovatelskaya Laboratoriya Okhotnichego Khozyaistva i Zapovednikov, Moscow, pp 21–29 (in Russian)
Kerechun SF (1975) Impact of predators on the Carpathian ungulate populations. In: Sokolov VE (ed) Kopytnye fauny SSSR. Nauka, Moscow, pp 197–198 (in Russian)
Kerechun SF (1980) Numerical increase and use of Carpathian ungulates. In: Sokolov VE (ed) Kopytnye fauny SSSR. Nauka, Moscow, pp 93–94 (in Russian)
Khakhin GV (1998) Recent status of reindeer in Novaya Zemlya archipelago. In: Primak AP (ed) GGAZU – agropromyshlennomu kompleksu. Rossiysky Gosudarstvennyi Agrarnyi Zaochnyi Universitet, Balashikha, pp 62–63 (in Russian)
Khakhin GV, Prisyazhnyuk VE (1985) Modern status and numbers of sika deer in the USSR. In: Flint VE (ed) Ekologicheskie osobennosti okhrany zhivotnogo mira. Vsesoyuznyi Nauchno-Issledovatelsky Institut Okhrany Prirody i Zapovednogo Dela, Moscow, pp 14–23 (in Russian)
Kheruvimov VD (1967) Rational method of moose hunting. In: Bannikov AG (ed) Biologiya i promysel losya, vol 2. Rosselkhozizdat, Moscow, pp 142–152 (in Russian)
Kheruvimov VD (1969) Moose. Comparative studies with example from the Tambov population. Tsentralno-Chernozemnoye Izdatelstvo, Voronezh (in Russian)
Khokhryakov SA (1997) Biocenological connections of goral in the reserve ecosystems. In: Kuleshova LV (ed) Nauchnye issledovaniya v zapovednikakh i natsionalnykh parkakh Rossii (federalnyi otchet za 1992–1993 gody). Vsesoyuznyi Nauchno-Issledovatelsky Institut Okhrany Prirody i Zapovednogo Dela, Moscow, pp 138–139 (in Russian)
Kholodova MV (1989) Consumption and assimilation of food by wild ungulates in the temperate zone. Dissertation. Institut Evolutsionnoy Morphologii i Ekologii Zhivotnykh, Moscow (in Russian)
Kholodova MV, Belousova IP (1989) Utilization and assimilation of nutrients and energy by European bison (*Bison bonasus*). Zool Zh 68(12):107–117 (in Russian with English summary)
Kholodova MV, Prikhodko VI (1984) Utilization and nutrients of food for musk deer (*Moschus moschiferus*). Zool Zh 63(6):923–928 (in Russian with English summary)
Kirikov SV (1959) Changes of animal world in natural zones of USSR (XIII–XIX). Forest steppe and steppe zones. Izdatelstvo Akademii Nauk SSSR, Moscow (in Russian)

Kirikov SV (1960) Changes of animal populations in natural zones of the USSR (13th–19th centuries). Forest zone and forest tundra. Izdatelstvo Akademii Nauk SSSR, Moscow (in Russian)

Kirikov SV (1966) Hunting animals, natural environment and man. Nauka, Moscow (in Russian)

Kirilyuk VE, Tseveenmyadag N (1999) Space-time structure and numbers of Mongolian gazelle population (*Procapra gutturosa* Pall. 1777) in lower reaches of river Uldsa. In: Aristov AA (ed) Redkie vidy mlekopitayushchikh Rossii i sopredelnykh territoriy. Teriologicheskoe obshchestvo, Moscow, pp 154–167 (in Russian with English summary)

Kiryukhin ST (1990) Use of moose resources in the Karakan game husbandry. In: Evsikov VI (ed) Resursy zhivotnogo mira Sibiri. Okhotniche-promyslovye zveri i ptitsy. Nauka, Novosibirsk, pp 189–191

Kiselev AA (1979) The roe deer of the Urals. Thesis. Uralsky Institut Ekologii Rasteniy i Zhivotnykh, Sverdlovsk (in Russian)

Kiselev AA (1992) Roe deer of middle and southern Urals. In: Pyastolova OA (ed) Ekologiya mlekopitayushchikh Uralskikh gor. Nauka, Ekaterinburg, pp 150–164 (in Russian)

Kiselev NP (1975) Factors influencing on boar numbers in the Volga delta. In: Sokolov VE (ed) Kopytnye fauny SSSR. Nauka, Moscow, pp 103–104 (in Russian)

Kiseleva EG, Priklonsky SG, Teplov VP (1965) Data of moose distribution, its fertility and sex ratio. Trudy Okskogo Gosudarstvennogo Zapovednika 6:305–329 (in Russian)

Kishchinsky AA (1967) Snow sheep in the Kolyma plateau. Byulleten Moskovskogo Obshchestva Ispytatelei Prirody Otdelenie Biologicheskoe 72(4):36–42 (in Russian with English summary)

Kishchinsky AA (1971) Modern status of wild reindeer population (*Rangifer tarandus* L.) in the New Siberian Islands. Zool Zh 50(1):117–125 (in Russian with English summary)

Kiyli Y (1985) An analysis of some causes determining the moose dynamics in Estonia. Izvestiya Akademii Nauk Estonian SSR. Biologiya 31(4):266–270 (in Russian)

Knorre AV, Zyryanov AN, Andreeva EB, Shtarker (1999) Stolby Nature Reserve. In: Pavlov DS, Sokolov VE, Syroechkovsky EE (eds) Zapovedniki Sibiri, vol 1. Logata, Moscow, pp 102–115 (in Russian)

Knorre EP (1959) Ecology of moose. Trudy Pechoro-Ilychskogo Gos. Zapovednika 7:5–122 (in Russian)

Knorre EP (1961) Results and perspectives of moose domestication. Trudy Pechoro-Ilychskogo Gosudarstvennogo Zapovednika 9:5–113 (in Russian)

Kolisnik BI (1990) Number dynamics and use of ungulates in hunting grounds of Volyn Polesye. Vestnik Zool 3:65–67 (in Russian)

Kolpashchikov LA (1979) Utilisation of winter pastures by wild reindeer. In: Zabrodin VA (ed) Problemy okhrany i khozyaistvennogo ispolsovaniya resursov dikikh zhivotnykh Eniseyskogo Severa. Naucho-Issledovatelsky Institut Krainego Severa, Novosibirsk, pp 19–28 (in Russian)

Kolpashchikov LA (1990) Numbers and distribution of wild reindeer on Chukotka. In: Evsikov VI (ed) Resursy zhivotnogo mira Sibiri. Okhotniche-promyslovye zveri i ptitsy. Nauka, Novosibirsk, pp 272–274 (in Russian)

Kolpashchikov LA (2000) The reindeer population of Taymyr (Biological basis of management and sustainable use). Dissertation (MS). Institute Problem Ekologii i Evolutsii, Moscow (in Russian)

Kolpashchikov LA, Bolonin VD (1990) Peculiaritites of distribution and migrations of wild reindeer in Taymyr during autumn. In: Pavlov BM (ed) Resursy, ekologiya i ratsionalnoe ispolzovanie dikikh severnykh oleney v SSSR. Naucho-Issledovatelsky Institut Krainego Severa, Novosibirsk, pp 36–44 (in Russian)

Kolpashchikov LA, Kuksov VA, Pavlov BM (1983) Ecological substantiation of maximum numbers of the Taymyr wild reindeer population. In: Solomakha AI (ed) Ekologiya i ratsionalnoe ispolsovanie nazemnykh pozvonochnykh severa sredney Sibiri. Naucho-Issledovatelsky Institut Krainego Severa, Novosibirsk, pp 3–14 (in Russian)

Kolpashchikov LA, Pavlov BM (1995) Specifics of distribution and productivity of the Taymyr reindeer population. In: Shilov IA (ed) Ekologiya populyatsii: struktura i dinamica, vol 2. Nauka, Moscow, pp 406–409 (in Russian)
Korkishko VG (1992) Species composition, numerical dynamics of predators and ungulates in the Kedrovaya pad Nature Reserve. In: Makarchenko EA (ed) Sovremennoe sostoyanie flory i fauny gosudarstvennogo zapovednika "Kedrovaya pad". Biologo-Pochvennyi Institut, Vladivostok, pp 119–142 (in Russian)
Kormilitsyn AA (1970) Ecology of red deer (*Cervus elaphus brauneri* Charlemagne 1920) in Crimea. Vestnik Zoologii 5:15–19 (in Russian)
Korochkina LN (1971) Breeding parameters of European bison in the Belowezhskaya Forest. Belovezhskaya pushcha. Issledovaniya 3:176–184 (in Russian)
Korochkina LN (1973) Range and biotopes of European bison in the Belowezhskaya Forest. Belovezhskaya pushcha. Issledovaniya 7:148–165 (in Russian)
Korochkina LN, Bogdanovich VI (1975) Impact of ungulates on undergrowth in pine-blueberry forests. Belovezhskaya pushcha. Issledovaniya 9:107–121 (in Russian)
Korshunov VM (1985) The hunting of leopard in Turkmenistan. Okhota i Okhotniche Khozyaistvo 8:10–11 (in Russian)
Korshunov VM (1986) Restoration of ungulate numbers in Kopet-Dag. In: Shcherbak NN (ed) Priroda Centralnogo Kopetdaga. Ilym, Ashkhabad, pp 183–192 (in Russian)
Korshunov VM (1988a) Herd structure and its dynamics of mountain ungulates in the Central Kopet-Dag. In: Sokolov VE (ed) Populyatsionnye issledovaniya zhivotnykh v zapovednikakh. Nauka, Moscow, pp 70–97 (in Russian)
Korshunov VM (1988b) Comparative ecology of Turkmenian mountain sheep and Turkmenian goat. Thesis. Institut Evolutsionnoy Morphologii i Ekologii Zhivotnykh, Moscow (in Russian)
Kostin VP (1955) Data on ecology and hunting impact on gazelle. Trudy Instituta Zoologii Akademii Nauk Kazakhskoi SSR 4:67–82 (in Russian)
Kostyuk VA, Khudoleev FI, Skrubert VS, Katkov PV (1979) Distribution and numbers of wild reindeer in the north of Evenky okrug. In: Zabrodin VA (ed) Problemy okhrany i khozyaistvennogo ispolsovaniya resursov dikikh zhivotnykh Eniseyskogo Severa. Naucho-Issledovatelsky Institut Krainego Severa, Novosibirsk, pp 11–14 (in Russian)
Kotov VA (1960) Review of chamois numbers in the Kavkazsky Nature Reserve. Trudy Kavkasskogo Gosudarstvennogo Zapovednika 4:185–191 (in Russian)
Kotov VA (1968) Kubansky goat, its ecology and practical significance. Trudy Kavkazskogo Gosudarstvennogo Zapovednika 10:201–293 (in Russian)
Kotov VA, Ryabov LS (1963) Game and valuable mammals of mountain skirts and mountains of Krasnodarsky kray. Trudy Kavkazskogo Gosudarstvennogo Zapovednika 7:1–239 (in Russian)
Kovalev DP (1991) Space organization of the musk ox population in the Wrangel Island, during 1985–1990 years. In: Ovsyannikov NG (ed) Zhivotnye populyatsii i biocenozy na ostrove Vrangelya, pp 51–74. Gosudarstvennyi Komitet Po Okhrane Prirody, Moscow (in Russian)
Kovshar AF, Kargopolov AD (1982) Perspectives of breeding in captivity and establisment of new populations of kulan in Kazakhstan. In: Rustamov AK (ed) Razvedenie i sozdanie novykh populyatsyi redkikh i tsennykh vidov zhivotnykh. Ilym, Ashkhabad, pp 147–149 (in Russian)
Kovshar AF, Yanushko PA (1965) New data on mammals of the Aksu-Dzhabagly Nature Reserve. Trudy Gosudarstvennogo Zapovednika Aksu-Dzhabagly 2:203–236 (in Russian)
Kozhechkin VV, Zyryanov AN, Kelberg GV (1990) Winter stay and mortality of maral. In: Smirnov MN (ed) Problemy okhotovedeniya v Sibiri. Krasnoyarsky Universitet, Krasnoyarsk, pp 148–155 (in Russian)
Kozhukhov MV (1973) Results of experiments during 20 years on moose domestication in the Pechoro-Ilychsky Nature Reserve. In: Sablina TB (ed) Odomashnivanie losya. Nauka, Moscow, pp 17–27 (in Russian)

Kozhukhov MV (1989) Reproductive potential of *Alces alces* under domestication. Zoologichesky Zhurnal 68(5):150–153 (in Russian)

Kozlo PG (1975) Wild boar. Urozhay, Minsk (in Russian)

Kozlo PG (1983a) Morphometric characteristic of red deer of the Belowezhskaya Forest. Bulleten Moskovskogo Obshchestva Ispytatelei Prirody Otdelenie Biologicheskoe 91(1):21–30 (in Russian with English summary)

Kozlo PG (1983b) Ecological-morphological analysis of moose population. Izdatelstvo Akademii Nauk Belorusskoi SSR, Minsk (in Russian)

Kozlo PG (1995) Ungulates of Belarus: structure-functional analysis as a strategy to increase the effectiveness of hunting husbandry and rational use of population resources. Ekologija 2:13–17 (in Russian)

Kozlov DP, Nazarova NS (1979) Helmintho-fauna of European bison and its establishing. In: Sokolov VE (ed) Zubr. Nauka, Moscow, pp 471–475 (in Russian with English summary)

Kozlov GS, Antropov EV, Bespalov GI, Dzhurovich VM, Davidov VA (1992) Anabolic and anti-ulcer properties of moose milk. Alces [Suppl] 1:229

Krainev ED (1975) Territorial distribution of roe deer in Ukraine. In: Dobrovolsky GV (ed) Priroda i selskokhosiaistvennoe rayonirovanie SSSR. Moskovsky Gosudarstvennyi Universitet, Moscow, pp 112–114 (in Russian)

Krashevsky OR (1989) Lynx ecology in Putoran Mountains. Nauchno-Issledovatelsky Byulleten Naucho-Issl. Instituta Krainego Severa 1–2:23–26 (in Russian)

Krasińska M, Krasiński ZA (1995) Composition, group size and spatial distribution of European bison bulls in Białowieża Forest. Acta Theriologica 40:1–21

Krasińska M, Caboń-Raczyńska K, Krasiński ZA (1987) Strategy of habitat utilization by European bison in Białowieża Forest. Acta Theriol 32:147–202

Krasińska M, Krasiński ZA, Bunevich AN (2000) Factors affecting the variability in home range size and distribution of European bison in the Polish and Belorussian parts of the Bialowieża Forest. Acta Theriol 45(3):321–334

Krasiński ZA, Bunevich AN, Krasińska M (1994) Characteristics of the European bison populations in the Polish and Belorussian parts of the Białowieża Forest. Parki Narodowe I Rezerwaty Przyrody 13:25–67 (In Polish with English summary)

Krasiński ZA, Krasińska M, Bunevich AN (1999) Free-ranging populations of lowland European bison in the Białowieża Forest. Parki Narodowe I Rezerwaty Przyrody 18.4:23–75

Krechmar AV (1966) Wild reindeer in the basin of the Pyasina river. Zool Zh 45(4):599–608 (in Russian with English summary)

Krivoshapkin AA (1995) Snow sheep in Upper Yana: morphology, ecology, resources. Thesis. Institute Problem Ekologii i Evolutsii, Moscow (in Russian)

Krivoshapkin AA, Lukovtsev YS, Yakovlev FG (1997) Ecology of snow sheep in Central Verkhoyansky Upland. In: Tembotov AK (ed) Ekologiya mlekopitayushchikh gornykh territoriy. Elfa, Nalchik, pp 151–152 (in Russian)

Krivoshcheev VG, Rotshild EV, Smirin VM (1960) Distribution and numbers of gazelle and saiga in the north of Kyzylkum Desert. Voprosy Geographii 48:55–70 (in Russian)

Kucherenko SP (1977) Impact of Amur tiger on prey populations. In: Sokolov VE (ed) Redkie vidy mlekopitayushchikh fauny SSSR i ikh okhrana. Nauka, Moscow, pp 133–134 (in Russian)

Kucherenko S, Shvets V (1977) Roe deer of Amur-Ussuri region. Okhota i Okhotniche Khozyaistvo 3:22–23 (in Russian)

Kudaktin AN (1981) Wolf in the western Caucasus: ecology, behaviour, biocenothic position. Dissertation (ms). Institute Problem Ekologii i Evolutsii, Moscow (in Russian)

Kudaktin AN (1986) Impact of wolf on ungulates in the Kavkazsky Nature Reserve. In: Fedosenko AK (ed) Rol krupnykh khishchnikov v zapovednykh biotsenosakh. Vsesoyuznyi Nauchno-Issledovatelsky Institut Okhrany Prirody i Zapovednogo Dela, Moscow, pp 21–35 (in Russian)

Kudaktin AN (1998) Large predators of Kavkazsky Nature Reserve and surroundings (ecology, conservation, population management). Dissertation (ms). Institute Problem Ekologii i Evolutsii, Moscow (in Russian)

Kudryashova LM (1980) Influence of unfortunate weather factors on fertility of moose females and survival of moose calves. Ekologiya 5:63–67 (in Russian with English summary)

Kudryavtsev IV, Belousova IP (1999) The basic approaches for evaluation of modern condition and viability prospects of European bison. In: Aristov AA (ed) Redkie vidy mlekopitayushchikh Rossii i sopredelnykh territoriy. Teriologicheskoe Obshchestvo, Moscow, pp 188–198 (in Russian with English summary)

Kuksov VA (1981) Distribution of wild reindeer in Taymyr during calving season. In: Solomakha AI (ed) Ekologiya i khozyaistvennoe ispolzovanie nazemnoi fauny Eniseiskogo Severa. Naucho-Issledovatelsky Institut Krainego Severa, Novosibirsk, pp 3–13 (in Russian)

Kuksov VA, Zyryanov VA, Kolesnikov AL, Semenovich NS, Khasanov GS (1989) Distribution and numbers of forest wild reindeer and moose in Evenky okrug. Nauchno-Issledovatelsky Byulleten Naucho-Issledovatelskogo Instituta Krainego Severa 1/2:16–19 (in Russian)

Kulagin NM (1919) Bisons of the Belowezhskaya Forest. Moskovsky Nauchnyi Institut v Pamyat 19 Fevralya 1861 Goda, Moscow (in Russian)

Kulagin NM (1932) Moose of USSR. Izdatelstvo Akademii Nauk, Leningrad (in Russian)

Kuliev SM (1981) *Capra hircus* and *Capra cylindricornis* in Azerbaijan (biology, conservation and use). Thesis. Vsesoyuznyi Nauchno-Issledovatelsky Institut Okhrany Prirody i Zapovednogo Dela, Moscow (in Russian)

Kuliev SM (1989) Distribution and numbers of Dagestan goat in the Zakataly Nature Reserve. In: Sokolov VE (ed) Abstracts Vsesoyuznogo soveshchaniya po problemam kadastra i ucheta zhivotnogo mira, vol 2. Ufa, Bashkirgosizdat, pp 232–233 (in Russian)

Kupriyanov AG (1990) Feeding of wild reindeer in the north-west of Siberia. In: Pavlov BM (ed) Resursy, ekologiya i ratsionalnoe ispolsovanie dikikh severnykh olenei v SSSR. Naucho-Issledovatelsky Institut Krainego Severa, Novosibirsk, pp 127–135 (in Russian)

Kupriyanov AG, Belikov SE (1986) Wild reindeer of high latitudes in the Arctic. Bulleten Moskovskogo Obshchestva Ispytatelei Prirody Otdelenie Biologicheskoe 91(1):338–343 (in Russian with English summary)

Kurochkina LY, Osmanova LT, Karibaeva KI (1986) Food resources of Kazakhstan deserts. Nauka, Alma-Ata (in Russian)

Kuznetsov GV (1969) On daily activity rhythm of wild ungulates. Byulleten Moskovskogo Obshchestva Ispytatelei Prirody Otdelenie Biologicheskoe 74(1):47–55 (in Russian with English summary)

Kuznetsov GV (1987) Habitats, movements and interactions of moose with forest vegetation in USSR. Swedish Wildlife Research "Viltrevy" [Suppl] 1(1):201–212

Kuznetsov VI (1986) Gazelle in the Badkhyz area. In: Flint VE (ed) Okhrana i perspektivy restavratsii chislennosti dzheiranov v SSSR. Vsesoyuznyi Nauchno-Issledovatelsky Institut Okhrany Prirody i Zapovednogo Dela, Moscow, pp 31–37 (in Russian)

Kuzyakin VA (1985) Present status of wild ungulate resourses. Zoologiya Pozvonochnykh 13:5–66 (in Russian)

Lanina LB (1959) Experiences of studies of pastures in the Pechoro-Ilychsky Nature Reserve. Trudy Pechoro-Ilychskogo Gosudarstvennogo Zapovednika 7:182–212 (in Russian)

Laptev IP (1958) Mammals of the taiga zone of the western Siberia. Tomsky Universitet, Tomsk (in Russian)

Larin VV (1983) On ecology of *Ovis nivicola* in the Putoran Mountains. In: Zabrodin VA (ed) Problemy okhrany i ekonomicheskogo ispolzovaniya resursov dikikh zhivotnykh Eniseiskogo Severa. Naucho-Issledovatelsky Institut Krainego Severa, Novosibirsk (in Russian)

Larin VV (1989) On reproduction and maternal behaviour of snow sheep in the Putoran Mountains. In: Pavlov BM (ed) Mlekopitayushchie i ptitsy severa srednei Sibiri. Naucho-Issledovatelsky Institut Krainego Severa, Novosibirsk, pp 61–68 (in Russian)

Larin VV (1990) Snow sheep (*Ovis nivicola borealis*) in the Putoran Mountains. Thesis. Vsesoyuznyi Nauchno-Issledovatelsky Institut Okhrany Prirody i Zapovednogo Dela, Moscow (in Russian)

Larin VV, Krashevsky OR (1989a) Predation of wolf in the Putoran Mountains. Nauchno-Issledovatelsky Byulleten Naucho-Issledovatelskogo Instituta Krainego Severa 1/2:40–42 (in Russian)

Larin VV, Krashevsky OR (1989b) Predatory behaviour of wolf and wolverine towards snow sheep in the Putoran Mountains. In: Pavlov BM (ed) Mlekopitayushchie i ptitsy severa srednei Sibiri. Naucho-Issledovatelsky Institut Krainego Severa, Novosibirsk, pp 68–74 (in Russian)

Larin VV, Shelkovnikova TA (1985) Feeding peculiaritites of wild reindeer in Central Putoran Mountains. In: Lineitsev SN (ed) Ekologiya, okrana i khozyaistvennoe ispolsovanie dikikh severnykh olenei. Naucho-Issledovatelsky Institut Krainego Severa, Novosibirsk, pp 117–138 (in Russian)

Larin VV, Shelkovnikova TA (1988) Season peculiarities of the feeding ecology of *Ovis nivicola* in the Delochi river basin. In: Solomakha AI (ed) Zhivotnyi mir plato Putorany, ego ispolzovanie i okhrana. Nauchno-Issledovatelsky Institut Krainego Severa, Novosibirsk, pp 17–26 (in Russian)

Larin VV, Sipko TP (1999) The status of the *Ovis nivicola borealis* population. In: Aristov AA (ed) Redkie vidy mlekopitayushchikh Rossii i sopredelnykh territoriy. Teriologicheskoe Obshchestvo, Moscow, pp 199–203 (in Russian with English summary)

Larin VV, Pikuleva IN, Krashevsky OR (1990) Feeding ecology of wild reindeer in Northern Taymyr. In: Pavlov BM (ed) Resursy, ekologiya i ratsionalnoe ispolsovanie dikikh severnykh olenei v SSSR. Naucho-Issledovatelsky Institut Krainego Severa, Novosibirsk, pp 56–62 (in Russian)

Lavov MA (1974) Density of ungulate populations in Pribaikal and Transbaikal. In: Silinskiy PP, Kozhova OM (eds) Voprosy zoogeographii Sibiri. Irkutsky Gosudarstvennyi Universitet, Irkutsk, pp 103–117 (in Russian)

Lavov MA (1976) Forest reindeer in Evenky okrug and Priangarye and perspectives of their exploitation. Nauchno-Issledovatelsky Byulleten Naucho-Issledovatelskogo Instituta Krainego Severa 12/13:36–37 (in Russian)

Lavov MA (1978) Roe deer. In: Kaletsky AA (ed) Krupnye khishchniki i kopytnye. Lesnaya Promyshlennost, Moscow, pp 191–220 (in Russian)

Lavrov NP, Pavlov MP (1975) Acclimatization – one of ways to enlarge area and increase numbers of deer. In: Sokolov VE (ed) Kopytnye fauny SSSR. Nauka, Moscow, pp 11–12 (in Russian)

Lavrovsky AA (1952) Wild boar in the Volga mouth. Trudy Astrakhanskogo Gosudarstvennogo Zapovednika 7:1–67 (in Russian)

Lebedeva LS (1956) Ecological peculiarities of wild boar in Belovezhskaya Forest. Uchenye Zapiski Moskovskogo Pedagogicheskogo Instituta 61(4/5):105–271 (in Russian)

Lebedeva LS (1959) Feeding of saiga on the right bank of the Volga. Byulleten Moskovskogo Obshchestva Ispytatelei Prirody Otdelenie Biologicheskoe 64(5):27–35 (in Russian with English summary)

Lebedeva LS (1960) Data from a study of spring forages and pastures of saiga at the right bank of the Volga. Zool Zh 39(9):1438–1442 (in Russian with English summary)

Leshchenko AP (1997) Distributional peculiarities of saiga in North-Western CisCaspean. Dissertation (ms). Rossiisky Gosudarstvennyi Agrarnyi Zaochnyi Universitet, Moscow (in Russian)

Likhachev AI (1955) Reproduction and embryo development of moose. Trudy Tomskogo Universiteta 131:282–296 (in Russian)

Lineytsev SN (1983) Seasonal migrations of reindeer on the Putoran Mountains. In: Solomakha AI (ed) Ekologiya i ratsionalnoe ispolsovanie nazemnykh pozvonochnykh severa sredney Sibiri. Naucho-Issledovatelsky Institut Krainego Severa, Novosibirsk, pp 14–21 (in Russian)

Lineytsev SN, Krashevsky O (1988) Some aspects of wolverine feeding and behaviour during hunting. In: Solomakha AI (ed) Zhivotnyi mir plato Putorany, ego ispolzovanie i okhrana. Nauchno-Issledovatelsky Institut Krainego Severa, Novosibirsk, pp 58–63 (in Russian)

Ling KI (1973) Dynamics of moose fecundity in Estonia. Ekologiya 4:82–88 (in Russian with English summary)

Litvinov VF (1980) Parasitic diseases of moose and boar. In: Sokolov VE (ed) Kopytnye fauny SSSR. Nauka, Moscow, pp 174–175 (in Russian)

Litvinov VF (1981) Investigations on parasites in the Berezinsky Nature Reserve. Parazitologiya 15(4):391–392 (in Russian)

Lobanov PN (1975) Density of musk deer population and harvest in north-eastern ridges of the Eastern Sayan. In: Sviridov NS (ed) Problemy okhotovedeniya i okhrany prirody. Irkutsky Selskokhozyaistvennyi Institut, Irkutsk, pp 106–107 (in Russian)

Lomanov IK (1988) Analysis of synchronous fluctuations of moose numbers in the European part of RSFSR. In: Amirkhanov AM (ed) Khronologicheskie izmeneniya chislennosti okhotnichikh zhivotnykh v RSFSR. Tsentralnaya Nauchno-Issledovatelskaya Laboratoriya Okhotnichego Khozyaistva i Zapovednikov, Moscow, pp 101–106 (in Russian)

Lomanov IK (1995) Regularities of dynamic fluctuations and distribution of moose in European Russia. Tsentralnaya Nauchno-Issledovatelskaya Laboratoriya Okhotnichego Khozyaistva i Zapovednikov, Moscow (in Russian)

Lomanov IK (2000) Wild boar. In: Lomanov IK (ed) Sostoyanie resursov okhotnichikh zhivotnykh v Rossiiskoy Federatsii. Tsentrokhotkontrol, Moscow, pp 24–31 (in Russian)

Lomanov IK, Lomanova NV (1996) Moose. In: Lomanov IK (ed) Resursy osnovnykh vidov okhotnichikh zhivotnykh i okhotnichi ugodya Rossii (1991–1995). Tsentralnaya Nauchno-Issledovatelskaya Laboratoriya Okhotnichego Khozyaistva i Zapovednikov, Moscow, pp 31–50 (in Russian)

Lomanov IK, Lomanova NV (2000) Moose. In: Lomanov IK (ed) Sostoyanie resursov okhotnichikh zhivotnykh v Rossiiskoy Federatsii. Tsentrokhotkontrol, Moscow, pp 13–23 (in Russian)

Lukarevsky VS (1995) Modern status of large predator populations of Western Kopet-Dag and surroundings. Institute Problem Ekologii i Evolutsii, Moscow. Dissertation (in Russian)

Lukarevsky VS (1999) Large mammals of Southern Turkmenistan and their conservation. In: Aristov AA (ed) Redkie vidy mlekopitayushchikh Rossii i sopredelnykh territoriy. Teriologicheskoe Obshchestvo, Moscow, pp 216–230 (in Russian with English summary)

Lunitsyn VG (1988) Diseases of deer bred for velvet antlers. Vserosiyskaya Nauchno-Issledovatelskaya Stantsia Pantovogo Olenevodstva, Novosibirsk (in Russian)

Lunkashu MI, Lozan AM, Uspensky GA, Fenogenov AV (1986) Population composition and specifics of breeding of wild ungulates. In: Muntyanu AI (ed) Mlekopitayushchie i ptitsy antropogennogo landshafta Moldavii i ikh prakticheskoe znachenie. Shtinnitsa, Kishinev, pp 117–124 (in Russian)

Lushekina A, Struchkov A (2001) The saiga antelope in Europe: once again on the brink. Open Country 3:11–24

Lyalin VG, Aushev AA (1989) Fecundity of hunted moose population (*Alces alces*) in Tomsk Pri-Ob area during numbers increase. In: Bioproductivnost i biotsenoticheskie svyazi nazemnykh pozvonochnykh yugo-vostoka zapadnoy Sibiri. Tomsky universitet, Tomsk, pp. 139–145 (in Russian)

Lyamkin VF, Sokolova LP (1992) Impact of Siberian roe deer on taiga communities in the Pribaikalsky Nature Park. Ekologiya 4:81–84 (in Russian with English summary)

Lyapunova EA, Banch TD, Vorontsov NN, Hofman RS (1997) Chromozome patterns and systematic position of *Ovis ammon sevetzovi*. Zool Zh 76(9):1083–1093 (in Russian with English summary)

Lydolph E (1977) Climates of the Soviet Union. World survey of climatology, vol 7. Elsevier, Amsterdam

Lysenko EV, Shelkovnikova TV (1989) Wild reindeer feeding in Eastern Taymyr in autumn. Nauchno-Issledovatelskii Bulleten Naucho-Issledovatelskogo Instituta Krainego Severa 1/2:42–46 (in Russian)

Lyubchenko OV, Mamaev YF, Fedorov FF (1997) Practical steps at actual decision of "Forest - moose" problem. In: Chuenkov VS (ed) Problemy organizatsii mnogotselevogo lesopolzovaniya. Vsesoyuzny Nauchno-Issledovatelsky Institut Lesovodstva I Mekhanizatcii, Moscow, pp 162–175 (in Russian)

Magomedov MRD, Akhmedov EG (1994) Regularities of spatial distribution and numbers of *Capra cylindricornis* in the eastern Caucasus. Zool Zh 73(10):120–129 (in Russian with English summary)

Magomedov MRD, Akhmedov EG (1999) Modern status of mountain ungulates in Dagestan. In: Orlov VN (ed) VI sezd Teriologicheskogo obshchestva. Teriologicheskoe Obshchestvo, Moscow, 154 pp (in Russian)

Magomedov MRD, Akhmedov EG, Yarovenko YA (2001) Tur of Dagestan. Nauka, Moscow (in Russian)

Makarova OA (1976) Morphological peculiarities of wild reindeer in Murmansk oblast. Nauchno-Issledovatelsky Bulleten Naucho-Issledovatelskogo Instituta Krainego Severa 12(3):38–42 (in Russian)

Makarova OA (1981) Moose in Murmansk oblast. In: Ivanter EV (ed) Ekologia nazemnykh pozvonochnykh Severo-Zapada SSSR. Karelskoe Knizhnoe Izdatelstvo, Petrozavodsk, pp 160–166 (in Russian)

Makarova OA (1988) Wild reindeer as an indicator of environmental quality in the North. In: Kryuchkov VV (ed) Antropogennoe vozdeistvie na ekosistemy Kolskogo Severa. Kolsky Filial Akademii Nauk SSSR, Apatity, pp 64–70 (in Russian)

Makarova OA (1990) Characteristics of the wild reindeer population in the western part of Murmansk oblast during the period of deep depression. In: Pavlov BM (ed) Resursy, ekologiya i ratsionalnoe ispolzovanie dikikh severnykh oleney v SSSR. Naucho-Issledovatelsky Institut Krainego Severa, Novosibirsk, pp 110–118 (in Russian)

Makarova OA, Khokhlov AM (1985) Status and ways to restore the numbers of wild reindeer on Kola Peninsula. In: Lineitsev SN (ed) Ekologiya, okrana i khozyaistvennoe ispolsovanie dikikh severnykh olenei. Naucho-Issledovatelsky Institut Krainego Severa, Novosibirsk, pp 22–27 (in Russian)

Makovkin LI (1997) Status of sika deer in the Lazovsky Nature Reserve and its impact on natural environment. In: Kuleshova LV (ed) Nauchnye issledovaniya v zapovednikakh i natsionalnykh parkakh Rossii (federalnyi otchet za 1992–1993 gody). Vsesoyuznyi Nauchno-Issledovatelsky Institut Okhrany Prirody i Zapovednogo Dela, Moscow, pp 140–141 (in Russian)

Makovkin LI (1999) The sika deer of the Lazovsky Nature Reserve and surrounding areas of the Russian Far East. Lazovsky Gosudarstvennyi Prirodnyi Zapovednik, Vladivostok (in Russian)

Makridin VP (1962) Routes and data of wild reindeer migrations in Taymyr okrug. Zool Zh 41(6):927–934 (in Russian with English summary)

Malinovskaya GM (1975) Digestion of forest herbs by deer. Belovezhskaya pushcha. Issledovaniya 9:133–137 (in Russian)

Malygin YA, Katsarsky OP, Melnikov VL (1978) Distribution and numbers of wild reindeer in Eastern Taymyr during summer. Nauchno-Issledovatelsky Byulleten Naucho-Issledovatelskogo Instituta Krainego Severa 15:13–17 (in Russian)

Mamaev YF, Fedorov FF, Kotova SE, Lyubchenko OV (1991) Evaluation of moose hunting grounds in European part of RSFSR. Vsesoyuzny Institut Lesnykh Resursov, Moscow (in Russian)

Mambetzhumaev AM (1968) On ecology of Ustyurt mountain sheep (*Ovis ammon arcal* Eversmann). In: Reimov R (ed) Pushnie okhotnichi Zhivotnye Karakalpakii. Fan, Tashkent, pp 116–132 (in Russian)

Mambetzhumaev AM (1970) Gazella. Fan, Tashkent (in Russian)

Mamukelashvili GV (1987) Diet and seasonal utilisation of pastures by Caucasian red deer in the Lagodekhi Nature Reserve. Zapovedniki Gruzii 6:215–233 (in Russian)

Mardonov BK (1997) Peculiarities of influence of goitred gazelle on desert vegetation in enclosure. In: Sokolov VE (ed) Redkie vidy mlekopitayushchikh Rossii i sopredelnykh stran. Institut Problem Ecologii i Evolutsii, Moscow, p 57 (in Russian)

Markov EL (1938) Game animals of the Lagodekhi Nature Reserve. Gruzinsky Filial Akademii Nauk SSSR, Tbilisi (in Russian)

Markovsky VA (1978) Modern status of moose in Karelia. In: Ivanter EV (ed) Fauna i ekologiya ptits i mlekopitayushchikh taezhnogo severo-zapada SSSR. Institut Biologii, Petrozavodsk, pp 182–186 (in Russian)

Markovsky VA (1981) Dynamics of moose numbers in Karelia and perspectives of its resources use. In: Ivanter EV (ed) Ekologiya nazemnykh pozvonochnykh severo-zapada SSSR. Karelskoe Knizhnoe Izdatelstvo, Petrozavodsk, pp 154–159 (in Russian)

Mashkovtsev AA (1940) Biological and physiological meaning of sexual dimorphism for vertebrate animals. Doklady Akademii Nauk SSSR 27(1):91–94 (in Russian with English summary)

Mashkovtsev AA (1949) Role of the Ivan Petrovich Pavlov theory of high nervous activity for biology. Uspekhi Sovremennoi Biologii 28(4):47–87 (in Russian with English summary)

Matyushkin EN (1974) Large predators and scavengers in Middle Sikhote-Alin. Byulleten Moskovskogo Obshchestva Ispytatelei Prirody Otdelenie Biologicheskoe 79(1):5–21 (in Russian with English summary)

Matyushkin EN (1992) Tiger and red deer on seaside slopes of Middle Sikhote-Alin. Byulleten Moskovskogo Obshchestva Ispytatelei Prirody Otdelenie Biologicheskoe 97(1):3–20 (in Russian with English summary)

Meklenburtsev RN (1948) Pamirs mountain sheep (*Ovis polii polii* Blyth.). Bulleten Moskovskogo Obshchestva Ispytatelei Prirody Otdelenie Biologicheskoe 53(5):65–84 (in Russian with English summary)

Meklenburtsev RN (1949) On ecology of Siberian mountain goat in Pamirs. Zool Zh 23(5):482–483 (in Russian with English summary)

Mertz PA (1953) Ecology of European deer in the Usmansky forest. Thesis. Voronezhsky Gosudarstvennyi Universitet, Voronezh (in Russian)

Mertz PA (1957) Role of snow cover for the life of the Voronezh deer population. Trudy Voronezhskogo Gosudarstvennogo Zapovednika 7:159–166 (in Russian)

Metelsky AP (1976) The distribution and numbers of roe deer in Siberia and the Far East. Sbornik Nauchnykh Trudov Moskovskoi Veterinarnoi Akademii 84:54–57 (in Russian)

Michurin LN (1965a) Wild reindeer of the Taymyr Peninsula and rational utilisation of its resources. Thesis. Vsesoyuznyi Selskokhozyaistvennyi Institut Zaochnogo Obucheniya, Moscow (in Russian)

Michurin LN (1965b) Some morphological features of wild reindeer of the Taymyr Peninsula. Zool Zh 44(9):1396–1404 (in Russian with English summary)

Michurin LN, Makhaeva LV (1962) Feeding of wild reindeer in Taymyr. Zool Zh 41(12): 1883–1888 (in Russian with English summary)

Michurin LN, Mironenko SN (1966) On *Ovis nivicola borealis* Sevetzov in the Putoran mountains. Zool Zh 45(2):1736–1738 (in Russian with English summary)

Michurin LN, Vakhtina TV (1968) Winter feeding of wild reindeer in the extreme north. Zool Zh 47(3):477–479 (in Russian with English summary)

Mikhailov AP (1973) Main tasks and first results of moose domestication at the Kostroma agricultural experiment station. In: Sablina TB (ed) Odomashnivanie losya. Nauka, Moscow (in Russian)

Mikhailov AP (1976) Taming and domestication of moose in the Kostroma agricultural experiment station. Trudy Vsesoyuznogo Selskokhozyaistvennogo Instituta Zaochnogo Obucheniya 119:47–60 (in Russian)

Mikhailov VV, Pavlov BM, Zyryanov VA, Kolpashchikov LA, Kuksov VA (1990) Study of the Taymyr wild reindeer population by mathematical modelling. In: Pavlov BM (ed)

Resursy, ekologiya i ratsionalnoe ispolzovanie dikikh severnykh oleney v SSSR. Naucho-Issledovatelsky Institut Krainego Severa, Novosibirsk, pp 14–26 (in Russian)

Milner-Gulland EJ, Kholodova MV, Bekenov A, Bukreeva OM, Grachev Y, Amgalan L, Lushchekina AA (2001) Dramatic declines in saiga antelope populations. Oryx 35(4):340–345

Minaev AN (1992) Behaviour of moose during a domestication experiment (biotelemetry study). Moscow, Institut Evolutsionnoy Morphologii i Ekologii Zhivotnykh. Dissertation (MS) (in Russian)

Minasyan LG, Gaspayan KM, Antonyan AS, Egnatosyan VA, Ovsepyan LA, Khrimyan AN (1986) Peculiaritites of growth and development of Armenian mouflons (asiatic mountain sheep) in captivity. In: Ayrumyan KA (ed) Ecologicheskaya kharacteristika mlekopitayushchikh i ptits Armenii. Izdatelstvo Akademii Nauk Armyanskoi SSR, Erevan, pp 48–79 (in Russian)

Mirolyubov II, Ryashchenko LP (1948) Sika deer. Dalnevostochny Zverovodchesky Trest, Vladivostok (in Russian)

Mironenko ON (1984) Food utilisation by wild reindeer in the Putoran mountains. In: Syroechkovsky EE (ed) Wild reindeer of the Soviet Union. Oxonian Press, New Delhi, pp 99–108

Mirutenko VS (1996) Red deer. Sika deer. Roe deer. In: Lomanov IK (ed) Resursy osnovnykh vidov okhotnichikh zhivotnykh i okhotnichi ugodya Rossii (1991–1995). Tsentralnaya Nauchno-Issledovatelskaya Laboratoriya Okhotnichego Khozyaistva i Zapovednikov, Moscow, pp 114–121 (in Russian)

Mirutenko VS (2000) Red deer. Roe deer. In: Lomanov IK (ed) Sostoyanie resursov okhotnichikh zhivotnykh v Rossiiskoy Federatsii. Tsentrokhotkontrol, Moscow, pp 32–35, 48–51 (in Russian)

Mordosov II, Santaev VG, Leverev EA (1990) Conservation and rational use of tundra populations of wild reindeer in Yakutiya. In: Evsikov VI (ed) Resursy zhivotnogo mira Sibiri. Okhotniche-promyslovye zveri i ptitsy. Nauka, Novosibirsk, pp 274–276 (in Russian)

Mosheva TS (2000) Musk deer. In: Lomanov IK (ed) Sostoyanie resursov okhotnichikh zhivotnykh v Rossiiskoy Federatsii. Tsentrokhotkontrol, Moscow, pp 36–40 (in Russian)

Moskvitin VP, Smirnov MN (1975) Area, resources, use and protection of wild ungulates in Buryatia. In: Sokolov VE (ed) Kopytnye fauny SSSR. Nauka, Moscow, pp 115–116 (in Russian)

Moskvitin VP, Kurbetev GE, Maturova RT, Shargaev MA (1976) Wild reindeer of Buryatia. Nauchno-Issledovatelsky Byulleten Naucho-Issledovatelskogo Instituta Krainego Severa 12–13:52–54 (in Russian)

Mosolov VI (1985) Numbers and distribution of wild reindeer in the East of Kamchatka. In: Lineitsev SN (ed) Ekologiya, okrana i khozyaistvennoe ispolsovanie dikikh severnykh olenei. Naucho-Issledovatelsky Institut Krainego Severa, Novosibirsk, pp 17–22 (in Russian)

Mosolov VI (1986) Ecology of ungulates in the montane volcanic areas of the East of Kamchatka. In: Amirkhanov AM (ed) Ecosistemy ekstremalnykh usloviy sredy v zapovednikakh RSFSR. Tsentralnaya Nauchno-Issledovatelskaya Laboratoriya Okhotnichego Khozyaistva i Zapovednikov, Moscow, pp 38–49 (in Russian)

Mosolov VI (1992) Snow sheep of the south-east coast of Kamchatka. Byulleten Moskovskogo Obshchestva Ispytatelei Prirody Otdelenie Biologicheskoe 97(2):10–181 (in Russian with English summary)

Mosolov VI (1997) Territory of natural reserves and conservation of reindeer in the East of Kamchatka. In: Zhirmunsky AV (ed) III Far East conference on nature reserves. Dalnauka, Vladivostok (in Russian)

Movsesyan O, Ayrumyan KA (eds) (1987) Red book of Armenia. Institut Zoologii AN Armyanskoi SSR, Erevan (in Russian)

Muntyanu AI, Lozan MN, Fenogenov AV (1980) Status of roe deer population in Moldavia. In: Sokolov VE (ed) Kopytnye fauny SSSR. Nauka, Moscow, pp 183–184 (in Russian)

Muntyanu AI, Chegorka PT, Eremin AP (1992) Demographic structure of wild ungulates in Moldavia. In: Ganya IM (ed) Ekologiya i okhrana ptits i mlekopitayushchikh v antropogennom landshafte. Shtinitsa, Kishinev, pp 120–132 (in Russian)

Myslenkov AI, Voloshina IV (1978) Behaviour of goral during rutting season. Byulleten Moskovskogo Obshchestva Ispytatelei Prirody Otdelenie biologicheskoe 83(1):30–39 (in Russian with English summary)

Myslenkov AI, Voloshina IV (1989) Ecology and behaviour of Amur goral. Nauka, Moscow (in Russian)

Myslenkov AI, Voloshina IV (1992) Distribution of moose in the Sikhote-Alinsky Nature Reserve. Alces [Suppl] 1:233

Nadler CF, Korobitsyna KV, Hoffmann RS, Vorontsov NN (1973) Cytogenetic differentiation, geographical distribution and domestication in Palearctic sheep (*Ovis*). Z Säugetierkunde 38:109–125

Naniev VI (1956) Caucasian bison-"dombay". Uchenye Zapiski Severo-Osetinskogo Gosudarstvennogo Pedagogicheskogo Instituta 20:249–276 (in Russian)

Nasimovich AA (1936) On some regularities of winter distribution of ungulates in mountains of the western Caucasus. Byulleten Moskovskogo Obshchestva Ispytatelei Prirody Otdelenie Biologicheskoe 45(1):3–9 (in Russian with English summary)

Nasimovich AA (1938) A study of mineral feeding of wild animals in the Kavkazsky Nature Reserve. Trudy Kavkazskogo Gosudarstvennogo Zapovednika 1:103–152 (in Russian)

Nasimovich AA (1939) Winter life of ungulates of the western Caucasus. Problemy Ekologii i Biotsenologii 3:3–91 (in Russian)

Nasimovich AA (1941) Census method of goat numbers. Nauchno-Metodicheskie Zapiski Glavnogo Upravleniya Po Gosudarstvennum Zapovednikam 8:173–177 (in Russian)

Nasimovich AA (1949a) Description of ecology of Western Caucasian goat. Trudy Kavkazskogo Gosudarstvennogo Zapovednika 3:5–38 (in Russian)

Nasimovich AA (1949b) New data on chamois biology in the western Caucasus. Trudy Kavkazskogo Gosudarstvennogo Zapovednika 3:51–64 (in Russian)

Nasimovich AA (1955) Role of snow-cover regime for the life of ungulates within USSR territory. Academy of Sciences, Moscow (in Russian)

Nasimovich AA (1989) Biology and economic importance of musk deer, red deer, sika deer, moose, roe deer, European bison, snow sheep, urial, arkhar, mouflon, caucasian goat, dagestan goat, screw-horned goat, beard goat, siberian goat. In: Heptner VG, Naumov NP (eds) Mammals of the Soviet Union, vol 1. Ungulates. Model Press, Leiden, pp 112–123, 141–158, 190–234, 280–301, 363–424, 463–518, 573–598, 713–719, 734–747, 768–773, 798–815, 826, 838–858, 870–873, 929–953, 969–980

Naumov NP, Baskin LM (1969) Leadership in herds of reindeer as group adaptation. Zhurnal Obshchei Biologii 30(2):147–156 (in Russian with English summary)

Nazarova NS, Prisyazhnyuk VE (1986) Profylaxis of helminths invasions for breeding and acclimatisation of gazelles. In: Flint VE (ed) Okhrana i perspektivy restavratsii chislennosti dzheiranov v SSSR. Vsesoyuznyi Nauchno-Issledovatelsky Institut Okhrany Prirody i Zapovednogo Dela, Moscow, pp 84–89 (in Russian)

Nebolsin PI (1855) Essayes of trade of Russia and Middle Asia. Zapiski Imperatorskogo Russkogo Geograficheskogo Obshchestva 10:1–373 (in Russian)

Nedzelsky EM (1994) Boar of the Baykal region (*Suc scrofa sibiricus* Straffe 1922). Thesis. Vsesoyuznyi Selskokhozyaistvennyi Institut Zaochnogo Obucheniya, Moscow (in Russian)

Neifeld ND (1992) The current status of the upper Pechora moose population. Alces [Suppl] 1:234

Nemtsev AS (1985) Dynamics of European bison population in the Kavkazsky Nature Reserve. In: Kurazhkovsky YN (ed) Ekologicheskie issledovaniya v Kavkazskom biosfernom zapovednike. Izdatel'stvo Rostovskogo Universiteta, Rostov, pp 49–63 (in Russian)

Nemtsev AS (1998) Spatial and ecological structure of European bison population as a basis of its adaptation. In: Tkhakushinov AK (ed) Sotsialno-ekologicheskie problemy yuga

Rossii. Maikopsky Gosudarstvennyi Tekhnologichesky Institut, Maykop, pp 89–93 (in Russian)
Nemtsev AS, Kiseleva EG (1987) Comparative study of mountain and plain European bisons. In: Bolshakov VN (ed) Ekologiya i okhrana gornykh vidov mlekopitayushchikh. Institut Ekologii i Morphologii Zhivotnykh, Moscow, pp 127–129 (in Russian)
Nesterov DA (1992) Distribution of goral in Primorsky kray. In: Myslenkov AI (ed) Amursky goral. Tsentralnaya Nauchno-Issledovatelskaya Laboratoriya Okhotnichego Khozyaistva i Zapovednikov, Moscow, pp 6–20 (in Russian)
Novikov BV (1983a) Modern status of island populations of reindeer in the Soviet sector of the Arctic. In: Zabrodin VA (ed) Dikii Severnyi olen. Tsentralnaya Nauchno-Issledovatelskaya Laboratoriya Okhotnichego Khozyaistva i Zapovednikov, Moscow, pp 101–107 (in Russian)
Novikov BV (1983b) Wild reindeer of Yakutia: numbers, conservation and scientific study. In: Zabrodin VA (ed) Dikii Severnyi olen. Tsentralnaya Nauchno-Issledovatelskaya Laboratoriya Okhotnichego Khozyaistva i Zapovednikov, Moscow, pp 94–100 (in Russian)
Novikov BV (1996) Reindeer. In: Lomanov IK (ed) Resursy osnovnykh vidov okhotnichikh zhivotnykh i okhotnichi ugodya Rossii (1991–1995). Tsentralnaya Nauchno-Issledovatelskaya Laboratoriya Okhotnichego Khozyaistva i Zapovednikov, Moscow, pp 51–76 (in Russian)
Novikov BV (1998) Ecology, systematics, conservation and rational use of wolverine (*Gulo gulo*). Dissertation. Tsentralnaya Nauchno-Issledovatelskaya Laboratoriya Okhotnichego Khozyaistva i Zapovednikov, Moscow (in Russian)
Novikov GA, Timofeeva EK (1964) Roe deer feeding and role for forest in forest-steppe zone. Byulleten Moskovskogo Obshchestva Ispytatelei Prirody Otdelenie Biologicheskoe 69(2):39–53 (in Russian with English summary)
Ognev SI (1951) Descriptions of mammalian ecology. Izdatel'stvo Moskovskogo Obshchestva Ispytatelei Prirody, Moscow (in Russian)
Okarma H, Jędrzejewska B, Jędrzejewski W, Krasiñski Z, Miłkowski L (1995) The roles of predation, snow cover, acorn crop, and man-related factors on ungulate mortality in Białowieża Primeval Forest, Poland. Acta Theriol 40(2):197–217
Orlova VA (1986) Ecology of Siberian roe deer in the Sokhondinsky Nature Reserve. In: Sokolov VE (ed) Tesisy IV sezda Vsesoyuznogo Teriologicheskogo obshchestva, vol 1. Institut Evolutsionnoi Morphologii i Ekologii Zhivotnykh, Moskow, pp 307–308 (in Russian)
Ostanin V (1994) Roe deer of Kurgan oblast. Okhota i Okhotniche Khozyaistvo 9:2–3 (in Russian)
Ovsyukova NI (1966) Helminths and basic helminth diseases of the Chukotka mammals. Thesis. Vesoyuznyi Institut Gelmintologii, Moscow (in Russian)
Ovsyukova NI (1996) Wild boar. In: Lomanov IK (ed) Resursy osnovnykh vidov okhotnichikh zhivotnykh i okhotnichi ugodya Rossii (1991–1995). Tsentralnaya Nauchno-Issledovatelskaya Laboratoriya Okhotnichego Khozyaistva i Zapovednikov, Moscow, pp 77–98 (in Russian)
Padaiga VI (1969) Hunting husbandry of Lithuania. Okhota i Okhotniche Khozyaistvo 3:2–3 (in Russian)
Padaiga VI (1975) Ökologie und Prinzipien der Wirtschaftlichen Nutzung des Rehwildes (*Capreolus capreolus*) in den Wäldern des südlichen. Beitr Jagt Wildforsch 9:276–287
Panichev AM (1998) Stone-eating of animals and humans (Autoregulation of mineral feeding in natural ecosystems). Thesis. Biologichesky Institut, Vladivostok (in Russian)
Paponov VA (2000) Wild reindeer. In: Lomanov IK (ed) Sostoyanie resursov okhotnichikh zhivotnykh v Rossiiskoy Federatsii. Tsentrokhotkontrol, Moscow, pp 41–47 (in Russian)
Parshintsev AV (1992) Modern status of European mouflon population in the Krymsky Nature Reserve. In: Fedosenko AK (ed) Okhrana i izuchenye redkikh i ischezayushchikh zhivotnykh v zapovednikakh. Tsentralnaya Nauchno-Issledovatelskaya Laboratoriya Okhotnichego Khozyaistva i Zapovednikov, Moscow, pp 56–63 (in Russian)

Pasternak FA (1955) Data for systematics and biology of roe deer. Nauchnye Soobshcheniya Moskovskogo Pedagogicheskogo Instituta 38(3):29–140 (in Russian)

Pavlov BM, Savelev VD, Yakushkin GD, Zyryanov VA (1971) Ecological structure of the wild reindeer population of Taymyr. Ekologiya 1:49–57 (in Russian with English summary)

Pavlov BM, Arsenteva NF, Borzhonov BB, Dorogov VF, Kuksov VA (1985a) Demography of the Taymyr population of wild reindeer. In: Lineitsev SN (ed) Ekologiya, okhrana i khozyaistvennoe ispolsovanie dikikh severnykh oleney. Naucho-Issledovatelsky Institut Krainego Severa, Novosibirsk, pp 71–80 (in Russian)

Pavlov BM, Mikhailov VV, Zyryanov VA, Kolpashikov LA (1985b) Compex analysis of balance (replenishment-mortality) of the Taymyr population of wild reindeer. In: Lineitsev SN (ed) Ekologiya, okrana i khozyaistvennoe ispolsovanie dikikh severnykh oleney. Naucho-Issledovatelsky Institut Krainego Severa, Novosibirsk, pp 54–63 (in Russian)

Pavlov BM, Kolpashchikov LA, Zyryanov VA, Kuksov VA, Alabugin SV, Arsenteva NF (1989) Morphological characterstics of the Taymyr population of wild reindeer. In: Pavlov BM (ed) Mlekopitayushchie i ptitsy severa srednei Sibiri. Naucho-Issledovatelsky Institut Krainego Severa, Novosibirsk, pp 17–26 (in Russian)

Pavlov BM, Kolpashchikov LA, Zyryanov VA (1993) The Taymyr wild reindeer populations: management experiment. Rangifer 9:381–384

Pavlov MP (1976) Biotic peculiarities of moose distribution in a period of their census in the Vyatka taiga. Byulleten Moskovskogo Obshchestva Ispytatelei Prirody Otdelenie Biologicheskoe 81(1):41–50 (in Russian with English summary)

Pavlov MP (1990) Wolf. Agropromizdat, Moscow (in Russian)

Pereladova OB (1981) Space-time dynamics of roaring of Caucasian and Bukhara deer. In: Naumov NP (ed) Ekologiya i struktura populyatsii i vnutrividovye kommunikativnye protsessy u mlekopitayushchikh. Nauka, Moscow, pp 182–244 (in Russian)

Pereladova OB (1988) Permanence of subspecies features of Bukhara deer behaviour during rut in different ecological conditions. Byulleten Moskovskogo Obshchestva Ispytatelei Prirody Otdelenie Biologicheskoe 93(2):25–34 (in Russian with English summary)

Pereladova OB, Pereladov SV (1986) Breeding of gazelles in enclosures of the Syunt-Khasardag Nature Reserve. In: Flint VE (ed) Okhrana i perspektivy restavratsii chislennosti dzheiranov v SSSR. Vsesoyuznyi nauchno-issledovatelsky institut okhrany prirody i zapovednogo dela, Moscow, pp 49–59 (in Russian)

Pereladova OB, Sempere AZ, Agryzkov EB (1997) Bukhara deer: modern status and real perspectives of survival. In: Sokolov VE (ed) Redkie vidy mlekopitayushchikh Rossii i sopredelnykh stran. Institut Problem Ecologii i Evolutsii, Moscow, pp 70 (in Russian)

Pereladova OB, Sempere A, Marmozinskaya N, Baidavletov R, Toyshibekov M (1999a) Establishment of experimental groups of Bukhara deer to restore the population. In: Orlov VN (ed) VI sezd Teriologicheskogo obshchestva. Teriologicheskoe Obshchestvo, Moscow, pp 193 (in Russian)

Pereladova OB, Soldatova NB, Sidorenko E, Mordonov BK (1999b) Space and interspecies relationship of goitred gazelle, Przewalski's horse and kulan in ecological centre "Dzheiran". In: Orlov VN (ed) VI sezd Teriologicheskogo obshchestva. Teriologicheskoe Obshchestvo, Moscow, pp 193 (in Russian)

Pererva VI, Kiseleva EG (1991) Changes of reproductive cycles of European bison breeding in captivity. In: Perovsky MD (ed) Dichefermy I zoopitomniki. Tsentralnaya Nauchno-Issledovatelskaya Laboratoriya Okhotnichego Khozyaistva i Zapovednikov, Moscow, pp 165–179 (in Russian)

Pererva VI (1997) Account on European bison status in Russia for 31.12.1996 (ms) (in Russian)

Perovsky MD (1984) Changes of wild reindeer range in the USSR. In: Syroechkovsky EE (ed) Wild reindeer of the Soviet Union. Oxonian Press, New Delhi, pp 113–138

Perovsky MD (1988) Role of different factors in losses of wild ungulate populations in RSFSR. In: Nazarov AA (ed) Chronologicheskie izmeneniya chislennosti okhonichikh zhivotnykh v RSFSR. Tsentralnaya Nauchno-Issledovatelskaya Laboratoriya Okhotnichego Khozyaistva i Zapovednikov, Moscow, pp 106–126 (in Russian)

Perzanowski K, Kozak I (1999) The Carpathian bison *Bison bonasus*: its past and future. Biosphere Conserv 2(2):75–81

Petrashov VV (1977) Morphological characteristics of sika deer of the Khopyorsky Nature Reserve. Izvestiya Timiryazevskoi Selskokhozyaistvennoi Akademii 5:162–169 (in Russian)

Petrishchev BI (1998) Natal development. In: Sokolov VE, Zhirnov LV (eds) Saiga. Filogeniya, sistematika, ekologiya, okhrana i ispolzovanie. Nauka, Moscow, pp 193–202 (in Russian)

Petrov VS (1983) Saiga (*Saiga tatarica*) helminths in Kazakhstan. Byulleten Vsesoyuznogo Instituta Gelmintologii 34:27–31 (in Russian)

Petrov VS (1985) Helminths of saiga and their impact on epizootology of sheep helminthosis. Thesis. Vsesoyuznyi Institut Gelmintologii, Moscow (in Russian)

Petrov VS, Islamov PZ, Dutov MM (1979) Necrobiosis of saiga. Veterinaria 3:48 (in Russian)

Petrovsky YT, Kolesnikova ZM, Lisnevskaya NA (1967) Moose in Vitebsk oblast of Belorussia. In: Bannikov AG (ed) Biologiya i promysel losya, vol 3. Rosselkhozizdat, Moscow, pp 57–62 (in Russian)

Pevtsov MV (1951) Travels to China and Mongolia. Geographgiz, Moscow (in Russian)

Pikunov DG, Abramov VK, Skripchinskiy AA (1973) Some peculiaritites of distribution and conservation of rare mammalian species of the southern Far East. In: Sokolov VE (ed) Redkie vidy mlekopitayushchikh fauny SSSR i ikh okhrana. Nauka, Moscow, pp 25–62 (in Russian)

Pinegin NV (1932) Data from ecological studies in the New Siberian Islands. Trudy Soveta Po Izucheniyu Proizvoditelnykh Sil Ser Yakutskaya 7:23–78 (in Russian)

Plakhov KN (1997) Kulan (*Equus hemionus onager)* in Kazakhstan. In: Sokolov VE (ed) Redkie vidy mlekopitayushchikh Rossii i sopredelnykh stran. Institut Problem Ecologii i Evolutsii, Moscow, p 73 (in Russian)

Plakhov KN (1999) Ustyurt mountain sheep. In: Orlov VN (ed) VI sezd Teriologicheskogo obshchestva. Teriologicheskoe Obshchestvo, Moscow, 197 pp (in Russian)

Pleshak TV (1980) Spring ecology of mountain sheep. In: Sokolov VE (ed) Kopytnye fauny SSSR. Nauka, Moscow, pp 181–182 (in Russian)

Pobedinsky GD (1997) Antropogenic impact on numerical fluctuations of ungulates in Central Chernozemnyi region (by example of Lipetsk oblast). Thesis. Voronezhsky Gosudarstvennyi Universitet, Voronezh (in Russian)

Pole VB (1966) Aerial census of roe deer and moose in Northern Kazakhstan. Trudy Instituta zoologii Akademii Nauk Kazakhskoi SSR 26:186–192 (in Russian)

Pole VB (1973) Breeding of roe deer in Kazakhstan. Trudy Instituta Zoologii Akademii Nauk Kazakhskoi SSR 34:135–144 (in Russian)

Pole VB (1974) Systematic position of roe deer (*Capreolus capreolus*) of Kazakhsan. Zoologichesky Zhurnal 53(12):1833–1839 (in Russian)

Polezhaev NM (1976) Ecology of reindeer in northern Urals. Nauchno-Issledovatelsky Byulleten Naucho-Issl. Instituta Krainego Severa 12/13:68–70 (in Russian)

Polezhaev NM, Popov AA (1990) Status of wild reindeer in Komi republic. In: Pavlov BM (ed) Resursy, ekologiya i ratsionalnoe ispolsovanie dikikh severnykh olenei v SSSR. Naucho-Issledovatelsky Institut Krainego Severa, Novosibirsk, pp 141–144 (in Russian)

Popkova IF (1967) Chamois on the southern slopes of the Great Caucasus Ridge. Trudy Teberdinskogo Gosudarstvennogo Zapovednika 7:160–211 (in Russian)

Popov MV, Solomonov NG, Mordosov II, Labutin YV (1980) Biology of game mammals of Yakutia. Nauka, Novosibirsk (in Russian)

Pospelova EB (1989) Musk ox pastures in the Bikada basin and their production. Mlekopitayushchie i ptitsy Severa Srednei Sibiri. Naucho-Issledovatelsky Institut Krainego Severa, Novosibirsk, pp 79–87 (in Russian)

Prikhodko VI (1992) Food and feeding behaviour of musk deer (*Moschus moschiferus*). In: Sokolov VE (ed) Povedenie i kommunikatsiya mlekopitayushchikh. Nauka, Moscow, pp 42–76 (in Russian)
Prikhodko VI (1997) Decline of musk deer numbers in Russia. In: Sokolov VE (ed) Redkie vidy mlekopitayushchikh Rossii i sopredelnykh stran. Institut Problem Ecologii i Evolutsii, Moscow, pp 79 (in Russian)
Prilutskaya LI, Pishvanov YV (1988) Distribution and numbers of *Capra hircus* in Dagestan. In: Khokhlov AN (ed) Resursy zhivotnogo mira Severnogo Kavkaza. Stavropolsky Pedagogichesky Institut, Stavropol, pp 136–140 (in Russian)
Prisyazhnyuk VE (1978) Morphological peculiarities of aboriginal sika deer and its conservation. Thesis. Vsesoyuznyi Nauchno-Issledovatelsky Institut Okhrany Prirody i Zapovednogo Dela, Moscow (in Russian)
Prisyazhnyuk VE (1981) Morphometric characteristics of aboriginal sika deer (*Cervus nippon hortulorum*) in Primorsky Kray. Zool Zh 60(12):1817–1828 (in Russian with English summary)
Prisyazhnyuk VE (1986) Gazelle farm in Bukhara oblast. Watering of gazelles. In: Flint VE (ed) Okhrana i perspektivy restavratsii chislennosti dzheiranov v SSSR. Vsesoyuznyi Vsesoyuznyi Nauchno-Issledovatelsky Institut Okhrany Prirody i Zapovednogo Dela, Moscow, pp 59–70 (in Russian)
Prisyazhnyuk VE, Pigolkin AU (1974) Helminths of wild and tame sika deer of Far East. In: Vorobev GG (ed) Pyatnistyi olen Yuzhnogo Primorya. Kyrgyzstan, Frunze, pp 78–107 (in Russian)
Prisyazhnyuk NE, Prisyazhnyuk VE (1974) Systematical groups, life forms and seasonal use of food plants by sika deer. In: Vorobev GG (ed) Pyatnistyi olen Yuzhnogo Primorya. Frunze, Kyrgyzstan, pp 3–62 (in Russian)
Prisyazhnyuk VE, Pronyaev AV (1981) Production of saiga game husbandry. In: Flint VE (ed) Nauchnye osnovy ohrany I razionalnogo ispolzovaniya zhivotnogo mira. Vsesoyuznyi Nauchno-Issledovatelsky Institut Okhrany Prirody i Zapovednogo Dela, Moscow, pp 57–68 (in Russian)
Prisyazhnyuk VE, Soldatova NV (1984) The Bukharsky breeding station. Comunication 1. Feeding of goitred gazelle in an enclosed plot on hilly sands in conditions of enclosure. In: Flint VE (ed) Nauchnye osnovy okhrany i ratsionalnogo ispolzovaniya zhivotnogo mira. Vsesoyuznyi Nauchno-Issledovatelsky Institut Okhrany Prirody i Zapovednogo Dela, Moscow, pp 28–43 (in Russian)
Prostakov NI (1990) Ecology and social life of European roe deer. Voronezhsky Gosudarstvennyi Universitet, Voronezh (in Russian)
Prostakov NI (1996) Ungulates of the Central Chernozemnyi region. Voronezhsky Universitet, Voronezh (in Russian)
Prostakov NI, Obtemperansky SI (1986) Living conditions and spatial distribution of roe deer in forest islands of the Chernozemic regions. In: Tikhomirov VN (ed) Izuchenie i okhrana prirody v malykh zapovednikakh. Voronezhsky Universitet, Voronezh, pp 160–164 (in Russian)
Protoclitova TB (1961) Plant diet of sika deer in the Khopyorsky Nature Reserve. Trudy Khoperskogo Gosudarstvennogo Zapovednika 5:139–172
Prusaite YaA, Blazhis AS, Baleisis RM (1977) Breeding and fecundity of European roe deer in Lithuania. Trudy Akademii Nauk Litovskoi SSR Ser V, 3(79):105–110 (in Russian with English summary)
Pryadko EI (1976) Helminths of deer. Nauka, Alma-Ata (in Russian)
Pucek Z (1991) History of the European bison and problems of its protection and management. In: Bobek B, Perzanowski K, Regelin W (eds) Transactions of the 18th International Union Game Biologists Congress, Krakow 1987. Swiat Press, Krakow Warszawa
Puzansky VN (1999) Numbers of wild ungulates in Chita oblast. In: Orlov VN (ed) VI sezd Teriologicheskogo obshchestva. Teriologicheskoe Obshchestvo, Moscow, pp 208 (in Russian)

Pyatkov LP, Pryadko EI (1971) Maral husbandry. Kaynar, Alma-Ata (in Russian)
Raabe IS, Solopov OK(1992) On the behaviour of warble flies under attacking red deer and reciprocal reactions of red deer. In: Narchuk EP (ed) Bloodsucking and zoophilous two-winged insects (Insecta: *Diptera*). Zoologichesky Institut, St Petersburg, pp 137–139 (in Russian)
Rakhimov AA (1975) Short characteristic of Kyzylkumy mountain sheep. In: Khaitov RK (ed) Voprosy zootekhnii i karakulevodstva. Tashkentsky Selskohosyaistvenniy Institut, Samarkand, pp 65–69 (in Russian)
Rakov NV (1956) Saiga in western Kazakhstan. Trudy Instituta zoologii Akademii Nauk Kazakhskoi SSR 6:28–60 (in Russian)
Rakov NV (1965) Distribution and numbers of roe deer in Amur oblast. In: Dementev GP (ed) Okhotniche-promyslovye zveri. Rosselkhozizdat, Moscow, pp 93–107 (in Russian)
Randveer T (1995) Spruce damages caused by moose: a serious problem in the Estonian forestry. Ekologija 2:84–85
Rapota VV (1981) Plants of the Bikada river range (Eastern Taymyr) and their food utilisation by musk ox. In: Solomakha AI (ed) Ekologiya i khozyaistvennoe ispolzovanie nazemnoi fauny Eniseiskogo Severa. Naucho-Issledovatelsky Institut Krainego Severa, Novosibirsk, pp 73–93 (in Russian)
Rashek VA (1964) Parasitic fauna of kulan on the Barsakelmes Island. Byulleten Moskovskogo Obshchestva Ispytatelei Prirody Otdelenie Biologicheskoe 59(3):25–29 (in Russian with English summary)
Rashek VA (1966) Ecology of kulan and its acclimatization on the Barsakelmes Island. Thesis. Moskovskaya Selskokhozyaistvennaya Akademiya, Moscow (in Russian)
Rashek VA (1967) Seasonal peculiarities of kulan feeding in the Barsakelmes Island. Byulleten Moskovskogo Obshchestva Ispytatelei Prirody Otdelenie Biologicheskoe 82:5–13 (in Russian with English summary)
Rashek VL (1963) Data on saiga reproduction at the Barsakelmes Island (the Aral Sea). Trudy Instituta Zoologii Akademii Nauk Kazakhskoi SSR 20:164–193 (in Russian)
Rashek VL (1965) Saiga ecology during winter on the Barsakelmes Island. In: Dementev GP (ed) Okhotniche-promyslovye zveri. Biologiya i khozyaistvennoe ispolzovanie. Kolos, Moscow, pp 57–81 (in Russian)
Rashek VL (1974) Biology of saiga at the Barsakelmes Island. Thesis. Moskovskaya Veterinarnaya Akademiya, Moscow (in Russian)
Rashek VL (1978) Peculiarities of winter season in northern desert as a factor in the ecology of mammals. Trudy Moskovskoi Veterinarnoi Akademii 97:69–79 (in Russian)
Rastegaev J (1992) About species composition of gasterophiliid and oestriid marble flies of the USSR fauna. In: Narchuk EP (ed) Bloodsucking and zoophilous two-winged insects (Insecta: Diptera). Zoologichesky Institut, St Petersburg, pp 139–141 (in Russian)
Reimov A, Reimov R (1997) Anthropogenic influence on the status of populations of rare species of mammals in Karakalpak Ustyurt and Kyzylkum. In: Sokolov VE (ed) Redkie vidy mlekopitayushchikh Rossii i sopredelnykh stran. Institut Problem Ecologii i Evolutsii, Moscow, 81 pp (in Russian)
Reimov PP (1981) Ecological-geographical characteristics of mammalian therio-fauna of the southern Ustyurt. Vestnik Kara-Kalpakskogo filiala Akademii Nauk Uzbekskoi SSR 4(86):14–19 (in Russian)
Rementsova MM (1962) Brucellosis of wild animals. Nauka, Alma-ata (in Russian)
Revin YV (1989) Mammals of southern Yakutia. Nauka, Novosibirsk (in Russian)
Revin YV, Sopin LV, Zhelesnov NK (1988) Snow sheep. Nauka, Novosibirsk (in Russian)
Rozhnov VV (ed) (2001) Class Mlekopitayushchie – Mammalia. In: The Red Data book of Russian Federation (Zhivornye). Institute Problem Ekologii i Evolutsii, Moscow (in Russian)
Rukhlyadev DP (1964) Helminths of wild ungulates of the Crimean Peninsula and the Caucasus from ecological-zoogeographical view. Izdatelstvo Saratovskogo Universiteta, Saratov (in Russian)

Rumyantsev VV (1976) Peculiarities of winter pasture utilization by Chukchi reindeer (khargin) in Yakutiya. Yakutskoe Knizhnoe Izdatelstvo, Yakutsk (in Russian)
Rusakov OS (1978) Range dynamics and numbers of roe deer in the north-west of the USSR. In: Ivanter EV (ed) Fauna i ekologiya ptits i mlekopitayushchikh taezhnogo severo-zapada SSSR. Institut Biologii, Petrozavodsk, pp 175–181 (in Russian)
Rusakov OS (1979) Modern status of natural resources, ecology and use of ungulates in the north-west of the USSR. In: Vereshchagin NK, Rusakov OS (eds) Kopytnye severo-zapada SSSR. Nauka, Leningrad, pp 63–308 (in Russian)
Rusakov OS, Timofeeva EK (1984) Wild boar (ecology, resources, economical importance in NW Russia). Leningrad University, Leningrad (in Russian)
Rustamov AK, Atamuradov KI, Sopyev OS, Kurbanov OR, Korshunov MA (1986) Gazelles of Turkmenistan: status of populations, distribution and breeding, conservation measures. In: Flint VE (ed) Okhrana i perspektivy restavratsii chislennosti dzheiranov v SSSR. Vsesoyuznyi Nauchno-Issledovatelsky Institut Okhrany Prirody i Zapovednogo Dela, Moscow, pp 17–22 (in Russian)
Ryabov LS, Ovchinnikova SL (1964) Regulation of numbers of wild ungulates in the Voronezhsky Nature Reserve. Okhrana Prirody Tsentralno-Chernozemnoy Oblasti 5:173–185 (in Russian)
Ryabov LS, Likhatsky YP, Nikitin NM (1994) Wolf relations to wild and domestic ungulates. Ekologiya 1:53–59 (in Russian with English summary)
Ryashchenko LP (1976) Velvet antler husbandry in Primorsky kray. Dalizdat, Vladivostok (in Russian)
Sabilaev AS, Koptsev LA (1986) Past and modern status of gazelle numbers in the Karakalpak part of Ustyurt and Kyzylkumy. In: Flint VE (ed) Okhrana i perspektivy restavratsii chislennosti dzheiranov v SSSR. Vsesoyuznyi Nauchno-Issledovatelsky Institut Okhrany Prirody i Zapovednogo Dela, Moscow, pp 12–17 (in Russian)
Sablina TB (1955) Ungulates of the Belovezhskaya Forest Nature Reserve. Trudy Instituta Morphologii Zhivotnykh Akademii Nauk SSSR 15:1–191 (in Russian with English summary)
Sablina TB (1970) Evolution of digestive systems of Cervidae. Nauka, Moscow (in Russian)
Sablina TB (1980) Role of snow cover for forest reindeer. In: Sokolov VE (ed) Adaptatsiya zhivotnykh k zimnim usloviyam. Nauka, Moscow, pp 87–92 (in Russian)
Safronov VM, Platonov PN, Arzhanov IN (1990) Ecology, numbers and optimisation of cropping of reindeer in the Yakutian north. In: Pavlov BM (ed) Resursy, ecologiya i ratsionalnoe ispolsovanie dikikh severnykh olenei v SSSR. Naucho-Issledovatelsky Institut Krainego Severa, Novosibirsk, pp 6–14 (in Russian)
Safronov VM, Reshetnikov IS, Akhremenko AK (1999) Reindeer of Yakutia. Ecology, morphology, use. Nauka, Novosibirsk (in Russian)
Salmin Y (1972) Amur tiger (*Felis tigris*) in the Sikhote-Alinsky Nature Reserve. Priroda i Socialisticheskoe Khozyaistvo 8(2):199–202 (in Russian)
Sapozhnikov GN (1967) Numbers of some animal species in tugai of the Tigrovaya Balka Nature Reserve. Izvestiya Akademii Nauk Tadzhikskoi SSR Otdelenie Biologicheskikh Nauk 4(29):114–116 (in Russian)
Sapozhnikov GN (1975) Wild ungulates of Tadzhikistan. In: Sokolov VE (ed) Redkie vidy mlekopitayushchikh fauny SSSR i ikh okhrana. Nauka, Moscow, pp 48–62 (in Russian)
Sapozhnikov GN (1976) Wild mountain sheep (*Ovis*) of Tadzhikistan. Donish, Dushanbe (in Russian)
Sapozhnikov GN (1984) Feeding of some species of wild ungulates in Tadzhikistan. In: Narzikulov MN (ed) Prirodnye resyrsy i zapovedniki Tadzhikistana. Donish, Dushanbe, pp 174–209 (in Russian)
Sarkisov AA (1944) Armenian mouflon by Nasonov. Trudy Erevanskogo Zooparka 1(2):89–91 (in Russian)
Sarkisov AA (1950) Bovidae of Armenian plateau. Thesis, Erevansky Gosudarstvennyi Universitet, Erevan (in Russian)

Sashko VK, Smaznov VA (1999) An unusual migration of Siberian mountain goat. In: Orlov VN (ed) VI sezd Teriologicheskogo obshchestva. Teriologicheskoe Obshchestvo, Moscow, pp 223 (in Russian)

Savelev VD (1977) Behaviour of wild reindeer when crossing rivers. In: Sokolov GA (ed) Ekologiya i ispolzovanie okhotnichikh zhivotnykh Krasnoyarskogo kraya. Institut Lesa i Drevesiny, Krasnoyarsk, pp 17–20 (in Russian)

Savinov EF (1962) Reproduction and growth of Siberian mountain goat in the Dzhungarian Ala-Tau. Trudy Instituta Zoologii Akademii Nauk Kazakhskoi SSR 17:167–182 (in Russian)

Savinov EF (1964) Migrations and daily activity of Siberian mountain goat in Kazakhstan. Trudy Instituta Zoologii Akademii Nauk Kazakhskoi SSR 22:197–207 (in Russian)

Savinov EF (1975) Siberian mountain goat and mountain sheep in Kazakhstan. Thesis. Institut Zoologii, Alma-Ata (in Russian)

Savinov EF, Bekenov A (1983) Asiatic mouflon – *Ovis orientalis* Gmelin, 1774. In: Gvozdev EV, Kapitonov VI (eds) Mlekopitayushchie Kazakhstana, vol 3(3). Nauka, Alma-ata, pp 209–233 (in Russian)

Savinov V (1961) Hunting grounds of the Altai territory. Altaiskoe Knizhnoe Izdatelstvo, Barnaul (in Russian)

Sdobnikov VM (1935) Data on winter feeding of reindeer. Trudy Arcticheskogo Instituta 24:137–141 (in Russian)

Sdobnikov VM (1958) Wild reindeer in Taymyr and regulation of its cropping. Problemy Severa 2:156–168 (in Russian)

Sedov VA, Vedernikov VA, Chernichenko SA (1992) The most important infections of ungulates. In: Kolonin GV (ed) Bolezni i parazity dikikh zhivotnykh, Vsesoyuznyi Nauchno-Issledovatelsky Institut Okhrany Prirody i Zapovednogo Dela, Moscow, pp 4–11 (in Russian)

Segal AN (1959) Reindeer and reindeer husbandry in Karelia. Gosudarstvennoe Izdatelstvo Karelskoi ASSR, Petrozavodsk (in Russian)

Semenov-Tien-Shansky OI (1948a) Moose at the Kola Penninsula. Trudy Laplandskogo Gosudarstvennogo Zapovednika 2:91–162 (in Russian)

Semenov-Tien-Shansky OI (1948b) Wild reindeer of the Kola Peninsula. Trudy Laplandskogo Gosudarstvennogo Zapovednika 2:1–90 (in Russian)

Semenov-Tien-Shansky OI (1977) Reindeer. Nauka, Moscow (in Russian)

Semenov-Tien-Shansky OI (1989) History of wild reindeer in Kola Peninsula. In: Danilov PI (ed) Lesnoy severnyi olen. Institut Biologii, Petrozavodsk, pp 15–18 (in Russian)

Semyanov, LP, Starovoitov AV (1985) Wild reindeer of the Turukhansky rayon and neighbouring areas. In: Lineitsev SN (ed) Ekologiya, okhrana i khozyaistvennoe ispolsovanie dikikh severnykh oleney. Naucho-Issledovatelsky Institut Krainego Severa, Novosibirsk, pp 39–46 (in Russian)

Serzhanin IN (1961) Mammals of Belarus. Izdatelstvo Akademii Nauk Belorusskoi SSR, Minsk (in Russian)

Serzhanin YI, Kurskov AN, Rozhdestvensky VV (1975) Distribution and dynamics of numbers of ungulates at Rechitskoe Polesye. In: Sokolov VE (ed) Kopytnye fauny SSSR. Nauka, Moscow, p 119 (in Russian)

Severtsov SA (1940) Species constants of the Belowezhskaya Forest bison breeding and its population dynamics. Trudy Instituta Evolutsionnoy Morphologii 3(1):3–32 (in Russian with English summary)

Severtsov SA, Sablina TB (1953) Deer, roe deer and wild boar in the Belovezhskaya Forest Nature Reserve. Trudy Instituta Morphologii Zhivotnykh 9:140–205 (in Russian with English summary)

Shakula VF (1989a) Land use by Severtsov mountain sheep. In: Sokolov VE (ed) Ekologiya, morphologiya, ispolzozvanie i okhrana dikikh kopytnykh, vol 2. Nauka, Moscow, p 248 (in Russian)

Shakula VF (1989b) Daily cycle and activity of Severtsov mountain sheep. In: Sokolov VE (ed) Ekologiya, morphologiya, ispolzozvanie i okhrana dikikh kopytnykh, vol 2. Nauka, Moscow, pp 249–250 (in Russian)

Shakula VF (1989c) Reproduction, growth and development of Severtsov mountain sheep. In: Sokolov VE (ed) Ekologiya, morphologiya, ispolzozvanie i okhrana dikikh kopytnykh, vol 2. Nauka, Moscow, p 251 (in Russian)

Shamardanov ZA, Ramazanov ZM (1980) Home-ranges of ungulates in Dagestan. In: Sokolov VE (ed) Kopytnye fauny SSSR. Nauka, Moscow, pp 113–114 (in Russian)

Shaposhikov FD (1955) Ecology and morphology of Altai reindeer. Zool Zh 34(1):191–207 (in Russian with English summary)

Shaposhnikov FD (1956) Data on ecology of musk deer in North-Eastern Altai. Zool Zh 35(7):1084–1093 (in Russian with English summary)

Shargaev MA (1976) Wild reindeer of Ob – Irtysh North. Nauchno-Issledovatelsky Byulleten Naucho-Issledovatelskogo Instituta Krainego Severa 12/13:91–93 (in Russian)

Shargaev MA, Maturova RT (1976) Ecology of reindeer in Transbaikal. Nauchno-Issledovatelsky Byulleten Naucho-Issledovatelskogo Instituta Krainego Severa 12/13:94–96 (in Russian)

Sharov V (2001) Mongolian gazelle selected wrong route. Nezavisimaya gazeta, 13 Febr 2001, p 6 (in Russian)

Shaulskaya NA (1980) Seasonal changes of plants utilisation by goral in the Sikhote-Alinsky Nature Reserve. Rastitelnie Resursy 16(2):177–186 (in Russian)

Shaulskaya NA (1992) Pastures of goral in the Abrek area and their productivity. In: Myslenkov AI (ed) Amursky goral. Tsentralnaya Nauchno-Issledovatelskaya Laboratoriya Okhotnichego Khozyaistva i Zapovednikov, Moscow, pp 35–57 (in Russian)

Shchadilov YM (1986) Gazelle in Azerbaijan. In: Flint VE (ed) Nauchnye osnovy okhrany I ratsionalnogo ispolzovaniya zhivotnogo mira. Vsesoyuznyi Nauchno-Issledovatelsky Institut Okhrany Prirody i Zapovednogo Dela, Moscow, pp 41–43 (in Russian)

Shchelkunova RP (1980) Vegetation and food resources of reindeer husbandry in Taymyr. Dissertation (ms). Naucho-Issledovatelsky Institut Krainego Severa, Novosibirsk (in Russian)

Shchelkunova RP (1982) Distribution of plant biomass in zones of Taymyr. Bot Zh 67(4):479–491 (in Russian)

Shiryaev VV (1975) Impact of boar on muskrat population in the Ili delta. In: Sokolov VE (ed) Kopytnye fauny SSSR. Nauka, Moscow, pp 187–188 (in Russian)

Shmit EV (1936) Determination of duration of pregnancy of tame reindeer. Sovetskoe Olenevodstvo 8:35–43 (in Russian)

Shnarevich ID (1948) Data on the biology of mountain goat. Trudy Alma-Atinskogo Zooparka 1:73–93 (in Russian)

Shnitnikov VN (1936) Mammals of the Semireche. Izdatelstvo Akademii Nauk SSSR, Moscow-Leningrad (in Russian)

Shostak SV (1975a) Sex and age composition and spatial structure of European deer population in Belowezhskaya Forest. Belowezhskaya Pushcha Issledovaniya 9:144–155 (in Russian)

Shostak SV (1975b) On fertility of European deer in Belowezhskaya Forest. Belowezhskaya Pushcha Issledovaniya 9:137–144 (in Russian)

Shostak SV (1976) Reproduction of European deer in Belowezhskaya Forest. Belowezhskaya Pushcha Issledovaniya 10:81–92 (in Russian)

Shostak SV, Vasilyuk IF (1976) Diseases of European deer and their prevention. Belowezhskaya Pushcha Issledovaniya 10:93–108 (in Russian)

Shtarev YF (1966) Results of acclimatisation of sika deer in Mordovskaya SSR. Trudy Mordovskogo Gosudarstvennogo Zapovednika 3:55–125 (in Russian)

Shubin PN, Efimtseva EA (1988) Biochemical and population genetics of reindeer. Nauka, Leningrad (in Russian)

Shubin PN, Ionova TA (1984) Genetic relationship between tame and wild reindeer. Zool Zh 63(11):1725–1730 (in Russian with English summary)
Shulpin LM (1948) Data on mammals and reptiles of the Talas Ala-Tau. Izvestiya Akademii Nauk Kazakhskoi SSR Ser Zoologicheskaya 51:34–78 (in Russian)
Shvetsov YG, Smirnov MN, Monakhov GI (1984) Mammals of the Baikal Lake basin. Nauka, Novosibirsk (in Russian)
Sidorov SV (2000) Sika deer. Bighorn sheep. Caucasian turs. Siberian ibex. Saiga (2000) In: Lomanov IK (ed) Sostoyanie resursov okhotnichikh zhivotnykh v Rossiiskoy Federatsii. Tsentrokhotkontrol, Moscow, pp 52–54, 55–57, 58–65 (in Russian)
Sidorov SV, Bukreeva OM (1999) Population cycles of saiga in the north-west of Pricaspian. In: Orlov VN (ed) VI sezd Teriologicheskogo obshchestva. Teriologicheskoe Obshchestvo, Moscow, p 232 (in Russian)
Silantev AA (1898) Review of professional hunting in Russia. St Petersburg (in Russian)
Sinelnikov AM(1964) Some results of acclimatization of Bukhara deer in the mountain the Ramit Nature Reserve. Izvestiya AN Tadzhikskoi SSR Otdelenie Biologicheskikh Nauk 1:74–91 (in Russian)
Sipko TP, Rautian GS, Kiseleva EG (1999) European bison and its populations in the European part of Russia. In: Aristov AA (ed) Redkie vidy mlekopitayushchikh Rossii i sopredelnykh territoriy. Teriologicheskoe Obshchestvo, Moscow, pp 403–418 (in Russian with English summary)
Skalon VN (1936) Data on the fauna of the southern borders of Siberia. Izvestiya Gosudarstvennogo Protivochumnogo Instituta Sibiri i Dalnego Vostoka 3:135–209 (in Russian)
Skalon VN (1961) Distribution and way of life of wild ungulates in the Olekma basin. In: Granina AN, Domaradsky IV (ed) Biologichesky sbornik. Protivochumnyi Institut Sibiri i Dalnego Vostoka, Irkutsk, pp 150–158 (in Russian)
Sleptsov AM (1975) Moose of the upper part of Kolyma river. In: Sviridov NS (ed) Problemy okhotovedeniya i okhrany prirody. Irkutsky Selskokhozyaistvennyi Institut, Irkutsk, pp 153–157 (in Russian)
Slonim AD (1962) Ecological physiology of mammals. Izdatelstvo Akademii Nauk SSSR, Moscow (in Russian)
Slovtsov I (1897) Daily remarks during travel in Kokchetavsky uezd of Akmolinsk oblast in 1878. Zapiski Zapadno-Sibirskogo Otdela Imperatorskogo Russkogo Geographicheskogo Obshchestva 21:1–198 (in Russian)
Sludsky AA (1955) Saiga in Kazakhstan. Trudy Instituta Zoologii Akademii Nauk Kazakhskoi SSR 4:18–55 (in Russian)
Sludsky AA (1956a) Boar (morphology, ecology, economic use, epizootic significance, hunting). Izdatelstvo Akademii Nauk Kazakhskoi SSR, Alma-Ata (in Russian)
Sludsky AA (1956b) Breeding of gazelle. Trudy Instituta Zoologii Akademii Nauk Kazakhskoi SSR 6:78–108 (in Russian)
Sludsky AA (1962) Interrelations of predators and prey. Trudy Instituta Zoologii Akademii Nauk Kazakhskoi SSR 17:24–143 (in Russian)
Sludsky AA (1963) "Dzhut" in Eurasian steppes and desert. Trudy Instituta Zoologii Akademii Nauk Kazakhskoi SSR 20:5–88 (in Russian)
Sludsky AA (1977) Gazelle. In: Kaletsky AA (ed) Kopytnye zveri. Lesnaya Promyshlennost, Moscow, pp 28–61 (in Russian)
Sludsky AA, Baidavletov R (1984) Roe deer. In: Gvozdev EV, Strautman EI (eds) Mlekopitayushchie Kazakhstana, vol 3(4). Nauka, Alma-Ata, (in Russian)
Smirnov KA (1987) Moose impact in southern taiga habitats. Nauka, Moscow (in Russian)
Smirnov MN (1976) Distribution, numbers and conservation of wild reindeer in Buryatiya. Nauchno-Issledovatelsky Byulleten Naucho-Issledovatelskogo Instituta Krainego Severa 12/13:85–87 (in Russian)
Smirnov MN (1978) The roe deer in the west of the Trans-Baikal area. Nauka, Novosibirsk (in Russian)

Smirnov MN (1980) Wild ungulates of Tuva republic. In: Sokolov VE (ed) Kopytnye fauny SSSR. Nauka, Moscow, pp 104–105 (in Russian)

Smirnov MN (1989a) Regional peculiarities of ecology and game management of red deer in Tuva ASSR. In: Amirkhanov AM (ed) Ekologicheskie issledovaniya v zapovednikakh yuzhnoy Sibiri. Glavnoe Upravlenie Okhotnichego Khozyaistva, Moscow, pp 5–16 (in Russian)

Smirnov MN (1989b) Wild ungulates of the Southern Siberia (historical changing of population, problems of use). In: Glushkov VM (ed) Upravlenie populyatsiyami dikikh kopytnykh. Vsesoyuznyi Naychno-Issledovatelsky Institut Okhotnichego Khozyaistva i Zverovodstva, Kirov, pp 59–75 (in Russian)

Smirnov MN (1990) Modern numbers and perspectives of utilisation of ungulates in Krasnoyarsky kray. Krasnoyarskoe Knizhnoe Izdatelstvo, Krasnoyarsk (in Russian)

Smirnov MN (1994) Large game mammals of southern Siberia. Thesis. Institut Problem Ecologii i Evolutsii, Moscow (in Russian)

Sobansky GG (1975) Snow and migrations of maral in the north-east of Altai. Ekologiya 4:99–100 (in Russian with English summary)

Sobansky GG (1986) Enlargement of range and increasing of ungulates numbers at Altai. Byulleten Moskovskogo Obshchestva Ispytatelei Prirody Otdelenie Biologicheskoe 94(1):25–29 (in Russian with English summary)

Sobansky GG (1989) Altai mountain sheep and its distribution, numbers, migrations. In: Sablina TB (ed) Ekologiya, morphologiya, ispolzozvanie i okhrana dikikh kopytnykh, vol 2. Nauka, Moscow, pp 245–246 (in Russian)

Sobansky GG (1990a) Dzeren. In: Ilyashenko VY (ed) Itogi izucheniya redkhikh zhivotnykh. Tsentralnaya Nauchno-Issledovatelskaya Laboratoriya Okhotnichego Khozyaistva i Zapovednikov, Moscow, pp 31–33 (in Russian)

Sobansky GG (1990b) Ungulate resources in Altai. In: Evsikov VI (ed) Resursy zhivotnogo mira Sibiri. Okhotniche-promyslovye zveri i ptitsy. Nauka, Novosibirsk, pp 254–257 (in Russian)

Sobansky GG (1992) Ungulates of Altai mountains. Nauka, Novosibirsk (in Russian)

Sokolov AY, Kushnir AV (1986) Bioenergetics of reindeer. Nauka, Novosibirsk (in Russian)

Sokolov GA (1983) Wild reindeer of the southern Siberian mountains. In: Zabrodin VA (ed) Dikii severnyi olen. Tsentralnaya nauchno-issledovatelskaya laboratoriya okhotnichego khozyaistva i zapovednikov, Moscow, pp 120–130 (in Russian)

Sokolov GA (1988) Actual population status of some rare and endangered mammal species in the south of Siberia. In: Shvetsov YG (ed) Redkie nazemnye mlekopitayushchie Sibiri. Nauka, Novosibirsk, pp 211–218 (in Russian)

Sokolov II (1959) Fauna of the USSR. Mammals (Orders Artiodactyla and Perissodactyla), vol 1(3). Akademiya Nauk SSSR, Moscow-Leningrad (in Russian)

Sokolov VE (1979) Systematics of mammals. Vyshshayaya shkola, Moscow (in Russian)

Sokolov VE, Baskin LM (1984) Aims of applied ecological and ethological research of game land animals in forest zone. In: Sokolov VE (ed) Zhivotnyi mir yuzhnoi taigi. Nauka, Moscow, pp 6–38 (in Russian)

Sokolov VE, Danilkin AA (1981) Siberian roe deer. Nauka, Moscow (in Russian)

Sokolov VE, Gromov VS (1993) Chemical communication of roe deer (*Capreolus*). Zool Zh 72(8):123–136 (in Russian with English summary)

Sokolov VE, Lebedeva NL (1989) Commercial hunting in the Soviet Union. In: Hudson RJ, Drew KR, Baskin LM (eds) Wildlife production systems. Cambridge Univ Press, Cambridge, pp 170–186

Sokolov VE, Prikhodko VI (1992) Breeding and sexual behaviour of musk deer (*Moschus moschiferus*). In: Sokolov VE (ed) Povedenie i kommunikatsiya mlekopitayushchikh. Nauka, Moscow, pp 19–42 (in Russian)

Sokolov VE, Tembotov AK (1993) Vertebrates of the Caucasus. Mammals. Ungulates. Nauka, Moscow (in Russian)

Sokolov VE, Chernova OF, Voloshina IV, Myslenkov AI (1992a) Skin and hair cover of goral. In: Myslenkov AI (ed) Amursky goral. Tsentralnaya Laboratoriya Okhotnichego Khozyaistva i Zapovednikov, Moscow, pp 83–105 (in Russian)
Sokolov VE, Danilkin AA, Markov GG (1992b) Taxonomy of *Capreolus* in the light of modern research. In: Sokolov VE (ed) Evropeiskaya i sibirskaya kosulya. Nauka, Moscow (in Russian)
Sokolsky SM (1984) Wild reindeer of the upper part of Pechora river. Trudy Pechoro-Ilychskogo Gosudarstvennogo Zapovednika 9:238–263 (in Russian)
Sokov AI (1987) Bukhara red deer. Donish, Dushanbe (in Russian)
Sokov AI (1993) Mammals - Artiodactyla. Fauna Tadzhikistana 20(5):1–257 (in Russian)
Solomatin AO (1973) Kulan. Nauka, Moscow (in Russian)
Solomatin AO (1975) European deer in Voronezh oblast. In: Sokolov VE (ed) Kopytnye fauny SSSR. Nauka, Moscow, pp 127–128 (in Russian)
Solomkina NV (1983) Breeding of goral. Byulleten Moskovskogo Obshchestva Ispytatelei Prirody Otdelenie Biologicheskoe 88(3):21–26 (in Russian with English summary)
Solomkina NV (1992) Growth and development of goral. In: Myslenkov AI (ed) Amursky goral. Tsentralnaya Nauchno-Issledovatelskaya Laboratoriya Okhotnichego Khozyaistva i Zapovednikov, Moscow, pp 70–75 (in Russian)
Sopin LV (1975a) Wild sheep of the south of Siberia. Thesis, Irkutsky Selskokhozyaistvennyi Institut, Irkutsk (in Russian)
Sopin LV (1975b) Herd size of mountain sheep in the south of Siberia. In: Sviridov NS (ed) Problemy okhotovedeniya i okhrany prirody. Irkutsky Selskokhozyaistvennyi Institut, Irkutsk, pp 157–162 (in Russian)
Sopin LV (1976) Some data on ecology of mountain sheep of Chikhachev ridge and the Talduair mountains. In: Sokolov VE (ed) Redkie vidy mlekopitayushchikh fauny SSSR i ikh okhrana. Nauka, Moscow, pp 233–234 (in Russian)
Sopin LV, Ermolin AB (1988) Actual status and perspectives of protection of Kodar snow sheep. In: Shvetsov YG (ed) Rare land mammals of Siberia. Nauka, Novosibirsk, pp 222–226 (in Russian)
Sopina NK (1983) Ungulates of the Sokhondinsky Nature Reserve. In: Yudin BS (ed) Ekologiya mlekopitayushchikh Vostochnoi Sibiri. Irkutsky Universitet, Irkutsk (in Russian)
Standley LA (1992) Taxonomic issues in rare species protection. Rhodora 94:218–242
Steklenev EP (1980) Particular features of roe deer breeding in Ukraine. In: Sokolov VE (ed) Kopytnye fauny SSSR. Nauka, Moscow, pp 202–203 (in Russian)
Stepanova VV (2002) Ecology of red deer (*Cervus elaphus* L., 1758) in Yakutia. Dissertation (ms). Institut Problem Ecologii i Evolutsii, Moscow (in Russian)
Stroganov SU (1959) Data on ecology of tugai deer. Trudy Biologicheskogo Instituta Sibirskogo Otdeleniya Akademii Nauk SSSR Zoologiya 5:15–84 (in Russian)
Stubbe C, Danilkin A (1992) Breeding. Economic importance, rational use of resources and conservation. In: Sokolov VE (ed) Evropeiskaya i sibirskaya kosuli. Nauka, Moscow, pp 140–159, 276–336 (in Russian)
Sultanov GS (1953) Screw-horned goat in Uzbekistan. Trudy Instituta Zoologii i Parazitologii Akademii Nauk Uzbekskoy SSR Zoologiya 2:3–27 (in Russian)
Suvorov AP (1989) Maral in the Stolby Nature Reserve and problems of its economic use in Krasnoyarsky kray. Trudy gosudarstvennogo zapovednika "Stolby" 16:35–66 (in Russian)
Sviridov NS (1978) Maral. In: Kaletsky AA (ed) Krupnye khishchniki i kopytnye. Lesnaya Promyshlennost, Moscow, pp 129–159 (in Russian)
Sviridov NS (1983) *Cervus elaphus xantophygus* of the south of the Baikal region. Proceedings of the 14th Congress of the International Union of Game Biologists, Vysokie Tatry, CSSR, pp 56–59 (in Russian)
Syroechkovsky EE (1986) Reindeer. Agropromizdat, Moscow (in Russian)

Tarasov MA (1975) Factors determining numbers fluctuations of ungulates in the Teberdinsky Nature Reserve and neighbouring areas. In: Sokolov VE (ed) Redkie vidy mlekopitayushchikh fauny SSSR i ikh okhrana. Nauka, Moscow, pp 156–166 (in Russian)
Tarasov PP (1956) Some peculiaritities of reindeer morphology as the tundra animal. Byulleten Moskovskogo Obshchestva Ispytatelei Prirody Otdelenie Biologicheskoe 61(4):80–82 (in Russian with English summary)
Teplov VA (1938) Wolf in the Kavkazsky Nature Reserve. Trudy Kavkazskogo Gosudarstvennogo Zapovednika 1:366–443 (in Russian)
Teplov VP (1960) Numerical fluctuations and annual changes in ecology of game animals in Pechora Taiga. Trudy Pechoro-Ilychskogo Gosudarstvennogo Zapovednika 8:5–221 (in Russian)
Tikhomolov GA (1982). Establishing new populations of Bukharsky deer in Uzbekistan. In: Rustamov AK (ed) Razvedenie i sozdanie novykh populyatsiy redkikh i tsennykh vidov zhivotnykh. Ilym, Ashabad, pp 138–147 (in Russian)
Timofeeva EK (1974) Moose (ecology, distribution, economical role). Leningradsky Universitet, Leningrad (in Russian)
Timofeeva EK (1975) Ecology of boar at the north-west of the European part of the USSR. In: Sokolov VE (ed) Kopytnye fauny SSSR. Nauka, Moscow, pp 134–135 (in Russian)
Timofeeva EK (1978) Impact of wild ungulates on vegetation of oak forests in the steppe-forest zone. Vestnik Leningradskogo Universiteta 21:39–46 (in Russian)
Timofeeva EK (1985) Roe deer. Izdatelstvo Leningradskogo Universiteta, Leningrad (in Russian)
Tochiev TY (1968) Game animals of Chechen-Ingush republic. Thesis. Institut Zoologii AN Azerbaijanskoi SSR, Baku (in Russian)
Tochiev TY (1973) Problems of ecology and protection of Caucasian deer and *Capra hircus* in Checheno-Ingushskaya ASSR. In: Sokolov VE (ed) Redkie vidy mlekopitayushchikh fauny SSSR i ikh okhrana. Nauka, Moscow, pp 150–151 (in Russian)
Tochiev TY, Batkhiev AM (1980) On feeding of *Capra hircus* in Chechen-Ingush republic. In: Tembotov AK (ed) Fauna, ekologiya i okhrana na Severnom Kavkaze. Kabardino-Balkarsky Universitet, Nalchik, pp 57–64 (in Russian)
Tönisson J (1997) The moose research and management in Estonia. Conference abstracts of the 4th International Moose Symposium, University of Alaska, Fairbanks, p 70
Tönisson J, Roht K, Laanetu N (2000) Hunting. In: Yearbook forest 2000. Metsakaitse-ja Metsauuenduskeskus, Tartu
Trepet SA (1997) Dynamic of numbers of the main protected species of the Kavkazsky Nature Reserve. Proceedings Vtoroy nauchno-prakticheskoy konferentsii Maikopskogo gosudarstvennogo tekhnologicheskogo instituta, Maikop, pp 106–111 (in Russian)
Treus MY, Vitakova PN, Dzhurovich VM (1992) A comparative characteristic of lactation of moose and eland. Alces [Suppl] 1:238
Troitsky GA (1972) Moose of Karelia. Kareliya, Petrozavodsk (in Russian)
Tsalkin VI (1944) To biology of deer reproduction (*Cervus elaphus*). Zool Zh 23(6):369–377 (in Russian with English summary)
Tsalkin VI (1945) Data on study of deer (*Cervus elaphus*) antlers. Zool Zh 24(4):224–235 (in Russian with English summary)
Tsalkin VI (1950) Siberian mountain goat. Moskovskoe Obshchestvo Ispytatelei Prirody, Moscow (in Russian)
Tsalkin VI (1951) Mountain sheep of Europe and Asia. Moskovskoe Obshchestvo Ispytateley Prirody, Moscow (in Russian)
Tsapluk OE (1977) Age-related and seasonal peculiarities of the reproductive biology of the roe deer (*Capreolus capreolus* L.) in Kazakhstan. Zool Zh 56(4):611–618 (in Russian)
Tsaplyuk OE (1962) Age of sexual maturation and process of spermatogenesis of saiga males. Trudy Instituta Zoologii Akademii Nauk Kazakhskoi SSR 17:222–240 (in Russian)
Tsaplyuk OE (1966) Breeding cycles of saiga. Trudy Instituta Zoologii Akademii Nauk Kazakhskoi SSR 26:193–211 (in Russian)

Tsaplyuk OE (1972) Age and seasonal dynamics of breeding activity of gazelle. Izvestiya Akademii Nauk Kazakhskoi SSR Ser Biologicheskay 3:39–46 (in Russian)
Tsaplyuk OE, Savinov EF (1973) Breeding rhythms of Siberian mountain goat in Dzhungarian Ala-Tau Range. Trudy Instituta Zoologii Akademii Nauk Kazakhskoi SSR 34:200–212 (in Russian)
Tsarev SA (1980) Intraspecies relations of boar at the northern limits of the range. In: Sokolov VE (ed) Kopytnye fauny SSSR. Nauka, Moscow, pp 321–322 (in Russian)
Tsarev SA (1989) Cyclic modifications of social structure of boar groups. In: Sablina TB (ed) Ekologiya, morphologiya, ispolzozvanie i okhrana dikikh kopytnykh, vol 2. Nauka, Moscow, pp 216–217 (in Russian)
Tsarev SA (1999) Strategy of restoration of historical range of musk ox in the extreme North of Russia. Tsentralnaya Laboratoriya Okhotnichego Khozyaistva i Zapovednikov, Moscow (in Russian)
Tsarev YS (1975) Habitats and migrations of moose at the area of middle Enisei. In: Sokolov VE (ed) Kopytnye fauny SSSR. Nauka, Moscow, pp 139–140 (in Russian)
Tsybulin SM (1990) Summer numbers and distribution of roe deer in Northern Altai. In: Evsikov VI (ed) Resursy zhivotnogo mira Sibiri. Okhotniche-promyslovye zveri i ptitsy. Nauka, Novosibirsk, pp 248–250 (in Russian)
Tsybulsky AI (1995) Utilisation of renewable resources of the Ust-Lensky region (wild reindeer, fishes). In: Tikhomirov VN (ed) Zapovednoe delo v novykh socialno-economicheskikh usloviyakh. Nauka, St Petersburg, pp 161–162 (in Russian)
Tupitsyna LF (1988) Numerical fluctuations and distribution of wild boar in the Darwinsky Nature Reserve. In: Sokolov VE (ed) Populyatsionnye issledovaniya zhivotnykh v zapovednikakh. Nauka, Moscow, pp 128–139 (in Russian)
Turkin NV, Satunin KA (1902) Mammals of Russia. Izdatelstvo Turkina, Moscow (in Russian)
Tyshkevich A (1999a) Bukhara mountain sheep (urial). Okhota 9/10:22
Tyshkevich VE (1999b) Ecological-ethological adaptation of roe deer (*Capreolus capreolus* L) in the Polesye National Reserve. Vestsi Natsyjanalnoy Akademii Navuk Belarusi Seria Bijalagichnykh Navuk 1:113–116 (in Russian)
Uchitkin VI (1990) Wintering of wild reindeer in the tudra zone of Taymyr. In: Smirnov MN (ed) Ekologicheskie i ekonomicheskie aspekty okhrany i ratsionalnogo ispolzovaniya okhotnichikh zhivotnykh i rastitelnykh pishchevykh resursov. Sayansky Zapovednik, Shushenskoe, pp 158–160 (in Russian)
Ushkov SL (1947) The game fauna of the Ilmen Nature Reserve. Trudy Ilmenskogo Gosudarstvennogo Zapovednika 3(1):87–159 (in Russian)
Usov SA (1865) European bison. Zapiski Russkogo Obshchestva Akklimatizatsii, Moscow (in Russian)
Uspensky SM (1976) Musk ox in the USSR. Priroda 1:58–63 (in Russian)
Uspensky SM (1984) Musk ox in the USSR: some results and perspectives of their introduction. Biol Pap Univ Alaska Spec Rep 4:12–14
Ustinov SK (1960) Distribution and habitats of moose in the Barguzinsky ridge. Trudy the Barguzinskogo Gosudarstvennogo Zapovednika 2:75–80 (in Russian)
Ustinov SK (1965) Biology of musk deer in Pribaikal and Transbaikal areas. Okhotniche-Promyslovye Zveri 1:82–92 (in Russian)
Ustinov SK (1969a) Some adaptations of musk deer to snow cover. Trudy Moskovskoi Veterinarnoi Akademii 55:138–152 (in Russian)
Ustinov SK (1969b) Winter feeding of musk deer (*Moschus moschiferus*) in Eastern Sayan. Zool Zh 48(10):1558–1563 (in Russian with English summary)
Ustinov SK (1971) Geographical features of some biological peculiarities of musk deer. Izvestiya Vostochno-Sibirskogo Otdeleniya Geographicheskogo Obshchestva SSSR 68:101–107 (in Russian)
Vakhtina TV (1964) Dynamics of yield and utilisation of some tundra food shrub leaves in reindeer husbandry. Problemy Severa 8:290–296 (in Russian)

Varnakov AP, Mosheva TS (1975) Role of boar in agriculture in Moscow oblast. In: Sokolov VE (ed) Kopytnye fauny SSSR. Nauka, Moscow, pp 161–162 (in Russian)

Vasenko EP (1950) Ecology and distribution of saiga. Trudy Gosudarstvennogo Zapovednika Barsa-Kelmes 1:37–115 (in Russian)

Vatolin BA (1980) Distribution, numbers and impact of ecological factors on boar in Bryansk oblast. In: Sokolov VE (ed) Kopytnye fauny SSSR. Nauka, Moscow, pp 131–132 (in Russian)

Veinberg PI (1979) Biology of large predators in the Severo-Ossetinsky Nature Reserve. In: Sokolov VE (ed) Ecologicheskie osnovy okhrany i ratsyonalnogo ispolsovaniya khishchnykh mlekopitayushchikh. Nauka, Moscow, pp 22–23 (in Russian)

Veinberg PI (1984) Dagestan goat (*Capra cylindricornis* Blyth). Moskovskoe Obshchestvo Ispytatelei Prirody, Moscow (in Russian)

Veinberg PI, Komarov YE (1999) State of European bison in Northern Ossetia-Alania. In: Orlov VN (ed) VI sezd Teriologicheskogo obshchestva. Teriologicheskoe Obshchestvo, Moscow, p 45 (in Russian)

Veinger GM (1963) Goral population of the Sikhote-Alinsky Nature Reserve. Trudy Sikhote-Alinskogo gosudarstvennogo zapovednika 3:269–290 (in Russian)

Vendland OV (1938) Food plants of wild and tame sika deer. Vestnik Dalnevostochnogo Filiala Akademii Nauk 28(1):134–140 (in Russian)

Vereshchagin NK (1938) Goat of Azerbaijan. Trudy Zoologicheskogo Instituta Akademii Nauk Azerbaijanskoi SSR 9(45):1–70 (in Russian)

Vereshchagin NK (1947) Game animals of the Caucasus. Institut Zoologii Azerbaijanskoi SSR, Baku (in Russian)

Vereshchagin NK (1959a) Mammals of the Caucasus. Akademiya Nauk SSSR, Moscow-Leningrad (in Russian)

Vereshchagin NK (1959b) Musk ox in the north of Siberia. Priroda 8:105–106 (in Russian)

Vershinin AA, Kleymenov AD, Vyatkin PS, Fil VI (1984) The Kamchatka wild reindeer. In: Syroechkovsky EE (ed) Wild reindeer of the Soviet Union. Oxonian Press, New Delhi, pp 209–216

Vinogradov AF (1870) European bison of the north-west of the Caucasus. Trudy Vtorogo Sezda Russkikh Estestvoispytatelei, pp 181–189 (in Russian)

Vinogradov MP (1936) Dates of calving and rut in reindeer husbandry. Sovetskoe Olenevodstvo 6:25–89 (in Russian)

Vodopyanov BG (1970) Wild reindeer in Transbaikal (distribution, morphology, conservation and rational use. Thesis. Irkutsky Selkhozyaistvennyi Institut, Irkutsk (in Russian)

Vodopyanov BG (1971) Feeding of reindeer in mountains and taiga of the Stanovoy ridge. Izvestiya Vostochno-Sibirskogo Otdeleniya Geographicheskogo Obshchestva SSSR 68:94–100 (in Russian)

Vodopyanov BG (1984) Systematic position and ecological featurs of wild reindeer of the Trans-Baikal Region. In: Syroechkovsky EE (ed) Wild reindeer of the Soviet Union. Oxonian Press, New Delhi, pp 199–202

Voloshina IV (1997) Dynamics of sika deer number in middle Sikhote-Alin. In: Sokolov VE (ed) Redkie vidy mlekopitayushchikh Rossii i sopredelnykh stran. Institut Problem Ecologii i Evolutsii, Moscow, p 21 (in Russian)

Voloshina IV, Myslenkov AI (1992) Theoretical and practical aspects of reintroduction of goral. In: Myslenkov AI (ed) Amursky goral. Tsentr. Lab. Okhotnichego Khozyaistva i Zapovednikov, Moscow, pp 123–131 (in Russian)

Voloshina IV, Nesterov DA (1992) Characteristics of goral habitats in Primorsky kray. In: Myslenkov AI (ed) Amursky goral. Tsentr. Lab. Okhotnichego Khozyaistva i Zapovednikov, Moscow, pp 21–35 (in Russian)

Voloshina IV, Solomkina NV (1992) Diseases of goral in nature and in captivity. In: Myslenkov AI (ed) Amursky Goral. Tsentralnaya Laboratoriya Okhotnichego Khozyaistva i Zapovednikov, Moscow, pp 105–118 (in Russian)

Voloshina IV, Myslenkov AI, Smirnov EN (1986) Conservation of goral. In: Sokolov VE (ed) Redkie vidy mlekopitayushchikh fauny SSSR i ikh okhrana. Nauka, Moscow, pp 125–133 (in Russian)

Voloshina IV, Myslenkov AI, Obvertkina TV (1992) Daily activity of goral. In: Myslenkov AI (ed) Amursky goral. Tsentralnaya Laboratoriya Okhotnichego Khozyaistva i Zapovednikov, Moscow, pp 58–70 (in Russian)

Voronin AA (1980) Daily activity of boar at the south of Nechernozemic regions. In: Sokolov VE (ed) Kopytnye fauny SSSR. Nauka, Moscow, pp 285–286 (in Russian)

Voronin AA, Zaikin SM (1989) Analysis of hunting samples of boar population in Kaluga oblast. In: Sablina TB (ed) Ekologiya, morphologiya, ispolzovanie i okhrana dikikh kopytnykh, vol 2. Nauka, Moscow, p 194 (in Russian)

Vshivtsev VP, Pavlov BM, Kolpashchikov LA, Novikov BV (1980) Numbers population structure, use and conservation of wild reindeer in Yakutian tundra. In: Sokolov VE (ed) Kopytnye fauny SSSR. Nauka, Moscow, pp 78–79 (in Russian)

Wroblewski K (1927) European bison of Białowieża Forest. Wydawanictwo Polskie, Poznan (in Polish)

Yakushkin GD (1976) Herd characteristics of the Taymyr reindeer population. Nauchno-Issledovatelsky Byulleten Naucho-Issledovatelskogo Instituta Krainego Severa 12–13: 106–109 (in Russian)

Yakushkin GD (1979) Comparisons of ecological characteristics of areas of trapping (Canada, Alaska) and release of musk ox (USSR). In: Zabrodin VA (ed) Problemy okhrany i ekonomicheskogo ispolzovaniya resursov dikikh zhivotnykh Eniseiskogo Severa. VASKHNIL, Novosibirsk, pp 63–81 (in Russian)

Yakushkin GD (1998) The musk ox in Taymyr. Naucho-Issledovatelsky Institut Krainego Severa, Novosibirsk (in Russian)

Yakushkin GD, Michurin LM, Pavlov BM, Zyryanov VA (1970) Number and migrations of wild reindeer in Taymyr. Proceedings of the 9th International Congress of Game Biologists, Moscow, pp 335–338

Yakushkin GD, Kolpashchikov LA, Kokorev (2001) Reindeer in Taymyr in 2001. Okhota i Okhotniche Khozyaistvo 5:20–21 (in Russian)

Yanushevich AI, Aizin BM, Kydyraliev AK, Umrikhina GS, Fedyanina TF, Shukurov ED, Grebenyuk RV, Tokobayev MM (1972) Mammals of Kyrgyzstan. Frunze, Ilym (in Russian)

Yarovenko YA (1997) Feeding and energetic connections of goats *Capra cylindricornis* with pasture ecosystems in high mountains of the eastern Caucasus. Dissertation (MS). Prikaspiyskiy Institut Biologicheskikh Resursov, Makhachkala (in Russian)

Yavruyan EG (1969) Distribution of the wild sheep *Ovis ammon* Gmelin in Transcaucasia. Byulleten Moskovskogo Obshchestva Ispytatelei Prirody Otdelenie Biologicheskoe 24(2):26–32 (in Russian with English summary)

Yavruyan EG (1970) Transcaucasian wild sheep. Thesis. Erevansky Gosudarstvennyi Universitet, Erevan (in Russian)

Yavruyan EG, Popkov AS, Stepanyan ZS (1971) Aerial counts of game animals in mountain areas (Armeniya). Byulleten Moskovskogo Obshchestva Ispytatelei Prirody Otdelenie Biologicheskoe 76(2):52–55 (in Russian with English summary)

Yazan YP (1961) Biological peculiarities and ways of economical use of the migratory moose population in the Pechora taiga. Trudy Pechoro-Ilychskogo Gosudarstvennogo Zapovesnika 9:114–201 (in Russian)

Yazan YP (1972) Game animals of the Pechora taiga. Volgo-Vyatskoe Knizhnoe Izdatelstvo, Kirov (in Russian)

Yudin BS, Belov SN (1976) Wild reindeer of Tomsk and Novosibirsk oblast. Nauchno-Issledovatelsky Byulleten Naucho-Issledovatelskogo Instituta Krainego Severa 12/13: 103–105 (in Russian)

Yurgenson PB (1935) Moose of central areas of the European part of the USSR. In: Jurgenson PB (ed) Los i ego promysel. Glavpushnina, Moscow, pp 5–102 (in Russian)

Yurgenson PB (1964) Structure and composition of moose populations in forest hunting grounds. In: Bannikov AG (ed) Biologiya i promysel losya, vol 1. Rosselkhozizdat, Moscow, pp 13–34 (in Russian)

Zablotskaya LV (1964) Experience of moose number regulation in the Prioksko-Terrasny Nature Reserve and its surrounding. In: Bannikov AG (ed) Biologiya i promysel losya, vol 1. Rosselkhozizdat, Moscow, pp 156–173 (in Russian)

Zablotsky MA (1949) It is necessary to study features of European bison and its restoration in the USSR. Nauchno-Metodicheskie Zapiski Glavnogo Upravleniya Po Zapovednikam 13:128–147 (in Russian)

Zablotsky MA (1957) Some biological features of European bison and its changes in captivity. Trudy Priioksko-Terrasnogo Gosudarstvennogo Zapovednika 10:5–65 (in Russian)

Zabrodin VA (1984) Data on diseases of the Taymyr population of wild reindeer. In: Syroechkovsky EE (ed) Wild Reindeer of the Soviet Union. Oxonian Press, New Delhi, pp 113–120

Zabrodin VA, Pavlov BM (1983) Status and rational use of Taymyr reindeer population. In: Zabrodin VA (ed) Dikii severnyi olen. Tsentralnaya nauchno-issledovatelskaya laboratoriya okhotnichego khozyaistva i zapovednikov, Moscow, pp 60–75 (in Russian)

Zagorodskikh EE, Reimers NF (1984) Wild reindeer of Sakhalin. In: Syroechkovsky EE (ed) Wild Reindeer of the Soviet Union. Oxonian Press, New Delhi, pp 217–224

Zaitcev VA (1981) The habitats and feeding behaviour of the musk deer (*Moschus moschiferus* L.). Byulleten Moskovskogo Obshchestva Ispytatelei Prirody Otdelenie Biologicheskoe 86(7):19–30 (in Russian with English summary)

Zaitcev VA (1983) Escape behaviour of musk deer (*Moschus moschiferus parviceps*) in middle Sikhote-Alin. Zool Zh 62(11):1718–1726 (in Russian with English summary)

Zaitcev VA (1991) Musk deer of Sikhote-Alin. Ecology and behaviour. Nauka, Moscow (in Russian)

Zakariev AY (1982) Ecology of *Capra hircus* in Dagestan. In: Sokolov VE (ed) Papers of the third meeting of the All-Union Theriological Society. Nauka, Moscow, pp 202–203 (in Russian)

Zalessky PM (1930) Maral husbandry in the north-west of Altai. In: Pankratov KN (ed) Maralovodstvo v Sibiri. Sibkrayizdat, Novosibirsk (in Russian)

Zalikhanov MC (1967) Goats of Kabardino-Balkar. Kabardino-Balkarskoe Knizhnoe Izdatelstvo, Nalchik (in Russian)

Zaripov PZ, Znamensky VA (1964) Numbers, distribution and perspectives of the moose population use in Tatarstan republic. In: Zaripov PZ (ed) Prirodnye zapovedniki Volzhsko-Kamskogo kraya. Nauka, Moscow, pp 16–29 (in Russian)

Zaumyslova OY (1989) Peculiarities of salt licks use by Ussuri moose. In: Sablina TB (ed) Ekologiya, morphologiya, ispolzovanie i okhrana dikikh kopytnykh, vol 2. Nauka, Moscow, pp 127–128 (in Russian)

Zaumyslova OY (1998) Wild boar. Letopis prirody Sikhote-Alinskogo gosudarstvennogo zapovednika (ms), pp 145–148 (in Russian)

Zavatsky BP (1975) Impact of brown bear on moose population in Enisey taiga. In: Sokolov VE (ed) Kopytnye fauny SSSR. Nauka, Moscow, pp 48–49 (in Russian)

Zavatsky BP (1986) Selectivity of wolf winter hunting in western Sayan. In: Sokolov VE (ed) Abstracts of the fourth meeting of the All-Union Theriological Society, vol 1. Institute of Animal Morphology and Ecology, Moscow, pp 212–213 (in Russian)

Zavyalova LF (1997) Biocenotic role of wild boar in the Darwinsky Nature Reserve and its impact on neighboring agriculture fields. In: Kuleshova LV (ed) Nauchnye issledovaniya v zapovednikakh i natsionalnykh parkakh Rossii (federalnyi otchet za 1992–1993 gody). Vsesoyuznyi Nauchno-Issledovatelsky Institut Okhrany Prirody i Zapovednogo Dela, Moscow, pp 99–101 (in Russian)

Zharkikh TL, Yasinetskaya NI, Samchuk NG (1999) Establishment of free ranging population of Przewalsky horse in Ukraine. In: Orlov VN (ed) VI sezd Teriologicheskogo obshchestva. Teriologicheskoe obshchestvo, Moscow, 88 pp (in Russian)

Zharkov IV (1940) Results of censuses in the Kavkazsky Nature Reserve in 1939. Nauchno-Metodicheskiye Zapiski Komiteta Po Zapovednikam 7:138–150 (in Russian)
Zharkov IV (1949) Methods of ungulates censuses in Nature Reserves of RSFSR. Nauchno-Metodicheskie Zapiski Glavn. Uprav. Po Zapovednikam 13:77–104 (in Russian)
Zheleznov NK (1980) Characteristics of snow sheep pastures in Chukotka. In: Kolosov AM (ed) Povyshenie productivnosti zverovodstva i okhotniche-promyslovoi fauny. Vsesoyuznyi Selskokhozyaistvennyi Institut Zaochnogo Obucheniya, Moscow, pp 30–35 (in Russian)
Zheleznov NK (1981) Snow sheep (*Ovis nivicola* Eshch) food and its seasonal variability in Chukotka. In: Kontrimavichus VL (ed) Ekologiya mlekopitayushchikh Severo-Vostchnoy Sibiri. Nauka, Moscow, pp 115–137 (in Russian)
Zheleznov NK (1983) Data on wolf ecology (*Canis lupus* L) in the north-east of the USSR. In: Solomonov NG (ed) Teriologicheskie issledovaniya v Yakutii. Yakutsky filial Akademii nauk SSSR, Yakutsk, pp 28–50 (in Russian)
Zheleznov NK (1990) Wild ungulates in the north-east of the USSR. Tikhookeansky Institut Geographii, Vladivostok (in Russian)
Zheleznov-Chukotsky NK (1994) Ecology of *Ovis nivicola* in Northern Asia. Nauka, Moscow (in Russian)
Zheleznyakov DF (1952) New data on *Capra falconeri* Wagner in Middle Asia. Trudy Sredneasiatskogo Universiteta Novaya Seriya 32(2):27–33 (in Russian)
Zhestkova IA (1999) Peculiaritities of mountain sheep (*Ovis vignei*) in Badkhyz. In: Aristov AA (ed) Redkie vidy mlekopitayushchikh Rossii i sopredelnykh territoriy. Teriologicheskoe Obshchestvo, Moscow, pp 128–138 (in Russian with English summary)
Zhevnerov VV (1984) Gazella of the Barsa-Kelmes Island. Nauka, Alma-Ata (in Russian)
Zhevnerov VV, Bekenov AB, Sludsky AA (1983) Gazella. In: Gvozdev EV, Kapitonov VI (ed) Mlekopitayushchie Kazakhstana, vol 3(3). Nauka, Alma-ata, pp 11–54 (in Russian)
Zhirnov LV (1977). Dzeren. Screw-horned goat. In: Kaletsky AA (ed) Redkie zhivotnye SSSR. Kopytnye zveri. Lesnaya Promyslennost, Moscow, pp 62–78 (in Russian)
Zhirnov LV (1982) Restored to life (ecology, protection and use of saiga). Lesnaya Promyshlennost, Moscow (in Russian)
Zhirnov LV, Bekenov AB, Grachev JuA, Dulamtseren S, Lushchekina AA (1998a) Range and its dynamics during the 20th century. In: Sokolov VE, Zhirnov LV (eds) Saiga. Filogeniya, sistematika, ekologiya, okhrana i ispolzovanie. Nauka, Moscow, pp 60–67 (in Russian)
Zhirnov LV, Maksimuk AV, Bekenov AB, Grachev JA (1998b) The harvest rate and types of hunting products of the saiga antelopes. In: Sokolov VE, Zhirnov LV (eds) Saiga. Filogeniya, sistematika, ekologiya, okhrana i ispolzovanie. Nauka, Moscow, pp 264–267 (in Russian)
Zhirnov LV, Pronyaev AV, Bekenov AB, Grachev JA, Gruzdev AR (1998c) Structure of populations (sex and age composition). In: Sokolov VE, Zhirnov LV (eds) Saiga. Filogeniya, sistematika, ekologiya, okhrana i ispolzovanie. Nauka, Moscow, pp 180–192 (in Russian)
Zhiryakov VA (1975) Rare ungulates of the Almaata Nature Reserve and their conservation. In: Sokolov VE (ed) Redkie vidy mlekopitayushchikh fauny SSSR i ikh okhrana. Nauka, Moscow, pp 141–155 (in Russian)
Zhitenko P (1974) Saiga meat. Okhota i okhotniche khozyaistvo 5:20–21 (in Russian)
Zhukov MA (2000) Biological and resource evaluation of wild reindeer populations in Enisey taiga. Dissertation (ms). Institute of ecology and evolution, Moscow (in Russian)
Zimina RP (1953) A short description of mammals and bird fauna in surroundings of the Tien-Shan physics-geographical station. Trudy Instituta geographii AN SSSR 56:54–87 (in Russian)
Zobov AI (1995) Sika deer in the Khopersky Nature Reserve. In: Zobov AI (ed) Problemy izucheniya I okhrany zapovednykh prirodnykh kompleksov. Izdanie Voronezhskogo Universiteta, Voronezh, pp 66–68 (in Russian)

Zubkov AI (1935) Wild reindeer of Novaya Zemlya, archipelago. Trudy Arcticheskogo Instituta 22:55–61 (in Russian)
Zykova LY (1964) Numerical fluctuations and some data on moose ecology in the Oksky Nature Reserve. In: Bannikov AG (ed) Biologiya i promysel losya, vol 1. Rosselkhozizdat, Moscow, pp 127–142 (in Russian)
Zyryanov AN (1975) Wild ungulates of the Stolby Nature Reserve and its surroundings. Trudy Gosudarstvennogo Zapovednika "Stolby" 10:224–338 (in Russian)
Zyryanov AN (1990) Observations on wild reindeer in Sayan. In: Pavlov BM (ed) Resursy, ecologiya i ratsionalnoe ispolsovanie dikikh severnykh olenei v SSSR. Naucho-Issledovatelsky Institut Krainego Severa, Novosibirsk, pp 135–40 (in Russian)
Zyryanov AN, Rozhkov IV (1988) Distribution and ecology of wild reindeer in Central Sayan. In: Syroechkovsky EE (ed) Rare land vertebrates of Siberia. Nauka, Novosibirsk, pp 87–91 (in Russian)
Zyryanov AN, Zyryanov VN (1976) Population status of wild reindeer and experience of professional hunting in the taiga of the Lena river uppers. In: Zabrodin VA (ed) Dikii Severnyi olen. Tsentralnaya Nauchno-Issledovatelskaya Laboratoriya Okhotnichego Khozyaistva i Zapovednikov, Moscow, pp 17–19 (in Russian)
Zyryanov VA, Kolpashchikov LA (1981) Plant resources and their utilisation by wild reindeer in Western Taymyr. In: Syroechkovsky EE (ed) Ekologiya i khozyaistvennoe ispolzovanie nazemnoy fauny Eniseiskogo Severa. Naucho-Issledovatelsky Institut Krainego Severa, Novosibirsk, pp 27–38 (in Russian)
Zyryanov VA, Vinokurov YI (1990) Comparative ecological-morphological characteristics of newborns of wild and domestic reindeer in Taymyr. In: Pavlov BM (ed) Resursy, ekologiya i ratsionalnoe ispolsovanie dikikh severnykh olenei v SSSR. Naucho-Issledovatelsky Institut Krainego Severa, Novosibirsk, pp 74–84 (in Russian)

List of Monographs and Articles Containing Reviews of Ungulate Ecology

Bannikov AG (ed) (1978) Red data book of the USSR. Lesnaya promyshlennost, Moscow (in Russian)
Bannikov AG (ed) (1964, 1965, 1967) Biology and harvest of moose, 1, 2, 3. Rosselkhozizdat, Moscow (in Russian)
Bannikov AG, Zhirnov LV, Lebedeva LS, Fandeev AA (1961) Biology of saiga. Selkhozizdat, Moscow (in Russian)
Baskin LM (1970) Reindeer. Ecology and behaviour. Nauka, Moscow (in Russian)
Baskin LM (1976) Behaviour of ungulates. Nauka, Moscow (in Russian)
Bromley GF (1964) Ussuri wild boar. Nauka, Moscow (in Russian)
Bromley GF, Kucherenko SP (1983) Ungulates of the southern part of the Far East. Nauka, Moscow (in Russian)
Chernyavsky FB, Domnich VI (1989) Moose of north-eastern Siberia. Nauka, Moscow (in Russian)
Danilkin AA (1996) Behavioral ecology of Siberian and European roe deer. Chapman and Hall, London
Danilkin AA (1999) Deer (Cervidae). GEOS, Moscow (in Russian)
Danilkin AA (2002) Pigs (Suidae). GEOS, Moscow (in Russian)
Dinnik NY (1914) Mammals of the Caucasus, vol 2. Zapiski Zakavkazskogo Otdeleniya Imperatorskogo Russkogo Geographicheskogo Obshchestva 27(1):1–246, 28(1):247–536 (in Russian)
Dunin VF, Voronetsky NN, Tishkevich VE, Eliashevich NV, Piskuniv VS, Pareiko OA, Kireenko KM (1999) Moose and roe deer. Belyi veter, Mozyr' (in Russian)

Egorov OV (1955) Ecology of Siberian mountain goat. Trudy Zoologicheskogo instituta 17:7–134 (in Russian with English summary)
Egorov OV (1965) Wild ungulates of Yakutiya. Nauka, Moscow (in Russian)
Fedosenko AK (2000) Argali in Russia and neighboring countries (population status, ecology, behaviour, conservation and use). Tsentrokhotkontrol, Moscow (in Russian)
Fedosenko AK (2002) Urial. Tsentrokhotkontrol, Moscow (in Russian)
Filonov KP (1983) Moose. Lesnaya promyshlennost, Moscow (in Russian)
Filonov KP (1989) Ungulates and large predators in Nature Reserve territories. Nauka, Moscow (in Russian)
Filonov KP (1993) Evaluation of the status of populations of Cervidae. Nauka, Moscow (in Russian)
Flerov KK (1952) Musk deer and deer. In: Pavlovsky EN (ed) Fauna SSSR, new series 55, Mlekopitayushchie, vol 1(2). Izdatelstvo Akademii Nauk SSSR, Moscow Leningrad (in Russian)
Flint VE (ed) (1986) Conservation and perspectives of restoration of goitered gazelle population in the USSR. Vsesoyuznyi Nauchno-Issledovatelsky Institut Okhrany Prirody i Zapovednogo Dela, Moscow (in Russian)
Gambaryan PP (1974) How mammals run. Anatomical adaptations. Wiley, New York
Gvozdev EV, Kapitonov VI (eds) (1983) Mammals of Kazakhstan, vol 3(3). Nauka, Alma-Ata (in Russian)
Gvozdev EV, Strautman EI (eds) (1981, 1982, 1984) Mammals of Kazakhstan, vol 3(1, 2, 4). Nauka, Alma-Ata (in Russian)
Heptner VG, Naumov NP (eds) (1989) Mammals of the Soviet Union, vol 1. Model Press, Leiden
Heptner VG, Sludsky AA (1972) Mammals of the Soviet Union (Predators), vol 2(2). Vysshaya Shkola, Moscow (in Russian)
Hudson RJ, Drew KR, Baskin LM (eds) (1989) Wildlife Production Systems. Cambridge Univ Press, Cambridge
Kaplanov LG (1948) Tiger. Deer. Moose. Moskovskoye Obshchestvo Ispytatelei Prirody, Moscow (in Russian)
Kirikov SV (1966) Hunting animals, natural environment and man. Nauka, Moscow (in Russian)
Knorre EP (1959) Ecology of moose. Trudy Pechoro-Ilychskogo Gosudarstvennogo Zapovednika 7:5–122 (in Russian)
Knorre EP (1961) Results and perspectives of moose domestication. Trudy Pechoro-Ilychskogo Gosudarstvennogo Zapovednika 9:5–113 (in Russian)
Kotov VA (1968) Kubansky goat, its ecology and practical significance. Trudy Kavkazskogo Gosudarstvennogo Zapovednika 10:201–293 (in Russian)
Kozlo PG (1975) Wild boar. Urozhay, Minsk (in Russian)
Kozlo PG (1983) Ecological-morphological analysis of moose population. Izdatelstvo Akademii nauk Belorusskoi SSR, Minsk (in Russian)
Kulagin NM (1932) Moose of USSR. Izdatelstvo Akademii Nauk, Leningrad (in Russian)
Makovkin LI (1999) The sika deer of the Lazovsky Nature Reserve and surrounding areas of the Russian Far East. Lazovsky Gosudarstvennyi Prirodnyi Zapovednik, Vladivostok (in Russian)
Mambetzhumaev AM (1970) Gazella. Fan, Tashkent (in Russian)
Myslenkov AI (ed) (1992) Amur goral. Tsentralnaya Nauchno-Issledovatel'skaya Laboratoriya Okhotnichego Khozyaistva i Zapovednikov, Moscow (in Russian)
Myslenkov AI, Voloshina IV (1989) Ecology and behaviour of Amur goral. Nauka, Moscow (in Russian)
Nasimovich AA (1955) Role of snow-cover regime for the life of ungulates within USSR territory. Academy of Sciences, Moscow (in Russian)
Pavlov BM (ed) (1990) Resources, ecology and use of wild reindeer in USSR. Naucho-Issledovatelsky Institut Krainego Severa, Novosibirsk (in Russian)

Prostakov NI (1996) Ungulates of the Central Chernozemnaya zone. Voronezhsky Universitet, Voronezh (in Russian)
Revin YV, Sopin LV, Zheleznov NK (1988) Snow sheep. Nauka, Novosibirsk (in Russian)
Rozhnov VV (ed) (2001) Class Mlekopitayushchie – Mammalia. In: The Red Data book of Russian Federation (Zhivotnye). Institute Problem Ekologii i Evolutsii, Moscow (in Russian)
Rusakov OS, Timofeeva EK (1984) Wild boar (ecology, resources, economical importance in NW Russia). Leningrad University, Leningrad (in Russian)
Sablina TB (1955) Ungulates of the Belovezhskaya Forest Nature Reserve. Trudy Instituta Morphologii Zhivotnykh Akademii Nauk SSSR 15:1–191 (in Russian with English summary)
Safronov VM, Reshetnikov IS, Akhremenko AK (1999) Reindeer of Yakutiya. Ecology, morphology, use. Nauka, Novosibirsk (in Russian)
Sapozhnikov GN (1976) Wild mountain sheep (*Ovis*) of the Tadzhikistan. Donish, Dushanbe (in Russian)
Semenov-Tien-Shansky OI (1977) Reindeer. Nauka, Moscow (in Russian)
Serzhanin IN (1961) Mammals of Belorus. Izdatelstvo Akademii Nauk Belorusskoi SSR, Minsk (in Russian)
Sludsky AA (1956) Boar (morphology, ecology, economic use, epizootic significance, hunting). Izdatel'stvo Akademii Nauk Kazakhskoi SSR, Alma-Ata (in Russian)
Smirnov MN (1990) Modern numbers and perspectives of utilisation of ungulates in Krasnoyarsky kray. Krasnoyarskoe Knizhnoe Izdatelstvo, Krasnoyarsk (in Russian)
Sobansky GG (1992) Ungulates of the Altai mountains. Nauka, Novosibirsk (in Russian)
Sokolov II (1959) Fauna of the USSR. Mammals (Orders Artiodactyla and Perissodactyla), vol 1(3). Akademiya Nauk SSSR, Moscow-Leningrad (in Russian)
Sokolov VE (ed) (1992) European and Siberian roe deer. Nauka, Moscow (in Russian)
Sokolov VE (ed) European bison. Morphology, systematics, evolution and ecology. Nauka, Moscow, pp 442–470 (in Russian)
Sokolov VE, Tembotov AK (1993) Vertebrates of the Caucasus. Mammals. Ungulates. Nauka, Moscow (in Russian)
Sokolov VE, Zhirnov LV (eds) (1998) Saiga. Phylogenic, systematic, ecology, conservation and use. Nauka, Moscow, pp 105–115 (in Russian)
Sokov AI (1993) Mammals – Artiodactyla. Fauna Tadzhikistana 20(5):1–257 (in Russian)
Solomatin AO (1973) Kulan. Nauka, Moscow (in Russian)
Syroechkovsky EE (ed) (1984) Wild Reindeer of the Soviet Union. Oxonian Press, New Delhi
Syroechkovsky EE (1986) Reindeer. Agropromizdat, Moscow (in Russian)
Tavrovsky VA (ed) (1971) Mammals of Yakutiya. Nauka, Moscow (in Russian)
Timofeeva EK (1974) Moose (ecology, distribution, economical role). Leningradsky Universitet, Leningrad (in Russian)
Timofeeva EK (1985) Roe deer. Izdatel'stvo Leningradskogo universiteta, Leningrad (in Russian)
Tsalkin VI (1950) Siberian mountain goat. Moskovskoe Obshchestvo Ispytatelei Prirody, Moscow (in Russian)
Tsalkin VI (1951) Mountain sheep of Europe and Asia. Moskovskoe Obshchestvo Ispytateley Prirody, Moscow (in Russian)
Veinberg PI (1984) Dagestan goat (*Capra cylindricornis* Blyth). Moskovskoe Obshchestvo Ispytatelei Prirody, Moscow (in Russian)
Vereshchagin NK, Rusakov OS (eds) (1979) Ungulates of north-west of USSR. Nauka, Leningrad (in Russian)
Vorobev GG (ed) (1974) Sika deer of southern Primore. Kyrgyzstan, Frunze (in Russian)
Yakushkin GD (1998) The musk ox in Taimyr. Naucho-Issledovatelsky Institut Krainego Severa, Novosibirsk (In Russian)

Yanushevich AI, Aizin BM, Kydyraliev AK, Umrikhina GS, Fedyanina TF, Shukurov ED, Grebenyuk RV, Tokobayev MM (1972) Mammals of Kirgiziya. Frunze, Ilym (in Russian)

Yazan YP (1972) Game animals of the Pechora taiga. Volgo-Vyatskoe Knizhnoe Izdatelstvo, Kirov (in Russian)

Zaitcev VA (1991) Musk deer of Sikhote-Alin. Ecology and behaviour. Nauka, Moscow (in Russian)

Zalikhanov MC (1967) Goats of Kabardino-Balkar. Kabardino-Balkarskoe Knizhnoe Izdatelstvo, Nalchik (in Russian)

Zheleznov NK (1990) Wild ungulates in the north-east of the USSR. Tikhookeansky Institut Geographii, Vladivostok (in Russian)

Zheleznov-Chukotsky NK (1994) Ecology of *Ovis nivicola* in Northern Asia. Nauka, Moscow (in Russian)

Zhevnerov VV (1984) Gazella of the Barsakelmes island. Nauka, Alma-Ata (in Russian)

MIX
Papier aus verantwortungsvollen Quellen
Paper from responsible sources
FSC® C105338

If you have any concerns about our products,
you can contact us on
ProductSafety@springernature.com

In case Publisher is established outside the EU,
the EU authorized representative is:
Springer Nature Customer Service Center GmbH
Europaplatz 3, 69115 Heidelberg, Germany

Printed by Libri Plureos GmbH
in Hamburg, Germany